ULTRASONIC TREATMENT OF LIGHT ALLOY MELTS

SECOND EDITION

Advances in Metallic Alloys

A series edited by **D.G. Eskin**, *Brunel Centre for Advanced Solidification Technology (BCAST), Brunel University*

ADVANCES IN METALLIC ALLOYS
A series edited by D.G. Eskin

ULTRASONIC TREATMENT OF LIGHT ALLOY MELTS

SECOND EDITION

Georgy I. Eskin
Dmitry G. Eskin

CRC Press
Taylor & Francis Group
Boca Raton London New York

CRC Press is an imprint of the
Taylor & Francis Group, an **informa** business

CRC Press
Taylor & Francis Group
6000 Broken Sound Parkway NW, Suite 300
Boca Raton, FL 33487-2742

First issued in paperback 2017

© 2015 by Taylor & Francis Group, LLC
CRC Press is an imprint of Taylor & Francis Group, an Informa business

No claim to original U.S. Government works

ISBN-13: 978-1-4665-7798-5 (hbk)
ISBN-13: 978-1-138-07597-9 (pbk)

Library of Congress Cataloging-in-Publication Data

Eskin, G. I. (Georgii Iosifovich)
    Ultrasonic treatment of light alloy melts / authors, Georgy I. Eskin, Dmitry G. Eskin.
-- Second edition.
        pages cm
    Includes bibliographical references and index.
    ISBN 978-1-4665-7798-5 (hardback)
    1. Ultrasonics in metallurgy. 2. Light metal alloys. I. Eskin, D. G. II. Title.

TN773.E74 2014
669'.72--dc23   2014007001

Visit the Taylor & Francis Web site at
http://www.taylorandfrancis.com

and the CRC Press Web site at
http://www.crcpress.com

To Nelly

Beloved wife and mama

Contents

Preface

The expanding applications of light alloys demand better quality along with environmentally friendly processing technologies. The quality of cast metal is the foundation of its performance in downstream processing and in final parts. That is why research centers and universities are focusing on improving traditional and developing new means of melt processing and casting. Specific attention is being given to physical means of treatment that do not involve expensive additions or substances that are hazardous to the environment or public health. New types of materials, e.g., nanocomposites, require unconventional technological approaches that also involve physical means such as electrical and magnetic fields, intensive shearing, microwaves, etc. Ultrasonic melt processing accompanied by developed cavitation and acoustic streaming is among those physical processes that can potentially influence different stages of solidification with a resultant cleaner metal with fine grain structure and better mechanical properties.

The ultrasonic melt processing of light alloys has a long history that can be traced back to the 1960s. The first monograph was published in 1965 (G.I. Eskin, *Ultrasonic Treatment of Molten Aluminum*, Moscow, Metallurgiya, 1965), followed by two other monographs in 1988 (G.I. Eskin, *Ultrasonic Treatment of Molten Aluminum*, 2nd revised and enlarged edition, Moscow, Metallurgiya, 1988) and 1998 (G.I. Eskin, *Ultrasonic Treatment of Light Alloy Melts*, 1st edition, Amsterdam, Gordon and Breach, 1998). These books covered the fundamentals and practical application of ultrasonic melt treatment to the main technologies of light metals, mostly based on the author's research results. These works created a solid foundation for advanced research and development, and they demonstrated the industrial feasibility of ultrasound-aided technology. For many researchers, the practical aspects and fundamental issues published in these monographs (and other works of the same author) were the starting point of their own research. Many, even recent, works are attempts (mostly successful) to reproduce the results using different alloys and process parameters.

The first addition of the current book was published more than 15 years ago. The intervening years have marked a period of exploding interest in ultrasonic processing, as evidenced by an exponential increase in the number of publications and citations (see Figure 1). The geography of scientific centers and individual scientists active in ultrasonic research spans the entire globe, from Europe and the United States to China, Japan, and South Korea.

These 15 years have brought about new and exciting results in the application of ultrasonic processing to casting technologies—new areas of application (especially related to new materials) and a deeper understanding of the mechanisms of ultrasonic processing and the physical and metallurgical phenomena affected by ultrasonic cavitation. The fundamental insights into the underlying phenomena of ultrasonic processing have been facilitated by the development of new characterization techniques (particle image velocimetry, high-speed imaging, high-resolution

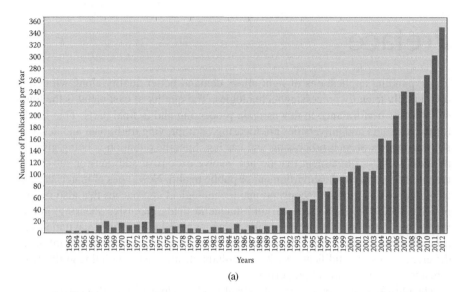

(a)

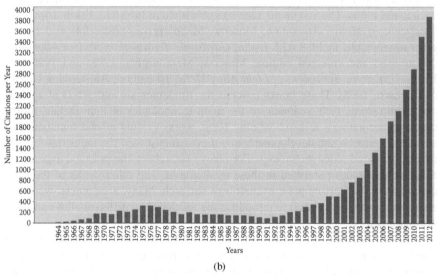

(b)

FIGURE 1 Publications (a) and citations (b) for ultrasonic processing of light metals in 1963–2012, according to Thomson–Reuters Web of Knowledge.

synchrotron radiation, a new generation of cavitometers) and advanced modeling capabilities (computational fluid dynamics and phase field).

Stringent requirements regarding the quality of degassing and grain refinement—including the elimination of harmful gases and excessive emissions, reduced amount of dross, and elimination of particle agglomerates—have created favorable conditions for reintroducing (albeit on a new technological level) ultrasonic technology in the melt processing and casting industry. Industrial application is further assisted by new developments in ultrasonic equipment that has become widely available, more compact, more reliable, and more controllable.

This new edition is revised and enlarged as compared with the first edition. It includes a historical overview, new practical and fundamental results obtained by the authors, as well as selected best examples of research performed by other scientists. New areas of application such as semisolid processing, immiscible alloys, and composite materials are thoroughly covered. The fundamentals of ultrasonic processing are also reviewed, taking into account recent developments.

Most of the results presented in this book have been obtained under the authors' supervision and with their direct participation in the All-Russia Institute of Light Alloys (Russia) (G. Eskin); and the Materials innovation institute (Netherlands) and the Brunel Centre of Advanced Solidification Technology at Brunel University (U.K.) (D. Eskin). The authors would like to thank all their collaborators over years; their names can be found in the References. Particular thanks go to Prof. V.I. Dobatkin, Drs. V.I. Slotin, P.N. Shvetsov, S.I. Borovikova, A.E. Ansyutina, S.G. Bochvar, T.V. Atamanenko, L. Zhang, and N. Alba-Baena.

About 40 years ago, broad and diverse applications of ultrasound to cleaning, degassing, casting, deformation, and postdeformation processing of metals and alloys were bravely forecast (G.I. Eskin, *Ultrasound Advanced to Metallurgy*. Moscow, Mashinostroenie, 1975). In the following years, the scientific foundation was laid for ultrasonic melt processing, and the advantages of ultrasound-aided technologies were clearly demonstrated in the main industrial processes that involve liquid metals.

The current book is not the end of the journey, but rather a continuation. As we write this preface, new projects are underway in many countries, projects aimed at more thorough understanding of the physics of the process; the modeling of very complex multiscale, multiphase, and multiphysics phenomena; and the upscaling of laboratory- and pilot-scale results to the industrial level.

The authors hope that this updated, enlarged, and revised edition will be useful for postgraduate students, postdoctoral researchers, and industrial development engineers, both as a starting point of their own research and as a textbook summarizing the state of the art as it is now.

Georgy Eskin

Dmitry Eskin
Moscow–London, December 2013

Authors

Georgy I. Eskin, Dr. Sc., Ph.D., graduated from Bauman Moscow State Technical University in 1956 as an engineer in metals science. He then worked in industry as a research scientist and received his PhD degree in 1963. Beginning in 1967, Dr. G.I. Eskin worked in the All-Russia Institute of Light Alloys (VILS) as a senior and principal scientist. He received his Dr. Sc. degree in 1982 and became a professor in the Moscow Institute of Steel and Alloys in 1990 and a full member of the Russian Academy of Natural Sciences in 1990. Prof. G.I. Eskin is a world-renowned specialist in the application of ultrasound in metals processing, more specifically in degassing, filtration, and grain refinement during casting of aluminum and magnesium alloys. He pioneered the industrial application of ultrasound to shape and continuous casting of light alloys and made major contributions to fundamental understanding of the underlying mechanisms. Prof. G.I. Eskin has authored and co-authored more than 400 scientific papers and ten monographs, reference books, and textbooks, including *Ultrasonic Treatment of Light Alloys Melts* (The First Edition, Gordon and Breach, 1998) and *Liquid Metal Processing* (Taylor and Francis, 2002). Prof. G.I. Eskin holds more than 100 patents in the area of ultrasonic melt processing.

Dmitry G. Eskin, Ph.D., graduated from the Moscow Institute of Steel and Alloys (Technical University) in 1985 as an engineer in metals science and heat treatment, received a PhD degree from the same university in 1988, and worked until 1999 in the Baikov Institute of Metallurgy (Russian Academy of Sciences) as a senior scientist. From 1999 to 2011, Dr. Eskin held various senior research positions in The Netherlands (Netherlands Institute for Metals Research, Materials innovation institute, Delft University of Technology). In 2011 he joined Brunel University (Brunel Centre for Advanced Solidification Technology) in the U.K. as a professor in solidification research. Prof. D.G. Eskin is a well-known specialist in light-metals processing. His main scientific contributions are to the processing of multicomponent alloys and to the formation of structure and defects during solidification of aluminum alloys. He has authored and co-authored more than 180 scientific papers, five patents, and five monographs, including *Multicomponent Phase Diagrams: Applications to Commercial Aluminum Alloys* (Elsevier, 2005), *Physical Metallurgy of Direct-Chill Casting of Aluminum Alloys* (CRC Press, 2008), and *Direct-Chill Casting of Light Alloys: Science and Technology* (Wiley, 2013).

1 Historical Overview of Ultrasonic Applications to Metallurgy

The application of ultrasound to the processing of liquids and slurries has a long history. This chapter gives a very sketchy overview of the plethora of experimental and theoretical work that has been done on ultrasound and its applications before the 1970s, which creates the foundation for the research reported in this book.

Physicists were the first to explore the realm of high-frequency oscillations. In 1842, Joule [1] discovered a phenomenon of magnetostriction, and J. Curie and P. Curie [2] described piezoelectricity in 1880. Both discoveries made possible the design and application of ultrasonic transducers later in the twentieth century. The theory of oscillations was developed by Lord Rayleigh [3], who laid the foundation for nonlinear acoustics. He also theoretically quantified the pressure pulse resulting from imploding cavitation bubbles. Altberg [4] was the first (1903) to experimentally measure sound pressure, confirming Rayleigh's thesis that sound pressure is directly related to wave energy and velocity.

Significant contribution to the theory of cavitation was made by Frenkel [5] and Harvey et al. [6], who suggested that the cavitation nuclei in real liquids are represented by stable, minute gas nuclei that exist in cavities at the solid surfaces and in crevices of suspended particles. This explained why the cavitation threshold in liquids is well below the theoretical tensile strength of the liquid phase. The pulsation of a cavitation bubble was described analytically by Nolting and Neppiras [7]. They introduced the resonance radius of the bubble. A bubble smaller than the resonance size will rapidly grow and then implode within the sound-wave cycle. The role of such collapsing bubbles is very important in cavitation. Each of the imploded bubbles generate a large pressure pulse and create many even smaller bubbles, starting a chain reaction of bubble multiplication. The bubble larger than the resonance size will not implode but, being relatively stable, will pulsate around its size. The product of the number of cavitation bubbles in the unit volume and the maximum volume of the single bubble is called the *cavitation index*. When this index approaches unity, the amount of bubbles in the unit volume becomes so large that they substitute the liquid phase, and the ultrasonic power transmitted to the liquid declines rapidly [8, 9]. Fundamentals of cavitation are discussed in more detail in Chapter 2.

With the physical background established, the practical aspects of ultrasonic cavitation started to attract the attention of physicists, chemists, and other applied scientists and researchers. Wood and Loomis [10] (1927) were pioneers in studying the effects of power ultrasound. They designed and constructed a sonoreactor with a very large quartz plate, generating a 200-kHz ultrasonic field from a 2-kW generator. They observed intensive acoustic streaming and fountaining, ultrasonic degassing, emulsification and atomization, cavitation damage of organic tissue, etc.

The direct observation of cavitation became possible with the development of high-speed film cameras, high-brilliance impulse lamps and, eventually, laser illumination in the 1950s and 1960s [11–16]. The images taken with the exposure 0.5–5 ms enabled the in-situ study of cavitation development, bubble collapse, and sonoluminescence.

The application of vibrations generally and ultrasonic-frequency vibrations in particular to treating metals is not a new idea. As early as the 1870s, Chernov [17], a prominent Russian metallurgist, stated that if steel solidifying in a mold was shaken so strongly that "all its particles come into motion," then the cooled ingot had very fine crystals. On the other hand, when cooled without shaking, the steel casting had large, well-developed crystals. Chernov considered dynamic effects on solidification to be of great importance.

The idea of dynamic solidification has been developed further as a result of the implementation of various techniques and devices for metal casting based on forced motion of a melt near the solidification front and in the bulk of liquid metal. Low-frequency vibration of a mold with liquid metal, insertion of vibrating tools into the solidifying melt, electromagnetic stirring and vibration of the melt during solidification, and eventually ultrasonic treatment of the solidifying metal (alloy) represent the methods that have been seeking industrial applications.

Good reviews of the influence of low-frequency vibration on liquid and solidifying metals were given by Balandin [18] and Campbell [19]. The method of shaking permanent molds containing melt with an eccentric vibrator has been in use since 1910. Periodical shaking of a permanent mold with a frequency of 1.5 Hz and an amplitude of 15–20 mm was shown to facilitate the removal of gases from the steel melt, favor grain refining, increase the density, and improve the mechanical properties of the casting [18]. In 1966, Southin [20] studied the effects of low-frequency (50 Hz) vibrations on solidification of metals and reported that dendritic grain structures were refined by fragmentation, whereas the cellular and plane-front structures required cavitation for any significant effect to be achieved. Mechanical vibrations are not the sole way of producing the favorable effects. Electromagnetic agitation was demonstrated to have benefits for the structure and segregation upon casting ingots and billets. Bondarev [21] reported in 1973 that electromagnetic stirring resulted in far greater temperature equilibration over the ingot liquid pool than that produced by mechanical agitation. For magnesium alloys, the temperature dropped below liquidus by 2–3 K for narrow-solidification-range alloys and by 10–20 K below liquidus for the alloys with a wide solidification range. Similar results were later obtained for aluminum alloys [22–25]. There are, however, conflicting reports on the effect of such an agitation on the structure. Dobatkin [26] considered that the temperature equilibration resulted in grain coarsening and formation of large primary intermetallics, while Balandin [18] reported grain refinement, but only in the case when the agitation and low-frequency vibrations caused the destruction of the solidification front and, therefore, fragmentation of grains.

Application of ultrasound to liquid and solidifying metals has been reviewed on different occasions and in different languages. The most notable reviews are those by

Hiedemann (1954) [27], Eskin (1961, 1965) [28, 29], Flynn (1964) [30], von Seemann, Staats, and Pretor (1967) [31], Abramov and Teumin (1970) [32], Kapustina (1970) [33], Abramov (1972) [34], Buxmann (1972) [35], and Campbell (1981) [19].

Quite a number of studies on crystal growth and nucleation were performed from the 1950s through the 1970s on so-called transparent analogues of liquid metals that show crystal morphologies similar to metals [36]. Salol (phenyl salicylate), camphene, naphthalene, timol (2-isopropil-5-metilfenol), piperine, and ammonium chloride were used extensively for observation of crystallization, solidification, and effects of ultrasound on these processes [29, 37]. Later, succinonitrile with camphor became popular as a metal analogue. These experiments helped a lot in understanding the effects of ultrasonic oscillations and cavitation. Some of these works are reviewed in Chapter 5.

When it comes to metals proper, Sokolov [38] is considered to be a pioneer. He was also one of the first who used piezoceramic transducers. As early as in 1935, Sokolov applied ultrasonic treatment at a frequency of 600–4500 kHz in solidifying zinc, tin, and aluminum. The vibrations were transmitted through a steel mold connected via a steel rod to the transducer. Sokolov reported formation of coarse grains with refined dendrite branches, which can be associated with low ultrasonic intensity transmitted through the mold bottom. He also experimentally observed the acceleration of solidification in the ultrasonic field.

Extensive studies on solidification of various metals and alloys under ultrasonic fields of different frequencies and intensities were performed by Seemann et al. [31, 39, 40,] and Schmid et al. [41, 42] in the 1930s and 1940s. Seemann et al. tried a variety of ways to introduce ultrasound into the melt, including through the mold and through a consumable or permanent waveguide. They also studied the effects of ultrasound on various metals from copper to tin, and on different materials from foundry alloys to composites. A significant refining of grain structure was achieved in these experiments, and mechanisms for the observed phenomena were suggested. In 1939, Schmid and Roll [42] showed that the grain-refining effect in Wood's alloy was the greatest at the highest ultrasonic intensity, and they speculated that crystal fragmentation assisted by viscous friction between the solid and the liquid phases was responsible for the refining.

At the same time, a group of scientists put forward an idea of the effect of ultrasonic waves on the nucleation of the solid phase [43–46]. In particular, Danilov and Teverovsky [45] advocated the idea of the effect of ultrasonic cavitation on heterogeneous nucleation through activation of insoluble impurities. Later, in the 1950s, using transparent analogues, Kapustin [37, 47] observed that the formation of new crystallization centers under sonication is proportional to the ultrasonic intensity and the degree of undercooling. He also reported that the nucleation was assisted by ultrasound even in very pure liquids; hence, something different from activation of impurities might have happened. One of the hypotheses was that the forced convection induced by ultrasonic streaming and cavitation caused high friction at the solid/liquid interface, thereby decreasing the surface tension and the size of critical nuclei [32]. Hunt and Jackson [48] considered three hypotheses that might explain the cavitation-aided nucleation, i.e., the nucleation on the expanding cavitation bubble due to cooling of its surface; a change of local equilibrium melting point due to the high-pressure spike produced by a collapsing bubble (according to the Clapeyron equation); and local undercooling of the liquid phase due to the high negative pressure associated with the bubble collapse. The calculations and dedicated experiments on water demonstrated that the collapse of cavitation bubbles might be responsible for the nucleation due to the local change of phase equilibria, the increase of the melting point (by tens of degrees!), and the effective local undercooling. Frawley and Childs [49] confirmed these findings by measuring undercooling in Bi and Bi–Sn alloys under various vibration frequencies.

The application of ultrasound to processing of commercial alloys started with the works of Seemann et al. (e.g., [40]), who demonstrated considerable refinement of the duralumin ingot grain structure and improvement of its mechanical properties. Bradfield [50] reported the works of Turner, who treated solidifying Al–4% Cu, Al–12% Si, Al–9% Si, and Al–0.5% Mn alloys and found that the structure was refined efficiently at a frequency of 26 kHz (ultrasound). Eskin [29] applied ultrasonic cavitation treatment to a variety of model and commercial foundry Al alloys (hypo- and hypereutectic Al–Si alloys), and demonstrated that the grain structure, intermetallics, and primary crystals were refined under cavitation conditions. The proper choice of the material for sonotrodes (horns, ultrasonic tips) was treated with special care, and Nb and its alloys were recommended for use in molten aluminum [29].

Chukhrov, Borovikova, and Sokolova [51] studied the solidification of magnesium alloys under ultrasonic processing and achieved significant grain refinement, better technological plasticity, and improved mechanical properties for a commercial Mg–Mn–Ce alloy. Later, ultrasonic processing was shown to be advantageous for receiving refined grain structure and improved mechanical properties of direct-chill (DC) cast wrought-magnesium alloys [52].

Considerable refinement of tin and zinc when treated by ultrasonic vibrations during solidification was reported by Seemann et al. [31]. They also studied higher-temperature alloys like brass and steel [31, 53]. The high melt temperature imposes severe restrictions on the materials for sonotrodes. In the case of copper alloys, alumina or molybdenum horns could be employed for relatively short laboratory-scale experiments [53]. Direct introduction of ultrasound to the steel melt is, however, almost impossible. Consumable steel horns were reported to be used [34]. Indirect ways of treatment were successfully tested, including transmission of the oscillations via mold, through solid substrate upon vacuum arc remelting, or through the layer of molten slag upon electro-slag remelting [54, 55]. Such indirect processing opens the way to treat virtually all metals.

This brief overview shows that by the middle of the 1960s, the possibility of refining the structure of metallic alloys was proved on the laboratory scale, and research efforts began to be focused on understanding the nature of this refinement and on its technological implementation. Chapters 5 through 8 are dedicated to these issues.

Another important effect of ultrasonic vibrations and cavitation that attracted the interest of metallurgists was degassing of the melt. Although ultrasonic and sonic degassing was known for water and other low-temperature liquids since the 1920s, Krüger [56] was probably the first who applied the ultrasonic degassing to melts, especially to molten glass. The main reported difficulty was the identification of a stable material for the sonotrode. Esmarch, Rommel, and Benther [57] studied the degassing of Al–Mg alloys by sonic vibrations induced by contactless electromagnetic stirring and vibrations in the crucible. Turner reported early results on ultrasonic degassing of Al–Cu alloys by direct introduction of ultrasonic oscillations into the melt [50].

The nature of ultrasonic degassing was first revealed on water. Lindström [58] investigated ultrasonic degassing and its relation to cavitation in water using oxidation of potassium iodide (so-called Wiessler reaction) as an indicator of cavitation activity. He suggested that the ultrasonic degassing of water is due to the diffusion of dissolved oxygen into the cavitation bubbles, which oscillate and grow and, finally, float to the surface. The roles of acoustic power (above the cavitation threshold), gas concentration, and the nature of the gas were reported to be unimportant. Kapustina [33] gave a thorough analysis of ultrasonic degassing mechanisms in water and concluded that the most important role is played by the oscillations of the bubbles in the acoustic field, while ultrasonic cavitation takes the supportive role in intensification of the bubble formation and acceleration of bubble/liquid interfacial diffusion. Eskin [29] argued that the cavitation is essential for ultrasonic degassing of metallic melts, where the natural gas bubbles are not typically present, unlike those in water. Therefore, the formation and multiplication of bubbles (essential for degassing) can only be achieved by cavitation. Mechanisms and practical issues of ultrasonic degassing are the subjects of Chapter 3.

Yet another important effect of ultrasound is atomization and dispersion of liquid and solid phases. This is widely used in the chemical and food industries for making emulsions and powders. For metallurgists, the obvious applications are production of powders/granules and composite materials as well as immiscible alloys. As early as 1936, Masing and Ritzau [59] reported that sonic vibrations of 500 Hz facilitated the fine distribution of lead in aluminum. A much more pronounced effect was observed next year by Schmid and Ehret [41], who used more powerful and higher frequency (10 kHz) oscillations. Later, Becker [60] described the Al–Pb and Zn–Pb alloys produced with ultrasonic melt processing as stable suspensions. Nonmetallic and solid particles can also be introduced into liquid metals, forming metal-matrix composites. Ultrasonic cavitation assists in cleaning the particle surface, improving its wetting, and distributing particles in the volume. One of the first reports on this subject was a paper by Pogodin-Alekseev and Zaboleev-Zotov [61] in 1958, where they reported introduction of particulate (2–20 μm) alumina, silicon carbide, and titanium nitride in liquid aluminum in quantities of 10 to 50 wt%. The mixture was then subjected to powerful ultrasonic processing, i.e., 21.5 kHz at 3 kW of generator power. Wood's alloy and Zn-, Sn- and Pb-based alloys were also used as the matrix. Seemann and Staats [62] published an important paper in 1968 where they summarized their earlier works on the dispersion of metallic (Ti, Fe) and ceramic (carbides, oxides) particles in molten aluminum using a 20-kHz magnetostrictive transducer and alumina sonotrode. Chapter 9 gives an overview of ultrasonic processing of composite materials and immiscible alloys.

The industrial application of ultrasonic melt treatment would not be possible without adequate sources of ultrasound, waveguiding systems, and a technological means to introduce the vibrations into the processed medium.

The technological schemes and equipment suitable for the industrial implementation of powerful ultrasonic treatment appeared only in the late 1950s and early 1960s. The main challenges in metallurgical application of ultrasonic processing can be summarized as follows:

- The source of ultrasound (generator and transducer) should provide enough power (frequency and amplitude) to overcome the cavitation threshold.
- The generator and transducer should be able to tune the resonance frequency of the entire system so that the variable conditions during processing (changing temperature, expansion of sonotrode, and active load) would not affect the efficiency of the processing.
- The waveguiding system should be designed in such a way as to diminish the losses of acoustic power at the joints and limit undesirable directions and modes of wave propagation.
- The tip (sonotrode) that is in contact with the processed media should stay in resonance and, hence, retain its acoustic and mechanical properties at elevated temperatures and not react with the molten metal while being in good acoustical contact, i.e., being wet.
- The transducer should be protected against heat and vapor coming from the processed molten metal, which typically means enclosures and efficient cooling systems.

A transducer is the essential part of any ultrasonic equipment. The function of a transducer is to convert the high-frequency electrical current coming from a generator to mechanical oscillations of the same frequency.

Most of the early installations used piezoelectric (e.g., quartz) and electrostrictive or piezoceramic (e.g., barium titanate or lead zirconate titanate) transducers operating in the MHz frequency range and transmitting relatively low acoustic power to the liquid metal (mainly low-melting metals such as zinc, tin, and antimony). Crawford [63] gives examples of powerful single-crystal quartz transducers working at frequencies 0.5 to 2 MHz and of matching 1.5–3-kW generators with variable frequency between 0.8 and 5 MHz. The reported maximum output power intensity was 60 W/cm^2, which is rather high. Single-crystalline quartz, however, requires very high voltage for excitation as compared to polycrystalline barium titanate; therefore, the latter requires much less powerful generators and has better power efficiency. For example, an ultrasonic output of 1 W/cm^2 at a frequency of 100 kHz would require 10 kV for quartz and only 100 V for barium titanate [63]. The polycrystalline barium titanate can also be shaped, which adds versatility to its applications. On the other hand, the stability of quartz output is greater.

The weak point of piezoceramic transducers is the use of adhesives in bonding crystals together, which makes the transducer assembly at the transmission end rather brittle and prevents the use of direct liquid cooling. Although there are examples of powerful piezoceramic transducers manufactured in the 1940s–1950s, the overall reliability and levels of acoustic power transmitted were low. This type of transducer could be used in lab-scale research, since the treated volumes did not exceed 100–200 g, and the researchers were mainly interested in the effects of ultrasound on liquid phase or on solidification. However, the upscaling of the discovered effects required reliability, robustness, and more power. The acoustic power intensity is proportional to $(fA)^2$, where f is the oscillation frequency and A is the amplitude of oscillation. The cavitation threshold in liquid depends on the frequency and remains relatively low and constant at frequencies below 20 kHz and then starts to increase rapidly with increasing frequency, especially above 50 kHz [64]. The high-frequency ultrasonic vibrations also attenuate more rapidly in the processed medium than the lower frequency oscillations, limiting the processed volume [65]. In general, the low-frequency ultrasonic vibrations at large amplitudes seem to be more suitable for high-power applications that involve cavitation of any appreciable volumes of liquid. With the addition of the requirement for robustness and efficient cooling, a magnetostrictive transducer looks like a suitable candidate for metallurgical applications.

Magnetostrictive transducers use a stack of metallic (e.g., permendur) plates that contract and expand in the alternating magnetic field produced by a coil excited by a generator. The frequency range is limited to 50 kHz, with 17–22 kHz being most typical. The power efficiency of the magnetostrictive transducers is less than 50%, but they can be easily automatically tuned to the variable resonance frequency of the waveguiding system, thereby maintaining the maximum output and amplitude. The magnetostrictive stack is mechanically bound together and then soldered to the transmitting piece. An advantage of magnetostrictive transducers for metallurgical applications is their water cooling, which prevents the transducer from overheating, maintaining the temperature below the Curie point even when the working sonotrode is dipped into the molten metal. Yet another advantage over piezoceramics lies in the recovery of magnetostrictive properties in the case of overheating above the Curie point. The recovery requires just cooling of the magnetostrictive stack, while piezoceramics would need ex-situ repolarization.

The early experience in metallurgical applications of ultrasound made the choice in favor of magnetostrictive transducers, and until now the only successful industrial-scale examples of ultrasonic metallic melt processing were done using these types of transducers [29, 34, 65, 66]. Figure 1.1 shows some historic photos.

In 1958, Herrmann [67] described many pilot installations for ultrasonic treatment of molten metal. The versatility of the treatment was illustrated by examples with ultrasonic processing conducted in the furnace, melt flow, feeders of castings, and in the molds with the aim of removing dissolved gases, refining structures, and improving the casting properties. For example, Dürener Metallwerke induced ultrasonic oscillations in the liquid pool of an ingot to degas the melt and refine grains during solidification. Oscillations were transmitted via a gas-tight vibrating slab connected to a transducer. Unfortunately, this method lacked sufficient power for casting large ingots. Another German company, Vereinigte Deutsche Metallwerke, developed a dynamic scheme in which oscillations were supplied to the melt by a sonotrode submerged into the molten part of the ingot, or by a pulsing gas jet blown onto the surface of the liquid pool.

The AFG company suggested that ultrasound could be used in continuous casting to degas the melt; however, the selection of a material resistant to the melt and vibration remained a challenge. Aktiebolaget Svenska Metall Werke tried to transfer oscillations to the liquid pool by submerged metal rings and discs connected to an

(a)

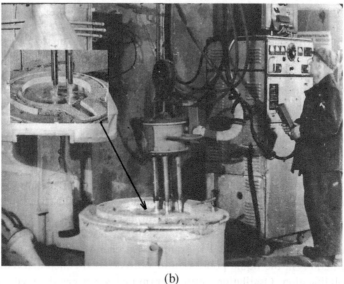

(b)

FIGURE 1.1 Industrial applications of magnetostrictive transducers in metallurgy: (a) steel melt processing through the bottom of a vacuum-arc furnace at Westinghouse Electric (adapted from Frederick [65]) and (b) aluminum degassing by submerged sonotrodes at Moscow Plant Nauka (adapted from Eskin [29]).

oscillation generator. Vogel [68] used a long tungsten rod tipped with a tungsten disk submerged in the melt.

Experimenting with a DC aluminum casting system with a sonotrode located in the liquid pool, Schoeler–Bleckmann Stahlwerke AG found that the amount of energy that could be transmitted from ultrasonic transducers in such a setup was sufficient for grain refinement, degassing, and diminishing macrosegregation. In continuous casting of steel with a much deeper sump, sonotrodes were placed at the water-cooled inner wall of the mold. Ingots obtained with this casting machine were fine grained and had a higher metal density over the entire cross section. Schoeler–Bleckmann also suggested degassing the melt by treating it with ultrasound in the feeder.

One of the early pilot-scale trials of ultrasonic melt processing of aluminum alloys during aluminum semicontinuous casting was described by Seemann and Menzel in 1947 [40]. Commercial-size (290 mm) billets from duralumin (AA2024) were cast with ultrasonic processing of the melt in the sump of the billet. The casting scheme is shown in Figure 1.2a. As one can see, the casting was performed in a movable water-cooled mold with four magnetostrictive transducers with connected sonotrodes submerged into the melt. Although this method was somewhat

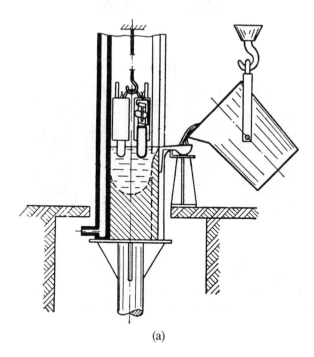

(a)

FIGURE 1.2 Early schemes of aluminum billet casting with ultrasonic melt processing: (a) casting in a moving water-cooled mold using four magnetostrictive transducers with sonotrodes submerged in the melt (adapted from Seemann and Menzel [40]). (*continued*)

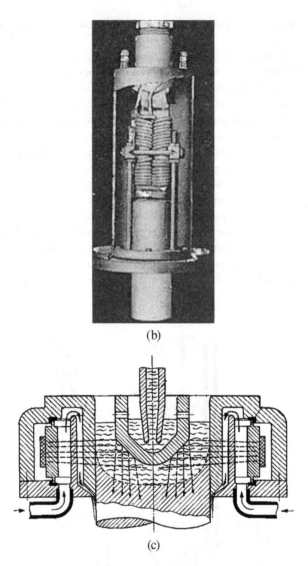

(b)

(c)

FIGURE 1.2 (*continued*) Early schemes of aluminum billet casting with ultrasonic melt processing: (b) an Atlas-Werke magnetostrictive transducer used in the setup shown in (a) (Adapted from Hiedemann [27]); and (c) direct-chill casting with ultrasonic oscillations transmitted to the mold through cooling water (Adapted from Seemann and Staats [69]).

inferior to direct-chill casting with a stationary water-cooled mold, it was demonstrated that the treatment was reasonably efficient. For the experimental purposes, a powerful ultrasonic generator (up to 25 kW) was manufactured that excited four 2-kW magnetostrictive transducers at 40 kHz. These magnetostrictive transducers, made by Atlas-Werke of Bremen, Germany, had a stack of nickel plates coupled to a ceramic radiator made from hard porcelain (Figure 1.2b). The efficiency of the entire assembly was only 14%, but it was able to deliver an intensity of 2.0 W/cm^2 to the cross-sectional area of the billet (660 cm^2), which was sufficient to achieve grain refinement, a reduced porosity, and an increase in ultimate strength. In 1955, Seemann and Staats [69] patented the ultrasonic processing of the melt during proper DC casting whereby the ultrasound was introduced through the mold by exciting the cooling water (Figure 1.2c).

A variety of technical solutions have been reported for conveying ultrasound into the flowing melt, both in the feeder and in the launder. Seemann and Staats [70] proposed a series of schemes where the melt was sonicated while passing along the radiators made of the same material as the melt and placed in the launder/feeder walls, as shown in Figure 1.3a. Part of the radiators protruding into the flowing melt might be initially dissolved, but the water-cooling system of the radiators ensured that a thermal balance was established during continuous operation, thus providing an efficient treatment. Another interesting solution was suggested in the 1960s for the combined ultrasonic and vacuum degassing [29]. In this case, a siphon scheme was used, as seen in Figure 1.3b.

The dispersion and emulsification applications of ultrasound led to some interesting technological solutions summarized by Seemann and Staats in 1968 [62]. Two ways of introducing particles or mixing phase were suggested, as illustrated in Figure 1.4. The first is the introduction of the mixing phase through a feeding tube into the melt that is subjected to ultrasonication either through the walls of the crucible/mold or by the sonotrode (Figure 1.4a). The latter can be combined with the feeding tube. The second is the introduction of the mixing phase via a solid master alloy (sintered alumina powder or sold metal were used) that is put in direct contact with the vibrating sonotrode, as shown in Figure 1.4b.

By the end of the 1950s, it was clear that robust and reliable solutions should be sought for industrial metallurgical applications of ultrasonic processing. The first industrial ultrasonic degassing installation (UZD-200) was developed in 1959 to treat 100–200 kg of melt in a crucible before casting [29]. The degassing unit was developed in two modifications: a mobile unit and an overhead stationary unit with an operating range of 20–30 m. The latter is shown in Figure 1.1b. The UZD-200 unit included a 10-kW tube ultrasonic generator and a special switching circuit allowing for the alternate operation of four magnetostrictive transducers with Ti or Nb sonotrodes. Similar installations were used for DC casting of aluminum alloys when the ultrasonic treatment was performed in the sump of a billet.

Ultrasound was also commercially used in the 1960s for precision investment casting of aluminum alloys using a modification of low-pressure die

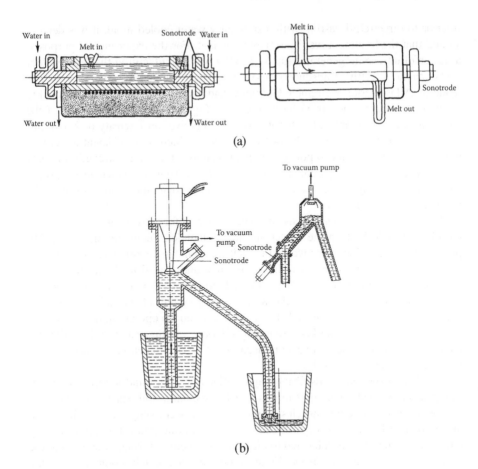

FIGURE 1.3 Examples of suggested schemes for ultrasonic melt processing in the melt flow: (a) treatment using sonotrodes from the same material as the melt (Adapted from Seemann and Staats [70]) and (b) treatment using siphon and submerged sonotrode from a nonreacting material (adapted from Eskin [29]).

casting [29, 66]. The principal scheme of the installation is given in Figure 1.5. The ultrasonic processing allowed for melt filling of very thin sections with high surface finish.

This brief historical overview shows that ultrasonic melt processing has a long tradition. It originated from advances in physics that led to the design of modern equipment, followed by technological developments and trials that started in the 1930s and continued through the 1960s, leading to the first industrial implementations. In the following chapters we will look at the fundamental and applied aspects of ultrasonic melt processing.

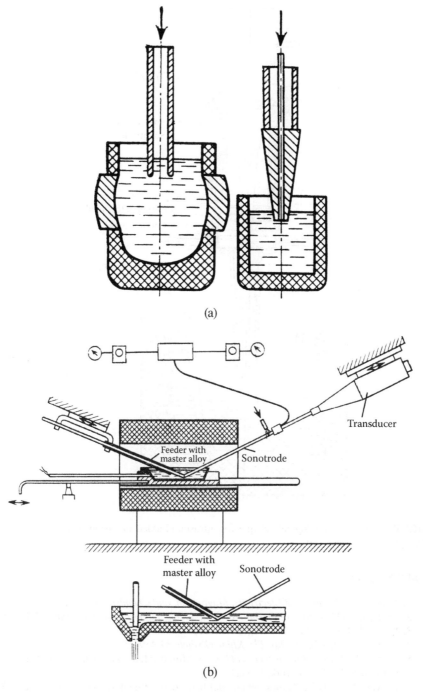

FIGURE 1.4 Examples of technological schemes suggested for ultrasound-assisted introduction of particles or mixing phases into the melt: (a) through a feeding tube and (b) via a solid master alloy rod. (Adapted from Seemann and Staats [62].)

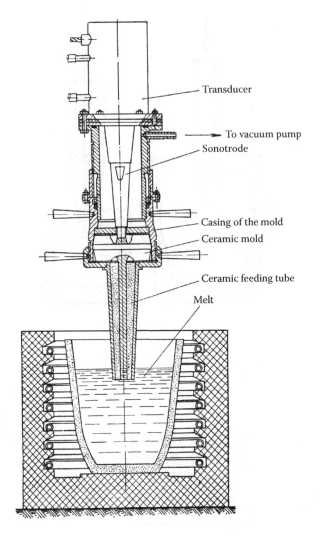

FIGURE 1.5   A technological scheme for ultrasound-aided precision investment casting. (Adapted from Eskin [29].)

REFERENCES

1. Joule, J. 1842. *Annals of Electricity, Magnetism, and Chemistry* 8:219–24.
2. Curie, J., and P. Curie. 1880. *Bulletin de la Société minéralogique de France* 3:90–93.
3. Rayleigh, Lord. 1917. *Phil. Mag.* 34:94–98.
4. Altberg, W.J. 1903. *Zh. Rus. Fiz. Khim. Obshch.* 34:459–60.
5. Frenkel. Ya.I. 1945. *Kinetic theory of liquids*. Moscow: Izd. Akad Nauk SSSR [Translated to English. 1955. New York: Dover].
6. Harvey, E.N., D.K. Barnes, W.D. McElroy, A.H. Whiteley, D.C. Pease, and K.W. Cooper. 1944. *J. Cell. Comp. Physiol.* 24:1–34.
7. Nolting, B.E., and E.A. Neppiras. 1950. *Proc. Royal Soc. London* 63B:674–85.
8. Kikuchi, Y., and H. Shimizu. 1959. *J. Acoust. Soc. Am.* 31:1385–86.

9. Rozenberg, L.D., and M.G. Sirotyuk. 1960. *Akust. Zh.* 6:478–81.
10. Wood, R.W., and A.L. Loomis. 1927. *Phil. Mag.* 4:417–25.
11. Knapp, R.T., and A. Hollander. 1948. *Trans. ASME* 70:419–35.
12. Ellis, A.T. 1956. Cavitation in hydrodynamics, paper 8. In *Proc. Nat. Phys. Lab. Symp.* London: Her Majesty's Stationery Office.
13. Ellis, A.T., and M.E. Fourney. 1963. *Proc. IEEE* 51:942–43.
14. Knapp, R.T. 1955. *Trans. ASME* 77:1045–54.
15. Schmid, J. 1959. *Acustica* 9:321–26.
16. Sirotyuk, M.G. 1962. *Akust. Zh.* 8:216–19.
17. Chernov, D.K. 1879. *Zapiski Imperat. Russ. Teckhnich. Obshch.*, no. 1:1–24.
18. Balandin, G.F. 1973. *Formation of crystal structure in castings*, 192–261. Moscow: Mashinostroenie.
19. Campbell, J. 1981. *Int. Met. Rev.*, no. 2:71–108.
20. Southin, R.T. 1966. *J. Inst. Met.* 94:401–7.
21. Bondarev, B.I. 1973. *Melting and casting of wrought magnesium alloys*, 214–23. Moscow: Metallurgiya.
22. Vives, Ch. 1989. *Metall. Mater. Trans. B* 20B:623–29.
23. Vives, Ch. 1993. *Mater. Sci. Eng. A* 173:169–72.
24. Zuo, Y., H. Nagaumi, and J. Cui. 2008. *J. Mater. Process. Technol.* 197:109–15.
25. Zuo, Y., J. Cui, Z. Zhao, H. Zhang, L. Li, and Q. Zhu. 2012. *J. Mater. Sci.* 47:5501–8.
26. Dobatkin, V.I. 1960. *Ingots of aluminum alloys*, 56–60; 90–108. Sverdlovsk: Metallurgizdat.
27. Hiedemann, E.A. 1954. *J. Acoust. Soc. Am.* 26:831–42.
28. Eskin, G.I. 1961. *Ultrasound in metallurgy*. Moscow: Metallurgizdat.
29. Eskin, G.I. 1965. *Ultrasonic treatment of molten aluminum*. Moscow: Metallurgiya.
30. Flynn, H.G. 1964. Physics of acoustic cavitation in liquids. In *Physical acoustics*, vol.1, part B, ed. W.P. Mason, 57–172. New York: Academic Press.
31. von Seemann, H.J., H. Staats, and K.G. Pretor. 1967. *Arch. Eisenhüttenwesen* 38:257–65.
32. Abramov, O.V., and I.I. Teumin. 1970. Crystallization of metals. In *Physical principles of ultrasonic technology*, ed. L.D. Rozenberg, 427–514. Moscow: Nauka [Translated to English. 1973. New York: Plenum].
33. Kapustina, O.A. 1970. Degassing of liquids. In *Physical principles of ultrasonic technology*, ed. L.D. Rozenberg, 253–336. Moscow: Nauka [Translated to English. 1973. New York: Plenum].
34. Abramov, O.V. 1972. *Crystallization of metals in an ultrasonic field*. Moscow: Metallurgiya.
35. Buxmann, K. 1972. *Z. Metallkde.* 68:516–21.
36. Jackson, K.A., and J.D. Hunt. 1965. *Acta Metall.* 13:1212–15.
37. Kapustin, A.P. 1962. *Effect of ultrasound on the kinetics of crystallization*. Moscow: Nauka.
38. Sokolov, S.Ya. 1935. *Acta Physicochim. URSS* 3:939–44; Sokolov, S.Ya. 1936. *Zh. Tekhn. Fiz.*, no. 3:81–85.
39. von Seemann, H.J. 1936. *Metallwirtschaft* 15:1067–69.
40. von Seemann, H.J., and H. Menzel. 1947. *Metall.* 1:39–46.
41. Schmid, G., and L. Ehret. 1937. *Z. Electrochem.* 43:869–74.
42. Schmid, G., and A. Roll. 1939. *Z. Electrochem.* 45:769–75.
43. Berlaga, R.Ya. 1939. *Zh. Eksperim. Teor. Fiz.* 9:1397–98.
44. Danilov, V.I., E.E. Pluzhnik, and B.M. Teverovsky. 1939. *Zh. Experim. Teor. Fiz.* 9:66–71.
45. Danilov, V.I., and B.M. Teverovsky. 1940. *Zh. Eksperim. Teor. Fiz.* 10:1305–10.
46. Altberg, V.J., and V. Lavrov. 1939. *Acta Physicochem. USSR* 11:287–90.
47. Kapustin, A.P. 1950. *Izv. Akad. Nauk SSSR* 14:357–65.

48. Hunt, J.D., and K.A. Jackson. 1966. *J. Appl. Phys.* 31:254–57.
49. Frawley, J.J., and W.J. Childs. 1968. *Trans. Metall. Soc. AIME* 242:256–63.
50. Bradfield, G. 1950. *Proc. Phys. Soc. B* 63 (5): 305–21.
51. Chukhrov, M.V., S.I. Borovikova, and A.I. Sokolova. 1963. *Research in alloys of nonferrous metals.* Issue 4, 141–56. Moscow: Izd. Akad. Nauk SSSR.
52. Ansyutina, A.E., I.I. Gur'ev, and G.I. Eskin. 1970. *Izv. Akad. Nauk SSSR*, no. 4:76–81.
53. von Seemann, H.J., and F. Wehner. 1973. *Metallwissehschaft und Technik* 27:971–77.
54. Lane, D.H., J.W. Cunningham, and W.A. Tiller. 1960. *Trans. Metal. Soc. AIME* 218:985–90.
55. Benua, F.F., I.V. Vologdin, and A.I. Katler. 1958. *Svarochn. Proizvod.*, no. 5:1–5.
56. Krüger, F. 1931. German patent no. 604486. *Glastech. Ber.* 1938 (7): 233–36.
57. Esmarch, W., T. Rommel, and K. Benther. 1940. *Werkstof Sonderheft*, 78–87. Berlin: W.V. Siemens Werke.
58. Lindström, O. 1955. *J. Acoust. Soc. Am.* 27:654–71.
59. Masing, G., and G. Ritzau. 1936. *Z. Metallkde.* 28:293–97.
60. Becker, V. 1941. *Novosti Tekhniki* 10 (4): 25–26.
61. Pogodin-Alekseev, G.I., and V.V. Zaboleev-Zotov. 1958. *Litein. Proizvod.*, no. 7:35–36.
62. von Seemann, H.J., and H. Staats. 1968. *Z. Metallde.* 59:347–56.
63. Crawford, A.E. 1955. *Ultrasonic engineering with particular reference to high power applications.* London: Butterworths.
64. Blitz, J. 1963. *Fundamentals of ultrasonics.* London: Butterworths.
65. Frederick, J.R. 1965. *Ultrasonic engineering.* New York: John Wiley & Sons.
66. Eskin, G.I., V.I. Slotin, and S.Sh. Katsman. 1967. *Precision casting of parts for aviation assemblies from aluminum alloys*, 56–76. Moscow: Mashinostroenie.
67. Herrmann, E. 1958. *Handbuch des Stranggiessens.* Dusseldorf, Germany: Aluminium Verlag.
68. Vogel, F. 1948. *Metall.* 2:229–30.
69. Seemann, H., and H. Staats. 1955. German patent 933779, Oct. 1955 (application Feb. 1952).
70. Seemann, H., and H. Staats. 1957. German patent 957533, Jan. 1957 (application Feb. 1953).

2 Fundamentals of Ultrasonic Melt Processing

This chapter is aimed at giving the reader some introduction to the fundamentals of ultrasound and ultrasonic metal processing. However, we do not aspire to cover all the aspects of acoustics and sonochemistry; instead, we refer the inquisitive reader to more suitable sources like Blitz [1], Rozenberg [2], Flynn [3], Suslick [4], Mason and Lorimer [5], Brennen [6], and Ensminger et al. [7, 8] as well as to numerous papers published over the years by specialized scientific journals like *Ultrasonics*, *Ultrasonics Sonochemistry*, *Journal of the Acoustical Society of America*, and *Akusticheskiy Zhurnal* (Russian Acoustics Journal).

2.1 DYNAMIC MEANS OF MELT PROCESSING

The idea of applying some dynamic action onto the liquid or solidifying metal with the aim to improve the structure dates back to D.K. Chernov, whom we cited in Chapter 1. The dynamic action can be delivered to the liquid phase or semiliquid slurry by various means that may conditionally be classified as follows: (1) stirring, (2) vibration, and (3) concentrated pulse [9]. The first group includes mechanical stirring using special mechanical impellers and mixers, electromagnetic stirring, and gas blasting through the melt. The second group comprises low-frequency vibration and shaking by mechanical, pneumatic, or electromagnetic sources and high-frequency oscillations at ultrasonic frequency induced by acoustic, mechanical, or electromagnetic sources. The third group covers electrical discharge and explosion action. There are different mechanisms associated with these means of dynamic action. It is important to note that all of the dynamic actions affect the solidifying (semiliquid) metals, while only some affect the liquid.

Stirring produces microscopic and macroscopic mass transport and acts in the same manner as convection. The moving masses of liquid and solid phases assist in transporting heat and chemical species, which has direct consequences for solidification and structure formation. The moving liquid phase intensifies the heat transfer at the solid/liquid interfaces, i.e., macroscopically at the solidification front and mold face and microscopically at the crystal/liquid interface. The intensive stirring with high Reynolds numbers may considerably increase the heat transfer between the liquid and solid phases. At the same time, active stirring equilibrates the temperature and decreases the temperature gradient over the entire affected volume. This has direct consequences for solidification, promoting spatial over progressive solidification. As a result, the columnar structure may be substituted with equiaxed morphology while the intermetallic phases grow large. The stirring may affect the viscosity of the solidifying melt. Liquid metals are considered to be Newtonian liquids with viscosity independent of the shear rate. The slurries or semiliquid mixtures are different. Intensive stirring/shearing of the semiliquid (up to 50% solid) mixtures lowers their viscosity and sustains their fluidity, which is the basis of so-called rheocasting of semisolid materials.

While the stirring of the liquid phase affects mostly the temperature and species distribution, the stirring of the semiliquid slurries results in a change of the grain morphology. With respect to the structure management, the stirring invokes two mechanisms: (a) fragmentation of dendrites by mechanical or thermosolutal mechanisms and (b) coarsening of dendrites up to the complete elimination of dendrite branches by thermosolutal mechanism. The mechanical breakup of dendrites by the melt flow was quite a popular idea in the 1960s–1980s; however, the exact mechanisms of such a breakup remain unclear, and the modeling of this process suffers from the unavailability of constitutive properties of the solid phase in the semisolid temperature range. Most probable is the combination of solid-phase deformation with liquid-metal embrittlement. The thermosolutal fragmentation and coarsening are well defined and confirmed both experimentally and by modeling. This mechanism involves solute accumulation at the solid/liquid interface, its redistribution to the roots of dendrite branches, change of local equilibrium temperature, and dissolution or remelting of the solid phase.

Low-frequency vibrations (less than 1000 Hz with amplitudes of several mm) are similar to stirring in their action on the liquid and semiliquid materials. The difference is in the periodicity of the pressure differences that are formed in the fluid medium by a vibration source and in the lesser mass transport compared to that achieved by stirring. As a result, the effects related to the liquid phase are minor, while the fragmentation of dendrites is frequently observed. Moderate degassing has been reported as well, provided that the liquid phase already contains cavities or bubbles. High-frequency or ultrasonic oscillations (above 16 kHz with amplitudes of several μm) produce some specific effects. The major one is cavitation, i.e., formation of cavities in the liquid phase and at liquid/solid/gas interfaces, with their pulsation, growth, and implosion. The relatively low input energy of an ultrasonic source is translated to high-temperature, -pressure, and -momentum (velocity) surges upon collapse of cavitation bubbles. These high-energy effects change locally the equilibrium (due to the temperature and pressure increase) and may affect the nucleation of the solid phase. At the same time, shock waves generated by imploded bubbles may have mechanical and physicochemical action on the solid phase, facilitating the penetration of the liquid into thin capillary channels, decreasing surface tension, and breaking up agglomerates. The cavitation zone created by an ultrasonic source acts like a pressure and momentum source in the liquid volume, creating powerful acoustic streams and secondary flows that promote rapid mixing of soluble and insoluble inclusions.

As we will show in this book, cavitation is very important for efficient ultrasonic melt processing, especially when it comes to structure modification, but also for degassing and mixing. In the precavitation regime, the acoustic source creates a standing wave or a traveling wave, with the inclusions in the liquid phase oscillating in resonance, which may promote degassing, provided that the gas bubbles already exist in the melt, e.g., in Ar-assisted degassing.

Concentrated pulses can be produced by electrical discharge or explosion and are represented by shock waves that propagate from the surface of the melt or the mold walls through the liquid. The potential effects include mixing, destruction of the solidification front, intensified mass and heat transport, and degassing. These means of solidification processing are, however, the least technologically advanced due to their hazardous nature and practical limitations.

We can summarize that the dynamic melt processing in the liquid state may affect the nucleation (dynamic nucleation) and release of dissolved gas from the liquid phase (degassing). The dynamic processing of the semiliquid (solidifying) mixture influences the structure formation by fragmentation of solid crystals and their subsequent distribution in the volume.

Dynamic nucleation was introduced in the 1960s by Kapustin [10], Walker [11], Chalmers [12], Hunt and Jackson [13], and Frawley and Childs [14], who based their works on earlier research cited in Chapter 1. Two main mechanisms were considered, i.e., undercooling of the cavitation bubble surface during the expansion phase of oscillations and undercooling of the liquid phase resulting from the instantaneous increase of pressure during cavitation bubble collapse (according to the Clapeyron equation). The latter mechanism seems most probable [13]. In addition to dynamic nucleation, multiplication of solidification nuclei by activation of heterogeneous substrates was suggested in the 1930s–1950s by Danilov et al. [15–17] and Kazachkovsky [18]. In this case, the dynamic action upon solid/liquid interface improves wetting, decreases surface tension, and promotes heterogeneous nucleation in the available insoluble substrates such as oxides, carbides, etc., being assisted by penetration of the liquid phase into discontinuities of the substrate surface and the formation of the adsorbed boundary layer at the substrate surface.

The fragmentation of the solid phase under dynamic action was adopted by many as the main mechanism of structure refinement [19–23]. Chvorinov [19] suggested that the dendrites growing in the two-phase transition region are separated from the solidification front by forced convection, with the resultant crystals moving to the bulk of the melt, where they act as nuclei for new grains, provided that they are not completely remelted. Balandin [20] enriched this idea with the thesis that the insoluble inclusions deactivated by alloy melting with high superheat reactivate once the solid phase is formed around them. After separation from the two-phase zone by forced convection, a solid crystal containing this activated inclusion is transported to the liquid phase, and the solid phase is melted away, leaving behind the active insoluble substrate. However, these solidification concepts have not explained the mechanisms of dendrite (crystal) separation from the two-phase zone or its fragmentation.

One of the earliest suggested mechanisms was the seemingly obvious fragmentation of dendrites by mechanical fracture caused by melt flow. This mechanical fracture assisted by bending deformation and formation of large-angle boundaries, and liquid metal embrittlement is still considered as one of the possibilities [22, 24], though the ductility of the solid phase and the velocities of the flow in the interdendritic space make the purely mechanical fracture unlikely [25].

Other mechanisms are currently considered as more plausible. On the mesoscopic scale, forced melt flow can bring hot melt from the liquid pool into the undercooled two-phase zone and cause its partial remelting, with subsequent washing-out of loose solid crystals. On the microscopic scale and in the absence of cavitation, the most realistic mechanism of fragmentation is dendrite arm separation by root remelting effects because of thermal, solute, or capillary effects. Solute accumulation at the solidification front causes the fluctuations in growth velocity that have a direct effect on the kinetics of the growth and coarsening of dendrite branches [26]. The coarsening of dendrite branches results in their necking [27, 28] and the accumulation of solute at their roots both by rejection from the solid phase and by convection in the interdendritic space [29]. The local solute enrichment causes local superheating of the solid phase and its melting. Along with the local change of equilibrium, capillary effects make dendrite roots more soluble than other regions. To summarize: Uneven propagation of a solidification front and disturbances (also by convection) of the solute and thermal fields in the liquid phase surrounding the roots can lead to remelting of the roots and the fragmentation of dendrites. The forced flow assists further by transporting the fragments to the solidification front and farther to the bulk of the liquid.

In the presence of cavitation accompanied by the implosion of bubbles, the destruction of dendrites has been demonstrated on transparent analogues and resembles the fracture by explosion [30–32]. The ultrasound-induced streaming flow does not play a significant role in the fragmentation of dendrites, but it can be effective in transporting cavitation bubbles toward the dendrites to promote continuous fragmentation of the growing dendrites and in transporting the fragments to the bulk of the melt.

Let us look more closely at the physics of vibrational processing.

2.2 LOW-FREQUENCY VIBRATIONS AND ULTRASOUND: BASIC EQUATIONS

The term *vibration* commonly refers to elastic oscillations of various frequencies ranging from low-frequency vibration to ultrasound. In order to implement ultrasonic techniques in industry, particularly in metallurgy, one should clearly understand the difference between low-frequency vibrations and ultrasound. This will help in explaining the effects produced by the ultrasonication of melts and will facilitate the application of ultrasound in industrial metallurgical processes.

Under low-frequency vibration, e.g., shaking, the object moves periodically with a certain angular frequency $\omega = 2\pi f$ or linear frequency f, velocity v, and acceleration j. When the direction of the motion changes to the opposite, the velocity and acceleration pass through zero when the vibrating object instantaneously stops and then continues to move in the opposite phase. The corresponding dynamic action (pressure) on the unit volume of the vibrating melt is described by

$$P = m(g \pm j), \tag{2.1}$$

where m is the mass of the unit volume and g is the gravity acceleration.

During one vibration period $T_v = 1/f$, the pressure on this unit volume varies from the minimum value $P_1 = m(g - j)$ to the maximum value $P_2 = m(g + j)$. Obviously by increasing j, one can increase the pressure in the phases of rarefaction and compression, respectively. The maximum acceleration is given by

$$j_{max} = 4\pi^2 f^2 A, \tag{2.2}$$

where A is the amplitude of vibrations.

This alternating pressure may induce wave propagation in the melt; however, for vibrations at low frequencies (50–500 Hz), their wavelength

$$\lambda = c/f, \tag{2.3}$$

where c is the velocity of wave propagation, which will be tens of meters and significantly exceeds the feasible size of common liquid metal containers.

The wavelength becomes shorter if the melt is subjected to continuous harmonic excitation at higher sonic and, especially, ultrasonic frequencies. In this case, one would expect a regular wave to be formed in the liquid or solidifying metal. When $j = g$, the pressure goes through zero and then becomes negative at some point in the liquid. This negative pressure is very important, as it is responsible for cavitation phenomena (provided that other energetic conditions are met).

Generally, acoustic phenomena are classified with respect to their frequencies as follows: infrasound ranges up to 16 Hz, audible sound from 16 Hz to 16 kHz, ultrasound from 16 kHz to 10^4 MHz, and hypersound above 10^4 MHz. Ultrasonic frequencies are, therefore, commonly bordered on the lower side by 16,000 Hz, in agreement with the term, meaning beyond the hearing (sonic) limit.

The range of elastic oscillations extends up to the GHz range, i.e., to the frequencies of thermal oscillations in atoms. At such a high frequency, we currently do not have technical means to produce significant acoustic power densities, i.e., above 10^4 W/m². On the other hand, commercial ultrasonic transducers are capable of generating power densities about 10^5–10^6 W/m² at frequencies of 18–20 kHz, which is sufficient for treatment of liquid and solidifying melts.

The physical nature of elastic oscillations is common regardless of frequency, and represents the alternating mechanic disturbances of the elastic medium. However, a number of new phenomena, such as acoustic cavitation or acoustic flows responsible for irreversible effects in the processed medium, require high-frequency and high-energy oscillations of high intensity.[*] As will be shown in this discussion, the ultrasonic intensity is proportional to the squared frequency and squared amplitude; therefore, the achievable ultrasonic intensity significantly exceeds that of audible

---

[*] Here and below, we understand the terms *sound* or *ultrasound intensity* to mean the density of acoustic energy unidirectionally propagating through the unit area, i.e., of a traveling acoustic wave.

frequencies.[*] Typical values of sound intensities (W/m²) in the vicinity of various sound (ultrasound) sources are listed here:

Voice	1.6×10^{-4}
Clarinet	6.3×10^{-4}
Grand piano	8.0×10^{-3}
Chamber orchestra	1.6×10^{-2}
Symphony orchestra	0.2
Air-raid siren	0.3
Airplane	1.0
Low-intensity ultrasound	1.0×10^{4}
Middle-intensity ultrasound	1.0×10^{5}
High-intensity ultrasound	1.0×10^{6}

When a source of ultrasonic oscillations is introduced into the liquid pool, it induces an ultrasonic field whose characteristics depend on the oscillation parameters and on the properties of the treated medium.

One of the basic parameters is the propagation velocity of elastic oscillations. This velocity is governed by the physical properties of the medium where the wave propagates. At a given temperature, the velocity (m/s) of sonic (ultrasonic) longitudinal waves in the solid phase with density ρ, Young's modulus E, and Poisson's ratio μ_P is determined by

$$c = \sqrt{\frac{E(1-\mu_P)}{[\rho(1+\mu_P)(1-2\mu_P)]}}.$$ (2.4)

In the liquid phase, where elastic properties depend on the compressibility, the velocity of an acoustic wave can be determined from

$$c = \sqrt{1/(\beta_{ad}\rho)},$$ (2.5)

where ρ is the liquid density and β_{ad} is the adiabatic compressibility. For gases, the molecular motion is related to the adiabatic index $\gamma = c_p/c_v$ (the ratio of specific heats at constant pressure and volume), gas pressure P_0, and density ρ:

$$c = \sqrt{\gamma P_0/\rho}.$$ (2.6)

---

[*] Ultrasonic testing and medical equipment use especially reduced levels of ultrasonic energy; otherwise, the tested material might suffer structural changes. Thus this method will no longer be considered nondestructive.

Generated at any point in the medium (solid, fluid, or gas), oscillating disturbances propagate through the medium as elastic waves of alternating compression and rarefaction stages. As it follows from Equations (2.4)–(2.6), the velocity of elastic waves in an unbound medium is independent of frequency and, up to certain magnitudes, of intensity (this relation is referred to as the *linear approximation*).

The product of propagation velocity c and density ρ (ρc) is called the wave, or acoustic, impedance of the given medium. It is equal to

$$\rho c = P/v = P/(2\pi f A), \tag{2.7}$$

where P is the sound pressure in the traveling wave and v is its oscillation velocity.

An important characteristic of an elastic wave is the distance between adjacent regions of compression or rarefaction, i.e., the distance that the wave passes in one period $T_v = 1/f$, where f is the frequency of sound. This distance is called the wavelength λ (see Equation 2.3).

We can state that ultrasonic oscillations excited in the medium are represented by alternating regions of positive and negative pressure (P), with wavelength λ as the distance between adjacent regions of compression and rarefaction. This pressure pattern propagates in the medium with the sound velocity c, whereas every particle of the medium oscillates with velocity v and displacement (amplitude) A around an equilibrium position.

In the one-dimensional case, the propagation of elastic disturbances is described by the wave equation

$$\frac{\partial^2 A}{\partial x^2} - \frac{1}{c^2}\frac{\partial^2 A}{\partial t^2} = 0 \tag{2.8}$$

where A is the displacement of particles of the medium and t is the time.

A specific solution to this wave equation can be represented as a harmonic wave

$$A = A_0 \sin\left(\omega t - \frac{\omega x}{c}\right), \tag{2.9}$$

where $\omega = 2\pi f$ is the angular frequency and x/c is the phase factor for harmonically oscillating particles with coordinate x.

Differentiate Equation (2.9) with respect to time and obtain the particle velocity:

$$v = A_0\omega \cos\left(\omega t - \frac{\omega x}{c}\right). \tag{2.10}$$

Differentiate Equation (2.10) again and obtain the oscillating acceleration:

$$j = -A_0\omega^2 \sin\left(\omega t - \frac{\omega x}{c}\right). \tag{2.11}$$

Multiplying the oscillating velocity (Equation 2.10) by the acoustic resistance ρc of the medium, we obtain the oscillating pressure P_A:

$$P_A = v\rho c = \rho c A_0 \omega \cos\left(\omega t - \frac{\omega x}{c}\right) \tag{2.12}$$

Solutions of Equations (2.10)–(2.12) are periodic in space and time. For the fixed moment of time, i.e., $t = 0$, they reduce to

$$v = A_0 \omega \cos\left(\frac{\omega x}{c}\right) \tag{2.13}$$

$$j = -A_0 \omega^2 \sin\left(\frac{\omega x}{c}\right) \tag{2.14}$$

$$P_A = \rho c A_0 \omega \cos\left(\frac{\omega x}{c}\right) \tag{2.15}$$

If we fix the position, i.e., $x = 0$, then we obtain

$$v = A_0 \omega \cos(\omega t) \tag{2.16}$$

$$j = -A_0 \omega^2 \sin(\omega t) \tag{2.17}$$

$$P_A = \rho c A_0 \omega \cos(\omega t) \tag{2.18}$$

The respective maximum values for the oscillating velocity (m/s), acceleration (m/s^2), and sound pressure (P_A) will be given by

$$v = A_0 \omega; \quad j = -A_0 \omega^2; \quad P_A = A_0 \rho c \omega.$$

A very important parameter of the ultrasonic field—one that determines to a great extent the efficiency of processing—is the ultrasonic intensity I, which is the power flux W_a normalized by area S. In the simplest case of a plane wave, the intensity (W/m^2) is given by

$$I = \frac{W_a}{S} = \frac{1}{2}\rho c v^2 = \frac{1}{2}\rho c (A_0 \omega)^2 = \frac{1}{2} P_A v \tag{2.19}$$

The acoustic intensity is proportional to the squared amplitude and frequency, which largely determines the selection of processing equipment and regimes.

Note that all of these relationships hold only for a traveling wave. They ignore the boundary conditions that give rise to diffraction, reflection, standing waves, and other phenomena complicating these expressions, though the main relationships between the oscillation parameters remain principally the same.

It is important to understand that the ultrasonic field should be considered while taking into account the ultrasonic source and the conditions of wave propagation and attenuation in the treated medium. The transducer and the waveguide not only transfer the energy, but also consume it. As a result, the energy is spent on mechanical and internal friction and on the active work of exciting the medium where the wave propagates. Without constant renewal, the oscillations will decay, and their amplitude will decrease with time as follows:

$$A = A_0 e^{-\delta t} \cos \sqrt{\omega_0^2 - \delta^2} t \qquad (2.20)$$

where $\omega_0 = 2\pi f_0$ is the system's natural angular frequency, A_0 is the initial amplitude, and δ is the attenuation showing how much the amplitude decreases with each unit of time.

A lossy system, once excited, oscillates with frequency f_L

$$f_L = \sqrt{f_0^2 - \left(\frac{\delta}{2\pi}\right)^2}. \qquad (2.21)$$

Consequently, the losses reduce not only the amplitude, but also the frequency of oscillation.

In order to maintain the amplitude and frequency and compensate for the losses, we should (1) constantly add power to the oscillation system from an external source and (2) maintain the resonance mode by tuning the source frequency to the natural frequency of the oscillating system.

In practical situation of our interest, the oscillating system is loaded by a melt, which plays the role of an active medium and consumes a considerable portion of the acoustic power transmitted by the system. When loaded, the oscillating system (transducer, waveguide, and sonotrode) may produce both standing and traveling waves. Their relative amplitudes are conventionally defined by a parameter called the standing-wave ratio. The quality of an oscillating system may be characterized, in some sense, by its ability to produce standing waves, i.e., oscillations of large amplitude occurring due to resonance amplification. Standing-wave amplitudes can be more than ten times higher than those of traveling waves, and the transducer or sonotrode radiating face should oscillate with the amplitude determined by standing waves.

In sonicated melts, however, the occurrence of standing waves is usually not an issue, unless the geometry of the container allows for that, being in resonance with the oscillating system. The sonotrode face excites predominantly traveling waves propagating through the melt. The amplitude, oscillation velocity, acceleration, and sound pressure of these waves are connected with sound intensity I and frequency f by the following relations:

$$A = \frac{1}{2\pi f}\sqrt{\frac{2I}{\rho c}} \tag{2.22}$$

$$v = \sqrt{\frac{2I}{\rho c}} \tag{2.23}$$

$$j = 2\pi f \sqrt{\frac{2I}{\rho c}} \tag{2.24}$$

$$P_A = \sqrt{2I\rho c} \tag{2.25}$$

From these formulae, it follows that in order to achieve the certain intensity in the given medium at different frequencies, the oscillation velocity and sound pressure may remain the same, while the amplitude and the acceleration should vary with the frequency.

2.3 PROPAGATION OF ACOUSTIC WAVES IN THE MELT

The foundation of successful ultrasonic processing is the acoustic power transferred to the treated medium, e.g., the melt. The relationships between the oscillation parameters and the properties of the sonicated melt determine the efficiency of processing.

The conditions for transferring a certain ultrasonic power to the liquid metal crucially depend on the acoustic impedance or, more generally, on the medium's mechanical resistance, which is defined in acoustics as the ratio of instantaneous values of force and velocity. This ratio has active and reactive components for harmonic oscillations with a phase shift: The *reactive component* of the impedance defines the shift of the natural frequency of the oscillating system effected by the load (melt), while the *active component* shows irreversible internal losses of energy (preferably small) as well as the useful radiation into the load.

The ultrasonic power transmitted to the melt is proportional, as Equation (2.19) shows, to the melt acoustic impedance and the squared oscillation velocity, i.e., it depends on both the resistance of the liquid metal and the oscillating velocity of the source.

When cavitation develops in the melt, the temporal characteristics of force and velocity at the sonotrode radiating face vary, so that Equation (2.19) may be used to describe the actual technological processes of melt sonication only in the first, or linear, approximation. However, as long as standing waves dominate traveling waves in the oscillating waveguide system proper, it retains the property of a resonance filter. As a result, the oscillation processes remain almost harmonic, and the acoustic impedance averaged over the period can still be a useful characteristic of the system.

In the presence of cavitation, the acoustic impedance of the melt is a function of oscillation amplitude or velocity and rapidly decreases, as the sound velocity and the

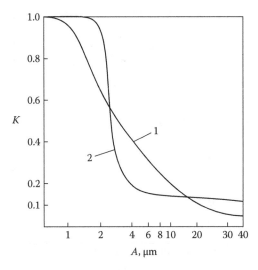

FIGURE 2.1 Relative acoustic impedance K vs. amplitude A of the sonotrode at 18 kHz: (1) water and (2) aluminum melt.

pressure in the cavitating liquid phase is no longer the same as in the noncavitating liquid. The intensity or transmitted power can still be considered proportional to the squared oscillation velocity, but with the acoustic impedance rapidly decreasing after the cavitation threshold.

Figure 2.1 shows the dimensionless parameter K, which is the ratio of acoustic impedance under cavitation to the acoustic impedance in the absence of cavitation, versus the oscillating amplitude of the sonotrode at 18 kHz for water [33] and aluminum melt [34]. When the null-to-peak amplitude exceeds 0.5 μm for water at 20°C and 2–3 μm for an aluminum melt, the relative acoustic impedance decreases to values 10 times smaller than for sonication without cavitation.

When acoustic cavitation begins, the acoustic power transferred to the fluid increases because the reactive component of the load falls to zero. Figure 2.2 gives the relation between the oscillation amplitude and the acoustic power generated by standard ultrasonic equipment and transmitted into an aluminum melt of commercial purity at a resonance frequency of 18 kHz for different surface areas of the sonotrode's radiating face.

As a rule, only longitudinal waves propagate in the liquid metal. This means that the liquid phase oscillates along the direction of wave propagation, thus generating alternate regions of compression and rarefaction. The profile and intensity distribution of the ultrasonic field in the melt is important for the processing of any appreciable volumes. In the usual case of pistonlike excitation of oscillations, the wave front deviates from planar geometry, in dependence on the oscillation-source geometry. The important parameter here is the ratio of the radiating face size (e.g., diameter of the sonotrode D_{rad}) to the wavelength of the sound λ. If the sonotrode diameter and the frequency are small (wavelength is large), i.e., if $D_{rad}/\lambda \ll 1$, then the oscillation source may be considered as the point source emitting spherical waves. With

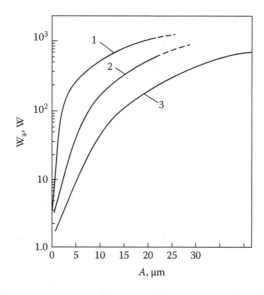

FIGURE 2.2   Power W_a transferred into the melt vs. amplitude A of the sonotrode at a resonance frequency of 18 kHz for three sonotrode diameters: (1) 65 mm, (2) 40 mm, and (3) 20 mm.

increasing sonotrode diameter and decreasing wavelength (increasing frequency), the wave front approaches the plane geometry, and the sound energy localizes in the direction of oscillations. The sound field (beam) starts to resemble a cone, but in reality it acquires a petal shape as the wave energy is progressively absorbed by the liquid. When $D_{rad}/\lambda > 1$, additional side "petals" appear at the radiating surface, as illustrated in Figure 2.3 [35]. Such directional patterns are well known in loudspeaker designs. In reality, the ratio D_{rad}/λ varies from 0.01 to 0.1; therefore, the ultrasonic beam can be considered as nearly cylindrical.

Generally, any propagating harmonic field has two characteristic regions: the near zone (also called the Fresnel zone), where the beam can be considered to be parallel, and the far zone (also called the Fraunhofer zone), where the beam diverges so that

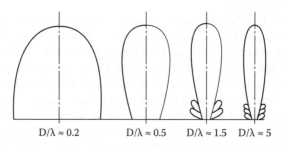

FIGURE 2.3   Sound-beam patterns in dependence on D_{rad}/λ ratio.

it appears to originate from the center of the source. The transition between those zones occurs at the distance:

$$N = D_{rad}^2/4\lambda \text{ or } N = D_{rad}^2 f/4c. \tag{2.26}$$

The near field is most important for ultrasonic processing of the melt, as the sound energy rapidly decays and disperses in the far field. Although being energy intensive, the near field is very nonuniform. Only a perfect sound source emits waves with a plane phase front and uniform amplitude. In reality, the radiating face of a sonotrode should be considered as a set of point sources that oscillate in phase but with different amplitudes. These sources emit elementary waves that produce a diffused sound field in which each point oscillates with different amplitudes resulting from superimposition of elementary waves emitted by the sonotrode.

Most of the ultrasonic measurements were performed on water and other low-melting materials. For low-melting metals, the (ultra)sound velocity was measured by pulse methods [36]. For higher-melting metals, the speed of sound can be calculated through interdependence with viscosity, density, surface tension, and atomic characteristics [37, 38]. It was found that the speed of sound in the melts near their melting points differed only slightly from that in solid samples at subsolidus temperatures. Table 2.1 gives the speed of sound and density for some molten metals and water for comparison. The temperature dependence of the sound velocity in liquid aluminum was estimated as

$$c = 4730 - 0.16\,(T - T_m), \tag{2.27}$$

where T_m is the melting point (933 K) [38].

TABLE 2.1
The Speed of Sound c (Different Values According to References), Melting Point T_m, Density ρ, and Viscosity μ for Some Molten Metals and Water

Metal	T_m, °C	Temperature, °C	c, m/s	ρ, g/cm³	μ, mPa·s
Cs	28.6	28.5	967; 983	1.84	0.68
Ga	29.8	30	2740; 2873	6.08	2.04
In	156.2	156	2215; 2320	7.02	1.89
Sn	231.9	300	2270; 2464	6.99	1.85
Bi	271	300	1635; 1640	9.84	1.80
Pb	327.4	327	1790; 1821	10.66	2.65
Zn	419.6	420	2790; 2850	6.57	3.85
Mg	649	650	4065	1.584	1.25
Al	660	666	4561; 4729	2.375	1.3
Cu	1083	1084	3440	8.02	4.0
Water	0.0	20	1482	1.0	1.002

Source: [36–39].

The instrumental facilities that allow the measurements of cavitation activity in liquid aluminum have been developed only recently [40], whereas the evaluation of flow patterns and detailed study of cavitation are still reserved to transparent liquids and, increasingly, to computer modeling and simulation.

The propagation of ultrasound is accompanied with losses of oscillation energy. The amplitude and intensity of a plane ultrasonic wave decrease exponentially with the propagation distance x:

$$A = A_0 e^{-ax} \tag{2.28}$$

$$I = I_0 e^{-2ax} \tag{2.29}$$

where α is the loss coefficient or sound absorption or attenuation factor.

The absorption of ultrasound in the liquid phase is related to the viscosity (see Table 2.1) and thermal conductivity of the melt, and changes with the ultrasound frequency. The attenuation factor depends on the squared frequency:

$$\alpha = \frac{\omega^2}{2\rho c^3} \left[\frac{4}{3}\mu + \mu' + a\left(\frac{1}{c_v} - \frac{1}{c_p} \right) \right], \tag{2.30}$$

where μ and μ' are the shear and volume viscosities, a is the thermal conductivity, and c_v and c_p are the specific heats at constant volume and pressure, respectively. This dependence demonstrates that very high ultrasonic frequencies would be impractical because of their strong attenuation.

In ideal liquids, the viscosity and thermal conductivity suffice to describe the absorption loss of ultrasonic energy, but in actual melts, the effect of impurities should be taken into account. The interfaces between the liquid phase and suspended particles (nonmetallic inclusions and crystals) may significantly affect absorption. Figure 2.4 illustrates this with kaolin particles suspended in water [41]. The attenuation factor increases with the amount of particles and with their fineness. A similar effect is produced by gas bubbles, whose interfaces with the melt act as scattering sources. As we will show in the following discussion, the very same interfaces of gaseous and solid inclusions act as cavitation nuclei and favor the development of cavitation, which absorbs additional ultrasonic energy.

In an unbound medium, ultrasonic oscillations may propagate over large distances. However, in practice, the sonicated liquid or solidifying alloys are confined within finite volumes such as crucibles, molds, launders, feeding systems, etc. In these cases, the reflections from the walls (reverberation) and, indeed, solidification front produce additional effects besides natural absorption along the wave propagation path.

High-frequency ultrasound in the megahertz range experiences reflection and refraction, while the contribution of diffraction becomes significant for low-frequency ultrasound used in metallurgy (18–25 kHz) when the ultrasonic wavelength becomes comparable with the casting dimensions.

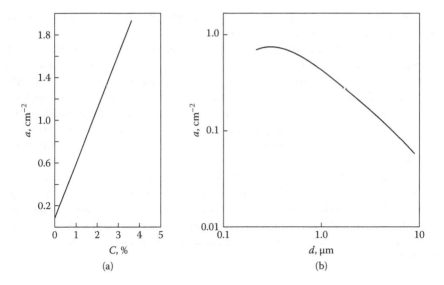

FIGURE 2.4 Attenuation factor a of ultrasound for kaolin suspension in water: (a) effect of 0.2–5-μm particle concentration C and (b) effect of particle diameter d.

Ultrasonic oscillations may excite standing (stationary) and traveling waves in the liquid medium. Standing waves occur as a result of the interference between the incident and reflected waves, provided that these are fully reflected at the interface (wall). In this case, there will be points or surfaces (nodes) in the sonicated medium where the medium particles (atoms, unit volumes) remain stable in their positions, while the sound pressure is doubled, similar to an oscillating string with fixed ends. There will also be points (surfaces) (antinodes) where the sound pressure is zero and the displacement amplitude doubles. The resonance condition of the sonicated volume is required for the formation of standing waves. For example, in a closed volume like a crucible with a sonotrode at the top, the ultrasonic wave path (distance between the sonotrode radiating face and the bottom of the crucible) should be exactly $(2n - 1)$ $\lambda/2$, where n is an integer.

If the propagation path does not consist of an integer number of half-wavelengths, or if it is very large compared to the wavelength, or if the condition of complete reflection is not satisfied, then the sound field has a traveling component. This is a typical situation in ultrasonic processing of melts.

We can describe the ultrasonic field using the wave rigidity $P_A/A = \rho c \omega$. When the ultrasound wave makes contact with a surface of a larger wave rigidity, e.g., melt–mold wall interface, the displacement phase changes by 180°. Therefore, the node of displacement and the antinode of pressure occur at this interface. Conversely, if the melt interfaces with a medium of a smaller wave rigidity, e.g., the melt–air interface, the pressure decreases to zero at the reflecting surface. In this case, the melt–air interface is in the antinode of displacement and the node of pressure. As a result, one can observe ripples and even eruptions or fountains at the free surface of liquid metal sonicated from the bottom. Parameters of the eruptions depend considerably on the liquid (melt) density and surface tension.

In practice, any kind of sonication, whether applied to a liquid metal in a crucible (furnace, launder) or to a solidifying melt in a mold, produces a combination of standing and traveling waves. Real interfaces are neither perfectly rigid nor ideally soft media. Therefore, a portion of the ultrasonic energy always passes through the interface into the adjacent media, thus decreasing the amplitude of the reflected wave below the level of the incident amplitude. As a result, a "nonideal" distribution of wave energy (amplitude) is established over the volume, with no point at which the amplitude vanishes.

Considering actual commercial metallic melts containing suspensions of insoluble nonmetallic inclusions and gaseous bubbles, we may assume that the ultrasonic field is diffuse. In other words, this field is characterized by spatially random and variable distributions of pressure and oscillatory displacements. The cavitation processes are an additional point in favor of the diffusion model for the sound field in the melt.

2.4 ACOUSTIC CAVITATION IN LIQUID METALS

Among the physical effects accompanying the propagation of power ultrasonic oscillations in melts, cavitation, or the formation of cavities filled mainly with gases dissolved in liquid metal, seems to be the most important. Formed by the tensile stresses characteristic of the half-period of rarefaction, these cavities continue to grow by inertia until they collapse (implode) under the action of compressing stresses during the compression half-period, thus producing high-intensity pressure pulses, temperature spikes, and high-velocity jets in the liquid [2–4, 42].

In real melts subjected to ultrasonic processing, cavities are produced at the weakest points of the melt during intervals of reduced pressure, and these collapse during intervals of increased pressure. Ultrasonic cavitation induces many physical and chemical processes that strongly affect the melt. Within the scope of our interest, these are the processes related to the melt degassing, filtration, wetting of solid inclusions, and structure modification.

In addition to the cavitation proper, the development of acoustic cavitation results in ultrasound energy adsorption (heating of the melt) and in the formation of acoustic streaming and secondary flows.

2.4.1 CAVITATION STRENGTH OF LIQUIDS

An intense alternating pressure applied to a liquid, for example by power ultrasound, results in disruption (fracture) of the liquid. The sound pressure must exceed a certain level in order to initiate cavitation in liquid. This cavitation threshold is a measure of the cavitation capability of liquids.

According to the kinetic theory of fluids, the strength of absolutely pure liquids is determined by the forces of molecular bonding, and therefore extremely high

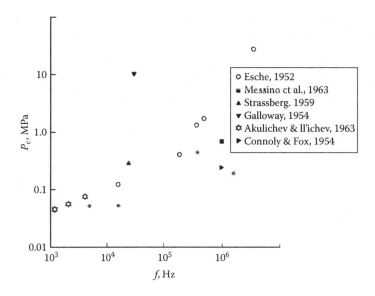

FIGURE 2.5 Cavitation threshold of water versus ultrasonic frequency: distilled water (filled symbols), tap water (open circles), and air-supersaturated water (stars).

cavitation thresholds corresponding to tensile stresses on the order of GPa magnitude are expected. As an example, the fracture strength of pure water is given by [2]:

$$P_c = \frac{2\sigma}{R_{im}} \tag{2.31}$$

where σ is the surface tension and R_{im} is the intermolecular distance. For water with $\sigma = 0.075$ N/m and $R_{im} = 20$ nm, this formula gives a fracture strength of about 1 GPa. This value decreases by an order of magnitude to about 100 MPa if we assume that vapor bubbles occur spontaneously, due to thermal fluctuations. This situation is described as [43]:

$$P_c = P_v - 44\sqrt{\frac{\sigma^2}{T}}, \tag{2.32}$$

where P_v is the vapor pressure in the bubble and T is the absolute temperature.

Even with this assumption, the calculated fracture strength remains much higher than the actual cavitation threshold obtained experimentally for water at various frequencies. Figure 2.5 shows that at frequencies up to 1.0 MHz, the distilled, tap, and air-supersaturated water has a cavitation threshold that seldom exceeds 10 MPa, which is three to four orders of magnitude below the calculated value [2].

According to well-adopted views on the cavitation threshold, the tensile-stress-induced disruptions in liquids are not governed by molecular forces, but rather by the presence of cavitation nuclei such as vapor and gas bubbles, solid gas-adsorbing suspensions, and hydrophobic inclusions. The formation of cavitation nuclei can be stimulated by disturbing the thermodynamic equilibrium with the surrounding medium, by fluctuations in the liquid phase, and by external radiation.

The cavitation strength is related (as shown by Equations 2.31 and 2.32) to the surface tension at the liquid–gas interface and the initial bubble radius. The viscosity μ also markedly influences the cavitation response of the liquid, increasing the cavitation threshold and the critical resonance radius of a cavitation bubble. The cavitation threshold or critical pressure is directly proportional to $\ln(\mu)$ [44] or to μ [45]. One can also relate the critical radius to the surface tension σ and viscosity μ of the liquid phase as

$$R_{cr} = \left(\frac{3\kappa-1}{3\kappa}\right)\frac{\sigma}{P_0}\left[-1+\sqrt{1+\frac{24\kappa\mu^2 P_0}{(3\kappa-1)^2\sigma^2\rho}}\right],\qquad(2.33)$$

where κ is the polytropic exponent varying from 1 to c_p/c_v [46], P_0 is the initial gas pressure, and ρ is the liquid density [47].

Neglecting the surface tension, the critical radius depends directly on the liquid viscosity [47]:

$$R_{cr} = \frac{2\sqrt{2\mu}}{\sqrt{3\kappa P_0\rho}}.\qquad(2.34)$$

Figure 2.6 illustrates the effects of surface tension and viscosity of water solutions of glycerin on the cavitation threshold at different ultrasound frequencies [48].

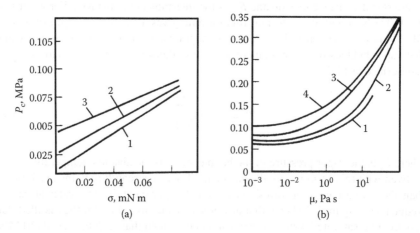

FIGURE 2.6 Cavitation threshold of water in dependence on: (a) surface tension and (b) viscosity at (1) 8 kHz, (2) 28 kHz, (3) 35 kHz, and (4) 61 kHz. (After Il'ichev [48].) (From previous edition.)

As the cavitation strength of real liquids is controlled by weak points in the form of inclusions (gas or solid), one should consider basic models for cavitation nuclei in order to choose an appropriate model applicable to molten metals of metallurgical purity. Soluble impurities can only affect the cavitation process indirectly, through variations in the surface tension of the liquid and changes in wettability of solid phases and solubility of gases. For example, soluble elements like Mg, Na, and Bi in Al can decrease the surface tension of liquid aluminum by as much as 30%, affecting the wettability of solid inclusions and increasing the cavitation threshold.

The energy of a bubble with curvature radius R on a surface contains a positive surface-energy term with liquid surface tension σ and a geometrical function $g(\theta)$ $= 2 + 3 \cos\theta - \cos^2\theta$ that depends on the contact angle θ (the smaller θ, the better wettability) and a volume term $E_v = -\pi/3R^3 g(\theta)P$ [49]. The pressure $P = P_A + P_g + P_v$ is composed of the acoustic pressure P_A (which is a function of ultrasonic processing parameters), the partial pressure P_g of a gas, and the vapor pressure of the liquid (P_v). Compared to P_A, other contributions may be negligible. The nucleation energy barrier is given by:

$$\Delta E_v = \frac{4\pi}{3} \frac{\sigma^3}{P^2} g(\theta). \qquad (2.35)$$

It can be easily shown and experimentally demonstrated that the nucleation of a cavitation bubble will not require much of nucleation energy in the case of hydrophobic surfaces (large θ, high surface tension) [49]. What could be the source of such hydrophobic interfaces in the liquid?

Several models of cavitation nuclei have been suggested [3]: electrostatic stabilization of free gas bubbles; adsorption of solute atoms on gas bubbles with their stabilization; nucleation of cavities by cosmic rays or other external radiation; and gas adsorbed on the surface of poorly wetted particles. The last model seems to be most applicable to metallic melts.

The real melt always contains a suspension of small particles that are not well wetted by the liquid phase, e.g., oxides, carbides, nitrides, etc. In addition, these particles have adsorbed gas on their surface and in the crevices of their, usually uneven, surface. Frenkel [50] suggested theoretically and Harvey et al. [51, 52] showed experimentally that very small gas bubbles can be stable inside the cracks of solid suspension particles, as shown in Figure 2.7. If the contact angle θ is smaller than 90°, then the solid phase is wettable by the liquid, and the gas pressure inside the crevice is larger by $2\sigma/R$ than the equilibrium gas pressure in the liquid phase. In this case, the gas bubble will be unstable and dissolve in the liquid. Figure 2.7 shows the other case, when the contact angle is larger than 90° and the liquid does not wet the solid particle. In this case, the gas pressure in the crevice is smaller than the equilibrium gas pressure in the liquid and the gas bubble is stable. Liquid aluminum contains both suspended particles, mostly alumina, and dissolved hydrogen. The equilibrium size of hydrogen bubbles in the aluminum melt under normal atmospheric pressure and a typical hydrogen concentration of 0.1–0.4 cm³/100 g varies between 14 and 30

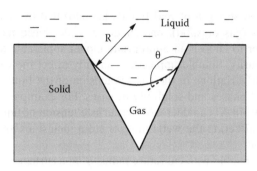

FIGURE 2.7 Model of a cavitation nucleus in the form of a nonwettable (hydrophobic) solid particle. (After Frenkel [50] and Harvey et al. [51, 52].)

μm [53]. This means that the crevices of smaller size at the surface of alumina particles will be necessarily filled with gaseous hydrogen that represents readily available cavitation nuclei. These nuclei will start to develop when the acoustic pressure in the liquid becomes negative (tensile).

The cavitation strength is, therefore, significantly affected by the concentration of dissolved gas and solid not-wetted particles with gas adsorbed at their surface. Both types of impurities ease the cavitation inception and decrease the cavitation threshold, as demonstrated in Figure 2.8 for water saturated with air and containing 20–30-μm solid, poorly wettable particles of aluminum oxide [48]. It is clear that solid impurities considerably vary the cavitation threshold only in the gas-free water (curve 1). With increasing gas concentration in water, the effect of the impurities becomes progressively negligible (curves 2–4).

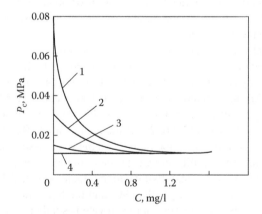

FIGURE 2.8 Cavitation threshold of water in dependence on solute gas concentration and solid alumina concentration C. Curves 1, 2, and 3 correspond to distilled water with gas contents of 16%, 20%, and 82%, respectively; and curve 4 corresponds to tap water with 86% dissolved gas. (After Il'ichev [48].)

2.4.2 CAVITY DYNAMICS

The behavior of cavitation bubbles has been extensively studied and is covered by many literature references [e.g., 2, 3, 46]. In this section, we will give only some essential information about the evolution of cavitating bubbles.

The dynamic behavior of a single vapor-gas cavity in an incompressible liquid is described (neglecting gas diffusion to the cavity) by the Nolting–Neppiras equation [54]:

$$\rho\left(\ddot{R}R+\frac{3}{2}\dot{R}^2\right)+4\mu\frac{\dot{R}}{R}+\frac{2\sigma}{R}-\left(P_0-P_v+\frac{2\sigma}{R_0}\right)\left(\frac{R_0}{R}\right)^3+P_0-P_v-P_A\sin(\omega t)=0$$

(2.36)

Here, R is the radius of the cavity, R_0 is the initial radius of the cavity, σ is the surface tension of the melt, μ is the viscosity of the melt, ρ is the melt density, P_v is the vapor pressure, P_A is the sound pressure, P_0 is the static pressure, and $\omega = 2\pi f$ is the angular frequency.

For correlation analysis, the dimensionless variables $\tau = \omega t$ and $r = R/R_0$ can be used to obtain the set of equations [55]:

$$\frac{du}{d\tau}=\frac{3v^2}{2r}-k_1\frac{v}{r^2}+k_2r^{-4}-\frac{k_3}{r^2}+\frac{AP_0\sin\tau-P_0}{k_4r}$$

(2.37)

$$\frac{dr}{d\tau}=v,$$

where

$$k_1=\frac{4\mu}{\rho\omega R_0^2};\ k_2=\frac{\left(P_0+\frac{2\sigma}{R_0}\right)}{\rho(\omega R_0)^2};\ k_3=\frac{2\sigma}{P_0R_0(\omega R_0)^2};\ k_4=\rho(\omega R_0)^2;\ \text{and}\ P_A=AP_0$$

This set of equations was solved with the initial conditions $v_0 = 0$ and $r_0 = 1$. Calculations were carried out for an aluminum melt at a temperature of 700°C and for water at 20°C. Table 2.2 summarizes the material properties used in these calculations.

TABLE 2.2

Material Properties of Liquid Aluminum and Water

Property	Aluminum (700°C)	Water (20°C)
Density (ρ), kg/m³	2350	1000
Surface tension (σ), N/m	0.860	0.079
Viscosity (μ), mPa·s	1.0	1.0
Pressure (P_0), MPa	0.1	0.1
Frequency (f), kHz	18	18
Vapor pressure (P_v), kPa	0	2.2

We investigated cavities with the initial radii R_0 varying between 1 μm and 100 μm, the minimum possible initial radius, i.e., critical radius R_{cr}, being determined from the stability condition

$$P_g + P_v - P_0 = 2\sigma/R, \tag{2.38}$$

where P_0 is the static ambient pressure and $(P_g + P_v)$ is the total pressure inside the bubble. This critical radius can also be described as [6]:

$$R_{cr} = \sqrt{\frac{9kT_b \mathbf{R} m_g}{8\pi\sigma}}, \tag{2.39}$$

where k is the Boltzmann constant, $\mathbf{R}$ is the gas constant, m_g is the mass of gas inside the bubble, and T_b is the temperature of the bubble. Bubbles smaller than the critical radius will be stable. The maximum initial radius (resonance radius) R_r has been defined from the Minnaert resonance condition [56]:

$$f = \frac{1}{2\pi R_r} \sqrt{\frac{3\kappa}{\rho}\left(P_0 + \frac{3\sigma}{R_r}\right)}. \tag{2.40}$$

Figures 2.9–2.11 show the relative radius R/R_0 and gas pressure in the bubble P_g calculated for various acoustic pressures. The families of curves differ in the initial radius and the intensity of the external sound field, and illustrate how the cavity evolves during 1–3 periods T_v of the sound wave.

From the curves in Figures 2.9 and 2.10, it follows that if the sound pressure is small enough ($P_A < P_C$, where $P_C = 0.6$ MPa is the cavitation threshold), the cavities pulsate and do not collapse during this time. The pressure in gaseous bubbles varies very little. As the sound pressure P_A increases to values above 1 MPa and exceeds P_C, the majority of cavities with $R_0 > R_{cr}$ behave like typical cavitation bubbles, collapsing at the end of the first or second period of oscillations. With a further increase in sound pressure, i.e., for $P_A \gg P_C$, cavitation becomes developed, and all cavities expand during one or two periods of the ultrasound wave and then collapse.

When a bubble expands tenfold or a hundredfold beyond its initial dimension, the pressure inside the bubble drops significantly, e.g., for $P_A > 2$ MPa, the pressure in the cavity falls to 100–133 Pa. The higher the acoustic pressure, the faster the cavities collapse.

This analysis is done with an assumption of spherical bubbles, which holds only for the first cycle of oscillations. In reality, the curved interface between the denser liquid and the less dense gas inside the bubble strongly accelerates inward, especially during the last stages of collapse [42]. This results in the distortion of an initially plane interface with the formation of kinks and folds. This phenomenon, as well as the collapse of the bubbles, resulting in the formation of a cloud of new, much smaller bubbles, has been observed experimentally [2, 42, 57]. The shape instability is counteracted by the smoothing effect of surface tension and energy dissipation by viscosity [42].

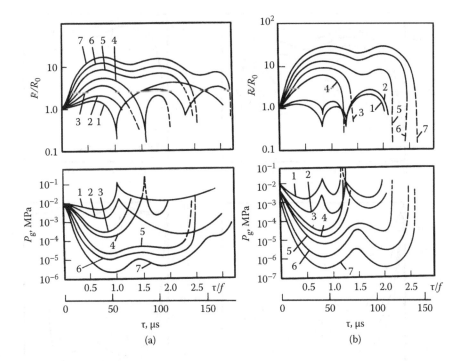

FIGURE 2.9 Evolution of cavities with initial radii (a) $R_0 = 100$ μm and (b) $R_0 = 50$ μm in aluminum melt for acoustic pressures P_A: (1) 0.1, (2) 0.2, (3) 0.6, (4) 1.0, (5) 2.0, (6) 3.0, and (7) 10.0 MPa.

The given results for bubble dynamics did not take into account the diffusion of gas dissolved in the liquid into the cavity. Allowing for this diffusion would increase the survival chances of the bubble due to gas diffusion in melts with low saturated vapor pressure or due to vaporization from the bubble walls in liquids with high saturated vapor pressure.

The effect of surface tension on the size and pressure in cavitation bubbles is illustrated in Figure 2.11 by comparison of liquid aluminum (σ = 0.860 N/m), liquid magnesium (σ = 0.50 N/m), and water (σ = 0.079 N/m) for $R_0 = 1$ μm and $P_A = 10$ MPa. Despite the significant difference in properties (the surface tension of molten aluminum is more than ten times that of water), the expansion of the bubble creates conditions (pressure drop) sufficient for the one-way gas diffusion from the liquid to the cavity in all considered systems.

The gas diffusion through bubble walls in the liquid during sonication is determined by the following phenomena. Firstly, in the rarefaction (expansion) phase, the surface area of a pulsating bubble is many times greater than its area in the compression phase; therefore, the gas diffusion influx during the rarefaction phase exceeds the gas outflux during the compression phase. Secondly, the diffusion is controlled by the diffusion barrier layer around the bubble. During the compression phase, its thickness increases and the concentration gradient decreases. Conversely, the layer

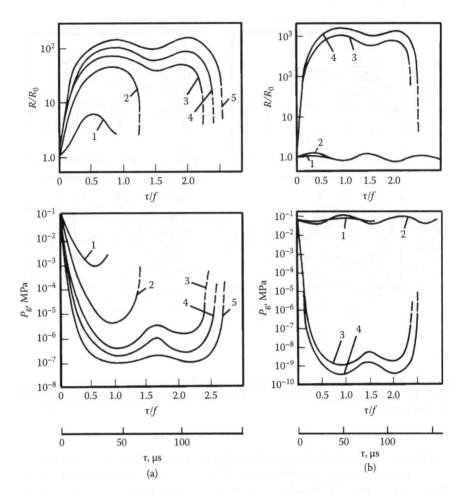

FIGURE 2.10 Evolution of cavities with initial radii (a) $R_0 = 10\ \mu m$ and (b) $R_0 = 1\ \mu m$ in the aluminum melt for acoustic pressures P_A: (1) 0.2, (2) 1.0, (3) 3.0, (4) 5.0, and (5) 10.0 MPa.

becomes thinner and the concentration gradient increases during the rarefaction phase. This additionally stipulates the gas influx into the bubble. As a result, the pulsating cavitation bubbles grow due to the one-way or "rectified" diffusion of gas from the liquid phase to the bubble.

The model of bubble evolution allowing for gas diffusion is a rather complex set of equations. The Nolting–Neppiras equation may also provide a solution of practical significance if one introduces time-dependent pressure $P(t)$ of gas in the bubble that replaces pressure-related terms in Equation (2.36), except for the sound pressure [58]:

$$\rho\left(R\ddot{R} + \frac{3}{2}\dot{R}^2\right) + 4\mu\frac{\dot{R}}{R} + \frac{2\sigma}{R} - P(t) - P_A\sin(\omega t) + P_0 = 0 \qquad (2.41)$$

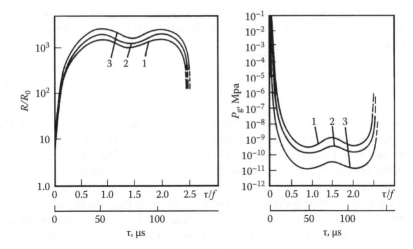

FIGURE 2.11 Effect of surface tension on the evolution of a cavity with initial radius $R_0 = 1$ μm at $P_A = 10$ MPa in (1) aluminum melt ($\sigma = 0.86$ N/m), (2) magnesium melt ($\sigma = 0.50$ N/m), and (3) water ($\sigma = 0.079$ N/m).

Obviously, the pressure $P(t)$ will not be a simple analytic function of radius if diffusion is taken into account.

The variation of gas mass in the bubble $m_g = (4/3)\pi R^3 M_g N_g$, where M_g is the molecular mass and N_g is the number of gas molecules, obeys the equation

$$\frac{4}{3}\pi \frac{d\left(N_g R^3\right)}{dt} = 4\pi R^2 i(t) \tag{2.42}$$

where $i(t)$ is the instantaneous gas flux through the bubble surface.

If we assume that the gas in the bubble is ideal, we can relate the gas pressure P to the temperature T and the number of gas molecules in the bubble N_g using $P = N_g kT$, where k is the Boltzmann constant. If we also assume that the temperature of the melt is kept constant during the cavitation (isothermal conditions), Equation (2.42) becomes

$$\frac{R}{3kT}\frac{dP}{dt} + \frac{P}{kT}\frac{dR}{dt} = i(t). \tag{2.43}$$

The gas flow to the bubble may be controlled by various processes, but most commonly by diffusion. The diffusion to the pulsating bubble can be described by the convective diffusion equation that takes the following form (assuming spherical symmetry) [59]:

$$\frac{\partial C}{\partial t} + v_r \frac{\partial C}{\partial r} = D\left(\frac{\partial^2 C}{\partial r^2} + \frac{2}{r}\frac{\partial C}{\partial r}\right), \tag{2.44}$$

where C is the gas concentration in the melt, r is the radial coordinate originated in the bubble center, v_r is the radial velocity of the liquid, and D is the diffusion coefficient.

The gas flux $i(t)$ is given by

$$i(t) = D\frac{\partial C}{\partial r}\bigg|_{r=R(t)}. \tag{2.45}$$

Thus, in order to describe the bubble evolution with gas diffused into the bubble, one should simultaneously solve Equations (2.41), (2.43), and (2.44) with the gas flux determined from Equation (2.45) and the radial velocity v_r dependent on the bubble expansion velocity $\dot{R}$. The solution of this set of equations is cumbersome even by numerical methods. The task can be simplified by using the relation for the gas flux to the cavitating bubble proposed by Boguslavsky [59]. Boguslavsky assumed that the convective gas diffusion occurs in a very thin liquid layer next to the bubble and that the bubble surface moves at a constant velocity. Under these conditions, the layer can be considered plane, and Equation (2.44) be written in a simpler form:

$$\frac{\partial C}{\partial t} - 2\frac{y}{t}\frac{\partial C}{\partial y} = D\frac{\partial^2 C}{\partial y^2}, \tag{2.46}$$

where y is the diffusion layer thickness $(r = R(t) + y)$.

Introducing new variables $\xi = t^2 y$ and $\tau = 1/5\ Dt^5$, the boundary conditions will be: $C = C_0$ at $\tau = 0$, $C = 0$ at $\xi = 0$ $(\tau > 0)$, and $C = C_0$ at $\xi \to \infty$ $(\tau \geq 0)$. Equation (2.46) can be solved analytically, which yields the gas flux through the bubble surface as

$$i(t) = D\frac{\partial C}{\partial y}\bigg|_{y=0} = C_0\sqrt{\frac{5D}{\pi t}}, \tag{2.47}$$

where C_0 is the equilibrium gas concentration (g/cm^3) independent of the applied tensile stress.

The total gas flow into the bubble in each time instant will be

$$I = \frac{8}{3}C_0\sqrt{5\pi Dt}\,\frac{z_0}{\rho_l}t^2, \tag{2.48}$$

where z_0 is the constant tensile stress that can be taken as $0.8P_A$, and ρ_g and ρ_l are the densities of gas and liquid, respectively.

By taking this into account and introducing the surface area of the bubble as an unknown $S = 4\pi R^2$, the set of equations describing the bubble dynamics, allowing for gas diffusion in the bubble, can be solved numerically [58].

$$\frac{\partial^2 S}{\partial t^2} = -\frac{1}{4S}\left(\frac{\partial S}{\partial t}\right)^2 + \frac{8\pi}{\rho}\left\{[P(t) - P_0 + P_A\sin(\omega t)] - \frac{4\sigma\sqrt{\pi}}{\sqrt{S}} - \frac{2\mu}{S}\frac{\partial S}{\partial t}\right\}$$

$$\frac{\sqrt{S}}{6kT\sqrt{\pi}}\frac{\partial P}{\partial t} + \frac{P}{4kT\sqrt{\pi S}}\frac{\partial S}{\partial t} = i(t) \tag{2.49}$$

This system yields a numerical solution for initial values $S(0) = S_0$, $P(0) = P_0$, and $\dot{S}(0) = 0$.

Some other approaches and solutions can be found in a review by Plesset and Prosperetti [46] and in the works of Fyrillas and Szeri [60] and Crum [61]. For example, the time evolution of a spherical bubble radius can be presented as [61]:

$$\frac{dR_0}{dt} = \frac{DRTC_0}{R_0 P_0}\left\{\frac{R}{R_0}\left(1+\frac{4\sigma}{3P_0R_0}\right)^{-1}\left(\frac{C_\infty}{C_0}-\frac{\left\langle\left(\frac{R}{R_0}\right)^4\left(\frac{P_g}{P_0}\right)\right\rangle}{\left\langle\left(\frac{R}{R_0}\right)^4\right\rangle}\right)\right\}, \qquad (2.50)$$

where D is the diffusion coefficient of gas, C_∞ is the concentration of dissolved gas far away from the bubble, C_0 is the saturation (equilibrium) concentration of gas, and the pointed braces imply time averaging.

Let us look at the results of numerical modeling performed for an aluminum melt with hydrogen concentration $C_0 = 0.2$ cm³/100 g and diffusion coefficient $D = 1$ cm²/s using the set of Equations (2.49) [58]. Other parameters were the same as in the simulations shown in Figures 2.9–2.11. The initial cavity radii R_0 were chosen to reflect the formation of 1–100-μm hydrogen lenses on 0.3–1.5-μm not-wettable alumina particles [62], which act as cavitation nuclei.

Figures 2.12 and 2.13 demonstrate the variation of relative radius R/R_0 and gas pressure P_g in cavitation bubbles according to Equations (2.36) and (2.49), respectively, with and without taking the gas diffusion into account.

At acoustic pressure $P_A \ll 1$ MPa, diffusion introduces only small variations in the bubble evolution, whereas for $P_A > 1$ MPa, it contributes significantly to the gas pressure in the bubble. Let us trace this process for a bubble with the initial radius $R_0 = 10$ μm. At $P_A = 0.2$ MPa, the bubble pulsates in a nonlinear mode. Gas diffusion to the bubble increases the gas pressure by about two orders of magnitude, with the relative radius being virtually unaffected (Figure 2.12a III). With increasing sound pressure to 1 MPa, gas diffusion drastically increases the gas pressure in the bubble by about four orders of magnitude (Figure 2.12b III). The gas pressure further increases at $P_A = 10$ MPa (Figure 2.13b II). A reduction of the bubble size to $R_0 = 1$ μm (Figure 2.13b III) also increases the effect.

It is possible to estimate the mass of gas (hydrogen) transferred from liquid aluminum to the cavitating bubble. The following expression can be used:

$$\Delta M = \frac{4\pi M_g}{3kT}\int_0^t R^3(P - P_1)dt \qquad (2.51)$$

with graphical integration of the curves like those in Figures 2.12 and 2.13. Here, P is the gas pressure with diffusion, P_1 is the gas pressure without diffusion, M_g is the molecular mass of the gas, and R is the average bubble radius during a rarefaction stage. The results of this calculation are shown in Table 2.3.

The behavior of a cavitating bubble and the sound-pressure-dependent rectified hydrogen diffusion into the cavity suggest that the cavitation threshold P_c of liquid

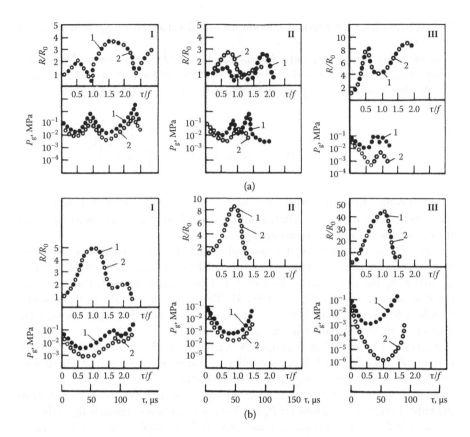

FIGURE 2.12   Effect of hydrogen diffusion on cavitation bubble dynamics (relative radius R/R_0 and pressure P in cavitation bubbles): (a) without cavitation ($P_A = 0.2$ MPa) and (b) at incipient cavitation ($P_A = 1.0$ MPa) for: (I) $R_0 = 100$ µm, (II) $R_0 = 50$ µm, and (III) $R_0 = 100$ µm—(1) with gas diffusion to the bubble and (2) without diffusion.

aluminum and magnesium is around 0.65–1.3 MPa at 18 kHz. This conclusion immediately follows from the given data: for $P_A > P_c$, the hydrogen mass in bubbles with $R_0 = 1$ µm increases by two orders of magnitude.

A correlation between the calculated data and actual tensile strength of a liquid metal can be established by measuring the cavitation threshold in aluminum and its alloys, as we will show later in this chapter.

2.4.3   Cavitation Region in Liquids

Cavitation in real liquids (also in melts) starts even at relatively low acoustic pressures forming a cavitation region or zone, typically close to the vibrating source. To characterize the evolution of cavitation, Rozenberg introduced the cavitation index

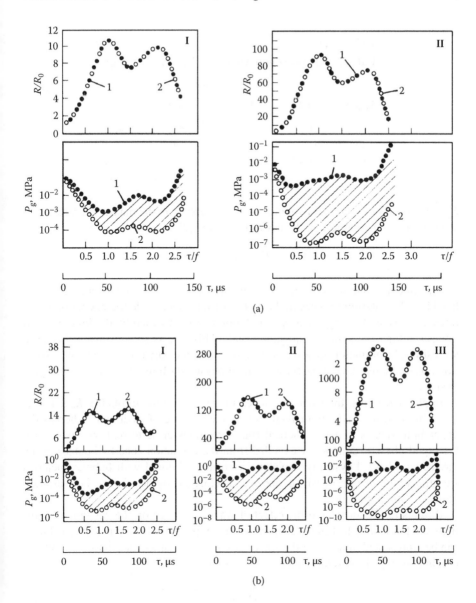

FIGURE 2.13 Effect of hydrogen diffusion on cavitation bubble dynamics (relative radius R/R_0 and pressure P in cavitation bubbles) at sound pressures corresponding to developed cavitation: (a) 5 MPa and (b) 10 MPa for: (I) $R_0 = 100$ μm, (II) $R_0 = 10$ μm, and (III) $R_0 = 1$ μm—(1) with gas diffusion to the bubble and (2) without diffusion.

TABLE 2.3

Calculated Hydrogen Mass in a Bubble (10^{-12} g)

R_0, μm	P_A, MPa			
	0.2	1.0	5.0	10.0
1	4.0	4.6	300	2000
10	0.4	4.6	300	1000
100	0.0004	0.7	200	650

K_c as the ratio of the integral volume ΔV of all bubbles in the maximum rarefaction stage to the total volume V of the cavitation region [2]:

$$K_c = \Delta V/V. \tag{2.52}$$

The integral bubble volume ΔV characterizes the potential energy accumulated by all cavitation bubbles in their rarefaction stage, while the cavitation index reflects the spatial energy density. Cavitation is accompanied by the loss of sound intensity along the sound-propagation path. This loss is so significant that even after several centimeters of wave propagation, the intensity drops below the cavitation threshold and cavitation ceases, which physically limits the cavitation region in space.

In physical terms, the cavitation index ranges within the limits $0 \leq K_c \leq 1$. The lower limit corresponds to a trivial case of the absence of cavitation, whereas the upper limit relates to a hypothetical (but sometimes close to reality) mode when the liquid phase is completely expelled from the cavitation zone by bubbles in the rarefaction phase. Physically, replacing a portion of liquid with bubbles means a variation of the average density ρ_c and sound velocity c_c of the mixture, i.e., variation of its acoustic impedance.

Rozenberg [2] considered the effect of wave resistance on the cavitation conditions and found that the time-averaged acoustic impedance $\overline{\rho_c c_c}$ of a cavitating liquid can be represented in terms of the wave resistance $\rho_0 c_0$ of a noncavitating liquid and the time-averaged cavitation index $\overline{K_c} \cong 0.1 K_c$ as

$$\overline{\rho_c c_c} = \rho_0 c_0 \sqrt{\frac{(1-\overline{K_c})+\overline{K_c}\rho_v/\rho_0}{(1-K_c)+\overline{K_c}\beta_v/\beta_0}}, \tag{2.53}$$

where ρ_v and β_v are the density and compressibility of the vapor–gas mixture in the bubble, and β_0 is the compressibility of the liquid. In water, $\beta_v/\beta_0 = 10^4$, $\rho_v/\rho_0 \ll 1$, and $K \ll 1$; then we can write

$$\overline{\rho_c c_c} = \rho_0 c_0 \sqrt{\frac{1}{1+\overline{K_c}\beta_v/\beta_0}} \tag{2.54}$$

For $0 < K \leq 0.5$, we can easily calculate the decrease in acoustic impedance with the increase in the cavitation index:

K_c	0.001	0.05	0.1	0.5
$\overline{\rho_c c_c} \big/ \rho_0 c_0$	0.71	0.14	0.10	0.04

The experimental measurements give values for the drop of acoustic impedance as 0.3 [2] and 0.25 [33] for water, 0.25 for liquid tin, and 0.1 for a liquid solder alloy [63].

Rozenberg [2] showed that any cavitating liquid behaves as a nonlinear medium, such that its average acoustic impedance is proportional to the ultrasonic power W_a transferred to the liquid and the angular velocity amplitude $(\omega A)^2$ at the radiating face of area S (see also Equation 2.19):

$$\overline{\rho_c c_c} = \frac{2W_a}{S(\omega A)^2} \quad (2.55)$$

Therefore, if we know the transferred ultrasonic power, the oscillation velocity ($\omega = 2\pi f$), and the area of the radiating face, we can estimate $\overline{\rho_c c_c}$.

The acoustic power W_a transferred to the melt was shown in Figure 2.2 as a function of the radiating-face amplitude A for three different radiator diameters at 18 kHz. These data were obtained by melt calorimetry (see Section 2.6, Table 2.5).

The size and geometry of the cavitation zone is not a very well-studied subject. Empirical observations show that the cavitation in a volume with dimensions commensurable with the wavelength of the sound originates on the solid/liquid interfaces (radiating face of the sonotrode, walls) as well as inside the melt volume, forming a concentrated region close to the ultrasound source and complicated, changing in time configurations at a distance (Figure 2.14). These configurations gradually transform to streams, jets, and flows.

The rule of thumb says that the average dimensions of the cavitation zone are on the same scale as the diameter of the sonotrode. A rough estimate of the dimensions of the cavitation zone can be obtained by observing cavitation erosion of a thin foil placed under the sonotrode (Figure 2.14b), or by measuring the loss of mass of special samples immersed into the liquid. When cavitation is established, the cavitation region has a volume with the cross section ranging approximately from $\lambda/4$ to $\lambda/2$. For example, this size is 20–40 mm for water and 50–100 mm for aluminum melts. When studying cavitation in molten aluminum and its alloys, a 20–500-μm thick foil of high-melting metals such as Ti, Nb, Mo, and W can be used. Foil samples are fixed in rigid frames and placed into the melt below the sonotrode. In the case of still melt, the foil can remain in the molten aluminum for an hour without even being wet. When sonication begins, a stable and strong diffusion layer is formed on the foil surface in 1–3 min, followed within 5–15 min by characteristic punctures,

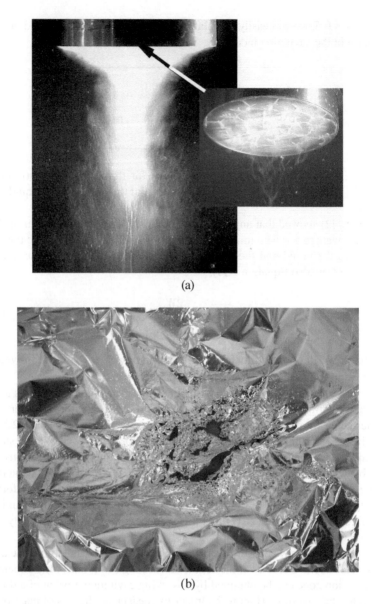

(a)

(b)

FIGURE 2.14 (a) Typical cavitation-zone configuration (adapted from Moussatov et al., 2003 [66]) and (b) simple cavitation test with a metallic foil.

which are evidence of cavitation erosion (Figure 2.14b). The degree of cavitation can be judged from the number of punctures in the foil. Our investigation showed that tungsten foil suffered more damage than molybdenum foil, and the latter undergoes cavitation erosion more readily than titanium and, especially, niobium foil. Similar cavitation sensitivity was observed in experiments on choosing material for the radiator [34, 64].

The real cavitation-zone pattern is a bit more complicated than can be estimated from foil erosion. Figure 2.14a illustrates a typical cavitation region generated by a cylindrical horn (the type that is frequently used in metallurgical applications). Evidently, there is a very densely cavitating zone close to the face of the sonotrode, which then develops into a conelike structure with loose boundaries. Dubus et al. [65] suggest that the thin and concentrated cavitation bubble layer close to the surface of the sonotrode grows in thickness with the amplitude of the source until it reaches the acoustic half wavelength (estimated to be on order of 1–2 mm at 20 kHz for a sound velocity in the bubbly medium of 60 m/s). As soon as the resonance thickness is reached, the pressure field inside and outside the layer dramatically increases, leading to the formation of the cone of bubbles. With further thickening of the dense cavitating layer, it goes out of the resonance conditions; the pressure and bubble density decrease; and the layer becomes thinner, going back to the resonance thickness. The process eventually stabilizes with the layer thickness oscillating around the resonance size. In addition, the dense bubble layer is not uniform across the surface of the radiating face of the sonotrode. It is thicker in the center, where the acoustic intensity is maximum, and thinner at the periphery of the sonotrode. As a result, the cavitation bubble layer forms a lenslike structure, focusing the acoustic field at a short distance.

Another interesting phenomenon that has consequences for the formation of the cavitation cone is the development of so-called streamers [66]. The insert in Figure 2.14a shows the netlike pattern at the surface of the sonotrode, reflecting the bubble trajectories before they leave the surface. The bubbles that are formed unevenly at the radiating surface self-organize in several streamers that propagate into the liquid volume. These streamers are not stable in time and space, and they tend to rotate around the axis of the sonotrode. Upon increasing the acoustic intensity, the streamers multiply and merge, forming the macroscopically stable cone structure. The experimental observations show that the cone structure is more pronounced with wide radiating faces, while narrow sonotrodes produce more turbulent and disturbed jetlike cavitation structure. Moussatov, Granger, and Dubus [66] relate the origin of cavitation cones to the formation of high-pressure repulsive zone[*] under the sonotrode, extended along its axis of symmetry. The bubbles cannot enter it and pass around it. With increasing distance under the sonotrode, the high-pressure zone becomes narrower and then disappears, transforming to the attractive zone. A cone of bubbles and then a narrow downward jet of bubbles are thus created. The entire cavitation cone is an active zone where the intensification of processes and liquid-phase treatment occur.

2.5 ACOUSTIC FLOWS

Sonication generates directed hydrodynamic flows in melts. These flows are represented by (1) acoustic streams that originate from the pressure wave caused by high-frequency vibration of the sonotrode and pulsation of the cavitation region and

---

[*] The sign of primary Bjerknes force is reversed at high sound amplitudes, and the high-pressure zone becomes repulsive.

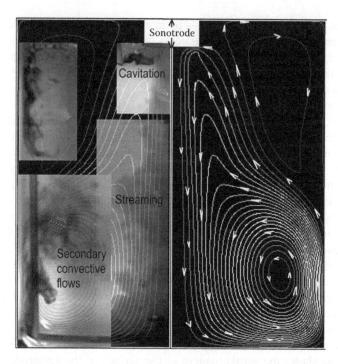

FIGURE 2.15 Typical flow pattern induced in a limited volume by an ultrasonic horn: (left) flow pattern visualized by dye injection; (right) streamlines obtained by computer modeling.

(2) secondary, forced convective flows. They occur both in the bulk of the liquid and near the walls, particles, and other objects within the volume subjected to the ultrasonic field. A typical flow pattern induced in a limited volume by an ultrasonic horn is shown in Figure 2.15.

The origin of streams relates to the momentum acquired by the liquid when it absorbs the wave. Therefore, the velocity of acoustic streams increases with the ultrasonic intensity and the sound absorption. Three types of acoustic streams are generally considered [2].

The first type is defined as flows in viscous boundary layers near the solid boundaries or at the phase interface. Flows of this type will be considered in more detail in Chapter 11, which is devoted to the zone refining of aluminum, where flows at the solidification front can be revealed and studied at ultrasonic intensities much below the cavitation threshold. A theory of streaming in boundary layers was developed by Schlichting, who stated that boundary layers in which turbulent boundary flows occur have a thickness of $\delta = (2 \, \nu/\omega)^{1/2}$, where ν is the kinematic viscosity and ω is the angular frequency of sound. Therefore, the characteristic scale of these boundary layers is small compared to the wavelength. The Schlichting flows can also be induced by oscillations of cavitation bubbles near a solid surface. They are significant in mass and heat transport, as well as faster than other types of flow of equal intensity.

The second type of acoustic streaming originates in the field of standing waves. The scale of this flow compares on the same scale to the acoustic wavelength and exceeds the thickness of the boundary layer. It is represented by vortex-type, turbulent movement.

Finally, the third type of acoustic streaming originates from the absorption of wave momentum of an inhomogeneous sound field in the bulk of the liquid. Here, the scale of mass transfer is larger than the wavelength, and this is determined by the geometry of the vessel where the processing occurs and by the viscosity of the liquid. This large-scale flow, also called Eckart flow, is relatively slow, i.e., its velocity is lower than the oscillation velocity on the ultrasound source.

Most of the experimental and modeling work on acoustic and induced convective flows has been done using water or transparent liquids. The properties of water and liquid aluminum are similar. Both liquids demonstrate Newtonian viscous behavior—the kinematic viscosity v of water at room temperature and liquid aluminum are of comparable magnitude ($v_{Al} = 0.5v_{H_2O}$)—and the minimum squared Reynolds number scales with a factor of 1.

Particle image velocimetry (PIV), laser Doppler anemometry (LDA), and high-speed filming of liquids with dye, tracers, or immiscible liquids have been widely used since the 1960s [2] and most intensely in the last fifteen years [67–70].

A combination of PIV and computer modeling nowadays allows the researcher to get accurate data on the velocity profile and magnitude outside the cavitation zone. Figures 2.16a–c demonstrate the velocity distribution in a water volume excited by an 18-kHz 4-kW magnetostrictive transducer with a Ti horn 20 mm in diameter. The linear flow velocity reaches approximately 1 m/s and then rapidly attenuates with the distance from the sonotrode tip. Secondary flow velocities are in the range of cm/s. These results agree well with earlier estimates of streaming velocities of 2–3 m/s close to the cavitation region [71]. The measurements of pressure in water ahead of a 20-mm horn working at 20-kHz frequency correspond with the velocity measurements and give the drop from 0.6 to 0.15 MPa over the distance of 40 mm from the horn surface [72]. Recent measurements of acoustic pressure in liquid aluminum give concurring results [73]. Figure 2.16d shows the results on the streaming velocity, recalculated from acoustic pressure. These data demonstrate very similar velocities to those measured and calculated in water, and also illustrate the dependence of the flow velocity on the ultrasound intensity (amplitude of vibrations).

The three-dimensional picture of the downstream flow is more complicated, as it includes radial and tangential components of the flow; in other words, the flow is not only directed downward, but also rotates around the vertical axis [68]. The direction of this rotational component may change with the distance from the radiating surface. The reason for the flow rotation may be the small asymmetry of the horn, a deviation from verticality, and the streamers that we have discussed in the previous section.

Abramov, Astashkin, and Stepanov [74] studied acoustic flows in an undercooled transparent analogue of melts, i.e., a 35% naphthalene–65% camphor eutectic alloy. The model material was chosen so as to satisfy its similarity in physical and chemical properties to metallic melts. According to Jackson and Hunt [75], $\alpha_s = \Delta S_m/R$ (where ΔS_m is the entropy of fusion and R is the gas constant) should be small for a good analogue. The tested material had $\alpha_s = 3.5$ and dynamic viscosity 1.5 mPa·s. The sonication was carried out at frequencies of 45–250 kHz in order to minimize the effects of cavitation. The observations were done using a high-speed camera and

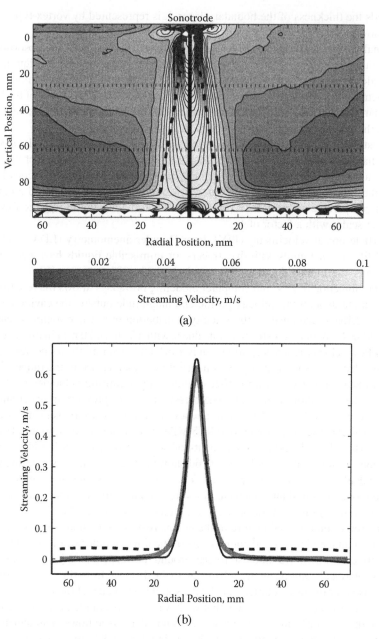

(a)

(b)

FIGURE 2.16 Acoustic stream velocity distribution: (a) velocity magnitude of the time-averaged flow; thin dashed horizontal lines in (a) represent downstream positions for velocity profiles in (b, c) are shown (sonotrode tip is located at the origin of the coordinate system) (adapted from Schenker et al. [70]); (d) experimentally measured velocities in liquid aluminum (null-to-peak amplitudes): (1) 12 μm; (2) 18 μm; (3) 24 μm (adapted from Ishiwata, Komarov, and Takeda [73]). (*continued*)

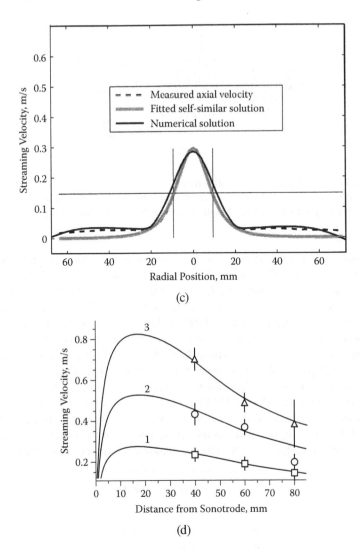

FIGURE 2.16 (*continued*) Acoustic stream velocity distribution: (a) velocity magnitude of the time-averaged flow; thin dashed horizontal lines in (a) represent downstream positions for velocity profiles in (b, c) are shown (sonotrode tip is located at the origin of the coordinate system) (adapted from Schenker et al. [70]); (d) experimentally measured velocities in liquid aluminum (null-to-peak amplitudes): (1) 12 μm; (2) 18 μm; (3) 24 μm) (adapted from Ishiwata, Komarov, and Takeda [73]).

using polarized light; in addition, the amplitude of the horn and the sound pressure in the liquid phase were measured.

Figure 2.17 shows schematically the streaming flows in the undercooled transparent melt. Within a few seconds of sonication with the amplitude above the cavitation threshold, solidification nuclei (crystals) start to form near cavitation bubbles. The analysis of the motion of bubbles and crystals allowed the observer to trace the

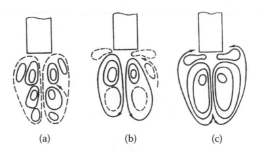

(a)   (b)   (c)

FIGURE 2.17   Development of acoustic streams in an undercooled ($\Delta T = 4°C$) transparent material: (a) 10 s, (b) 20 s, and (c) 30 s after onset of sonication at 139 kHz with amplitude of oscillation velocity = 5 cm/s. (After Abramov, Astashkin, and Stepanov [74].)

development of the acoustic flow. At the initial stage of sonication and when the number of formed crystals is small, flows with the scale $L \approx \lambda$ (wavelength) prevail (a), with larger flows with the scale $L = (2.5–3.0)\lambda$ appearing later. Upon continuation of sonication and with an increase in the solid-phase fraction, the small-scale flows are suppressed and give way to large-scale flows (b, c). Later, additional flows appear close to the radiating face (c). The flow development pattern remains the same in the range of exciting frequencies from 45 to 250 kHz. The superheating of the melt makes the development of acoustic flows more difficult, especially at low ultrasound intensity. This was related to the decreased viscosity of the melt [74].

For frequencies of 45 kHz and 139 kHz, the maximum velocity v_{max} of the flow, its relative scale L/λ, the acoustical Mach number Ma = v_{osc}/c (where v_{osc} is the oscillation velocity of the radiating face and c is the sound velocity in the liquid phase), and the ratio of the maximum flow velocity to the oscillation velocity were all estimated as [74]:

	45 kHz	139 kHz
v_{max}, cm/s	2.3	2.8
L/λ	1–2	2.5–5
Ma	3×10^{-4}	0.6×10^{-4}
v_{max}/v_{osc}	0.03	0.3

These results indicate that very high velocities are not characteristic of acoustic flows. The threshold for flow development almost coincides with the cavitation threshold. The acoustic streams do not interact with convective flows at low oscillation amplitudes, especially below the cavitation threshold. When the amplitude increases beyond the cavitation threshold, the acoustic flows start to interact with convective flows, and the finally established flows have velocities 5–10 times greater than those of natural convection. Abramov, Astashkin, and Stepanov [74] also showed that (a) the acoustic flows actively interact with a solidification front, especially when the transition zone is wide, and (b) the cavitation is accompanied by dispersion of growing solid crystals. This interaction equalizes the temperature field in the liquid volume.

These investigations show that the crucial factors determining the nature, velocity, and scale of streaming are the ultrasonic intensity and the melt temperature governing its viscosity.

2.6 EXPERIMENTAL STUDIES OF CAVITATION IN LIQUID METALS

Experimental studies of cavitation were mainly performed for water and other transparent liquids for obvious reasons of transparency, low temperature, and relative simplicity of the experimental techniques. For liquid metals, e.g., aluminum and magnesium alloys, the observations and especially quantitative measurements are difficult because of the opaque nature of metallic melts and due to high reactivity of melts with immersed solid tools and instruments, aggravated by high temperature and high-energy oscillations.

There are various ways to overcome these difficulties. The cavitation threshold can be determined using cavitation-noise monitoring. When cavitation begins, numerous bubbles of various sizes are formed and collapse in the melt. This adds to the main frequency of the acoustic signal, its harmonics, subharmonics, and incoherent noise. The beginning of the distortion of the main frequency signal can be taken as the onset of cavitation (see Figure 2.18, spectrograms). Figure 2.18 gives a principal diagram of a setup for studying cavitation in metallic melts. Major parameters of the process are controlled: frequency, amplitude and power of the transducer, melt temperature, and the acoustic pressure and noise generated by ultrasound.

The *acoustic power* transferred to the liquid metal can be measured directly, using calorimetry, and indirectly from the response of oscillating systems to loading. The oldest method used to measure ultrasonic intensity is calorimetry. If we know the temperature increment ΔT brought about by ultrasound in the liquid, then the acoustic intensity I in terms of power W_a injected by a radiating face of area S is determined as

$$I = \frac{W_a}{S} = \frac{c_1 m \Delta T}{tS}, \qquad (2.56)$$

where c_1 is the specific heat of the melt (J/kg K), m is the liquid-phase mass (kg), and t is the sonication time (s). Since all the ultrasonic oscillation energy spent on acoustic cavitation is eventually converted into heat, the temperature increment in this heating—other conditions remaining equal—may represent the acoustic power introduced into the melt. According to the measurement procedure [34], a melt specimen is prepared in a crucible inside an electrical-resistance furnace that can be kept at a constant temperature T_1 for a long time. Under these conditions, the furnace consumes a certain power W_1 that can be assessed by measuring the current and voltage. Ultrasonic treatment with oscillation amplitude A applied to the melt increases its temperature to T_2. When the ultrasonic treatment is completed, to keep the melt at the new temperature, the furnace is adjusted to a new power W_2 with new current and voltage values. The difference of powers $W_2 - W_1 = W_a$ gives the desired amount of acoustic power.

The results are given in Figure 2.2, and Table 2.4 shows the levels of acoustic power introduced into a melt of 99.7% pure aluminum using commercially produced

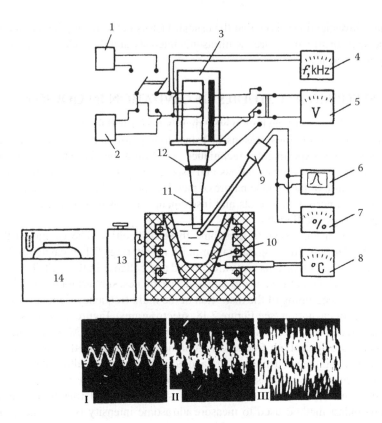

FIGURE 2.18 Setup for studying cavitation in liquid metals and typical spectrograms for main frequency signal (I) before cavitation, (II) at the cavitation threshold, and (III) at developed cavitation: (1) ultrasonic generator, (2) unit for pulse ultrasonic treatment, (3) magnetostrictive transducer, (4) frequency meter, (5) voltmeter for power assessment, (6) oscilloscope, (7) cavitometer, (8) thermocouple, (9) cavitometer high-temperature sensor, (10) crucible with melt, (11) ultrasonic sonotrode, (12) amplitude sensor, (13) furnace control, (14) hydrogen analyzer.

ultrasonic equipment (4-kW generator and 4-kW, 18-kHz magnetostrictive transducer) and Nb sonotrodes of different diameters. If the power consumed by the transducer is 4 kW, then the amplitude of sonotrode null-to-peak displacement exceeds 30 μm (20 mm diameter) or 20 μm (40 mm diameter), and the transducer converts up to 25% of consumed electrical power into the useful acoustic power transmitted into the melt.

It is instructive to note that the acoustic power measured by the heat-balance method may be viewed as the upper limit for the conversion in the ideal oscillatory system, as this power includes heat released into the furnace space and heat extracted by the cooling system. Real values of acoustic power conveyed into the melt can be obtained by correcting the data of Table 2.4 for these losses.

The direct method of measuring the acoustic power transmitted to the melt uses a measurement element that is part of the waveguiding system shown in Figure 2.19a [76].

TABLE 2.4

Acoustic Power Introduced in Commercially Pure Aluminum Melt

	1st Heating Regime			2nd Heating Regime				
A, μm	Voltage, V	Current, A	Power, W	Voltage, V	Current, A	Power, W	ΔT, °C	W_a, W
			Sonotrode 20 mm in Diameter					
5	168	4.25	714	168	4.3	722	2.5	8.4
12.5	160	4.3	680	168	4.5	756	11.8	68
26	154	4.5	697	186	4.75	883	42.0	260
38	165	3.9	643	265	6.5	1722	159	600
			Sonotrode 40 mm in Diameter					
15	65	8.5	552	141	9.2	1302	70	300
20	65	8.6	559	150	9.8	1560	250	1000

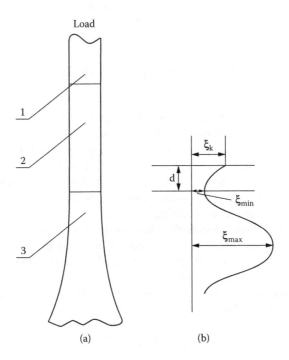

FIGURE 2.19 Measurement of acoustic power transmitted to the melt: (a) waveguide–sonotrode system with a measuring element: (1) sonotrode, (2) measuring element, (3) concentrator; and (b) diagram to determine displacement amplitude. (After Teumin [76].)

This method is convenient, as it enables one to estimate the acoustic flux transmitted to the load (e.g., melt) while taking into account the losses in the sonotrode and those due to the acoustic contact of the sonotrode with the melt. In other words, this method yields a real estimate of the power conveyed by the oscillating system from the transducer to the melt.

The measurement element, made of a low-acoustic-absorption material (titanium, aluminum, iron alloy with 6% silicon), is designed to sense the response of the load to variations of the displacement amplitude and is calculated to be a half-wavelength in size.

The acoustic power introduced into the melt may be defined as [76]:

$$W_a = \frac{1}{2}\omega^2 \xi_{max}\xi_{min}R_{load},\tag{2.57}$$

where ξ_{max} and ξ_{min} are the measured displacement amplitudes at the node and antinode of oscillations, respectively, and R_{load} is the active component of the load (Figure 2.19b).

In turn, R_{load} can be determined by the traveling-wave coefficient $k_\delta = \xi_{min}/\xi_{max}$, the wave resistance of the measurement element ρc, the resonance wavelength of the measurement elements λ, and the distance d from the end of the measurement element to the displacement node (Figure 2.19b):

$$R_{load} = \rho c \frac{k_\delta}{k_\delta^2 \cos^2\left(\frac{2\pi d}{\lambda}\right) + \sin^2\left(\frac{2\pi d}{\lambda}\right)}.\tag{2.58}$$

The *displacement amplitude* can be measured by a variety of methods, from direct observation of preliminary marks in a microscope to contactless measurements using capacitive, inductive, electrodynamic, or photonic sensors; interferometers; or lasers (see Section 13.4). The same methods can be used to measure the oscillation amplitude of the radiating face—the major characteristic that defines the conditions of ultrasonic treatment. However, the measurements are typically possible in air, without the melt loading. The correlation between thusly measured amplitude and the acoustic feedback voltage of the transducer makes it possible to estimate the actual amplitude of the loaded sonotrode by measuring the acoustic feedback. The measured acoustic power data can be related to the measured ultrasonic amplitudes as shown in Figure 2.2.

The sound pressure P_A before the onset of cavitation can be calculated from Equations (2.25) and (2.56) as

$$P_A = \sqrt{\frac{2W_a\rho_0 c_0}{S}},\tag{2.59}$$

where W_a is the measured acoustic power, $\rho_0 c_0$ is the acoustic impedance of the precavitation liquid (melt), and S is the sonotrode face area. The calculated acoustic power versus the sound-pressure amplitude in commercial aluminum melt before cavitation is given here:

W_a, W	1.5	3.3	5.0	6.5	7.7	8.4
P_A, MPa	0.37	0.55	0.68	0.77	0.88	0.96

It is obviously not easy to experimentally measure the sound pressure in liquid aluminum. A relatively simple scheme was used by Ishiwata, Komarov, and Takeda [73], who measured the dynamic pressure of the acoustic stream on a disk immersed into the aluminum melt. The disk was connected by a lever to a load on an electronic balance. The change in the weight of the load was recalculated to the pressure on the disk. These results were used to estimate the stream velocity shown in Figure 2.16d.

In modeling acoustic cavitation, it is important to estimate the pressure that arises in the melt when the cavity collapses. Abramov and Astashkin [63] estimated the pressure in the shock wave for several liquid metals by numerically solving the Kirkwood–Bethe–Gilmor equation [2] at a frequency of 20 kHz for various initial radii of bubbles (from 5 to 10 μm) and at a sound pressure amplitude of 0.3 MPa. The data for $R_0 = 6.5$ μm are given here:

	Sn	Bi	Cd	Pb	In
R_{max}/R_0	11.9	10.5	10.41	9.9	8.8
R_{min}/R_0	0.0226	0.02332	0.025	0.0247	0.031
V_{max}, km/s	5.8	4.07	4.38	3.62	3.54
P_{max}, GPa	11.6	10.7	9.87	9.5	6.5

In order to estimate the effect of the physical and chemical properties of the melt on the shock-wave pressure, the calculations were carried out for various viscosities from 0.7 to 7 mPa·s, keeping the surface tension constant. The estimated shock-wave pressure varied very little, e.g., for bismuth, from 10.7 to 10.2 GPa. Variations in surface tension had a more significant effect. As an example, an increase of the surface tension by a factor of 5 decreased the shock-wave pressure by an order of magnitude.

The *cavitation threshold* can be estimated from measuring the sound spectrum and intensity. Specially designed hydrophones are used in water, while solid-rod probes connected to piezoceramic receivers have been developed for liquid metals (see Section 13.4). The probe is made of Ti or W and has a resonance frequency significantly higher than that of the ultrasonic transducer, e.g., 200 kHz versus 18 kHz. Its signal is displayed on an oscillograph (or a computer using a data-acquisition system), so that cavitation could be observed on the screen as distortions around the main signal. Oscillograms of developed cavitation differ markedly from oscillograms of the threshold mode when cavitation starts.

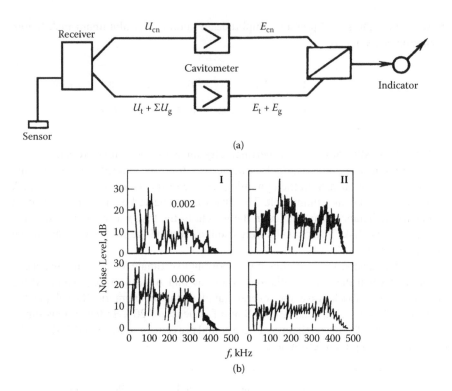

FIGURE 2.20 Diagram of (a) cavitometer and (b) spectrograms of the cavitation noise in various aluminum melts upon sonication at 18 kHz with amplitude $A = 20$ μm at 720°C: (I) effect of Al_2O_3 concentration in aluminum from 0.002 to 0.006 wt% on the noise spectrum; (II) a commercial 7XXX series alloy before and after fine filtration.

Spectral cavitometers that characterize the relation between the main carrier signal, its harmonics, subharmonics, and white noise as dimensionless numbers provide important information about the cavitation development. The work of cavitometers is based on the assumption that the conversion of the main-frequency sound into all other oscillations is governed by cavitation.

Figure 2.20a shows a schematic diagram of a cavitometer. The total noise received by the solid-rod sensor is separated into two signals: incoherent cavitation noise U_{cn} and the sum of the main frequency signal U_t and its harmonics U_g. Squared and integrated over the measurement time, these signals are transformed into electrical parameters proportional to their energies. Then the energy of the incoherent cavitation noise E_{cn} is divided by $E_g + E_t$ and the result

$$\Gamma = \frac{E_{cn}}{E_t + E_g} \tag{2.60}$$

is displayed by an indicator.

Examples of spectrograms obtained by a cavitometer in a wide frequency range are given in Figure 2.20b. These data demonstrate that the increased purity of the melt results in lesser cavitation development, manifested by lower noise levels.

The cavitation threshold in liquid metals is larger than in water. The reason is that most of the properties, like viscosity, surface tension, and density, have higher values in molten metals. Gaseous and nonmetallic solid inclusions existing in specific forms in the metals may be an additional cause. For example, there are no free hydrogen bubbles in the aluminum melt as opposed to free oxygen bubbles in the water.

Abramov [23, 63] studied the cavitation threshold of low-melting metals (Bi, Sn, In, Pb, and Cd). For these metals, the cavitation threshold at a frequency of 20 kHz was numerically estimated as 0.5–1.0 MPa, assuming a bubble radius $R_0 = 0.01$–100 µm. The experimentally determined cavitation threshold in liquid tin was 0.3–0.6 MPa [63], which agrees well with numerical calculations.

The experimental estimates obtained by Abramov and Astashkin [63] demonstrate that for all investigated melts with close acoustic impedance $\rho_0 c_0$, the cavitation threshold depends linearly on surface tension, similar to the experimental results for water shown in Figure 2.6a. It is obvious that for aluminum, magnesium, steel, and cast-iron melts that have stronger surface tensions, a further increase in the cavitation threshold could be expected. It should be noted, however, that theoretical estimates do not take into account the nonmetallic inclusions that were shown to considerably affect the formation of cavitation nuclei (see Section 2.4.1).

To estimate the cavitation threshold using Equation (2.59), which reflects the ultrasonic energy transfer to a noncavitating liquid, one should simultaneously measure the sonotrode amplitude and the generated noise spectrum to see whether or not the cavitation has started. Under experimental conditions, a 10–15-mm gap exists between the sonotrode face and the cavitometer probe. Accordingly, the cavitation conditions near the probe surface differ from those near the sonotrode face. Therefore, thresholds calculated with Equation (2.59) should be considered as the upper limit. For the lower limit of cavitation threshold, it is recommended to reduce the acoustic impedance by a factor of 3–4 [23]. In this case, the cavitation threshold increases by a factor of 1.5–2.0.

The experimentally measured data reflecting the development of cavitation in aluminum melts sonicated at 18 kHz are shown in Table 2.5. One can see that the acoustic impedance drops significantly with the onset of cavitation, while it changes only slightly with further cavitation development.

Figure 2.21a shows the temperature dependence of cavitation threshold and melt viscosity for an Al–6% Mg alloy. Near the liquidus temperature, at 655°C, the cavitation threshold is 0.8–1.0 MPa, whereas for overheated melts, the cavitation threshold decreases to 0.6–0.65 MPa. The correlation between the melt viscosity and the cavitation threshold is obvious. The effect, similar to that of temperature, has the increase in the Mg concentration in an aluminum alloy (Figure. 2.21b). In this case, the surface tension is the driving force for the decreasing cavitation threshold. Al–Mg alloys are characterized by enhanced hydrogen solubility: up to 0.6 cm³/100 g for an Al–6 wt% Mg alloy. However, the results shown in Figure 2.21c demonstrate that in this alloy, as in pure aluminum, the cavitation threshold weakly depends on the hydrogen content.

The cavitation threshold in aluminum alloys was recently estimated by using a high-temperature cavitometer [40] and acoustic emission [77]. These results

TABLE 2.5

**Average Acoustic Impedance and Transferred Acoustic
Power for Aluminum Melt Under Sonication at 18 kHz**

A, µm [a]	A_ω, m/s	W_a, W	$\overline{\rho_c c_c}$, kg/m²·s	Cavitation
2	0.228	2.2	17.7	absent
5	0.56	8.6	2.6	incipient
10	1.29	40	2.4	developed
15	1.69	100	2.7	developed
20	2.26	200	3.0	developed
30	3.39	400	2.7	developed
40	5.52	1000	2.6	developed

Note: See Equation (2.55) for reference.

[a] Null-to-peak amplitude.

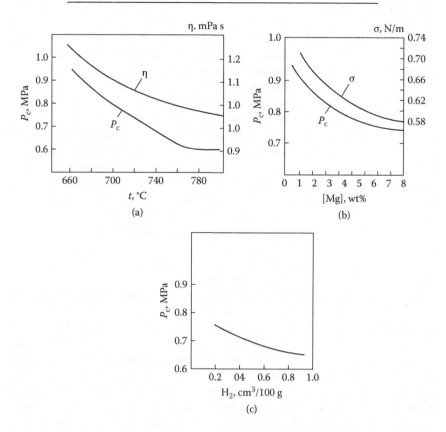

(a)

(b)

(c)

FIGURE 2.21 Experimentally measured cavitation threshold in Al–Mg alloys: (a) effect of temperature for Al–6% Mg melt, (b) effect of Mg concentration at 720°C, and (c) effect of hydrogen concentration in Al–6% Mg melt.

confirmed our earlier measurements of the cavitation threshold in aluminum alloys, giving the developed cavitation onset in an Al–17% Si alloy at about 10 μm peak-to-peak amplitude at 20 kHz [40] (Figure 2.22a). The effect of surface tension (varying from 0.851 N/m for pure Al to 0.75 N/m for Al–2.4% Mg and to 0.432 N/m for Al–10% Mg) is illustrated in Figure 2.22b [77]. This set of data also

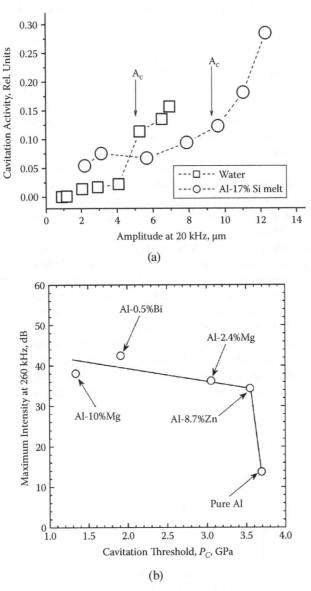

(a)

(b)

FIGURE 2.22 Experimentally measured cavitation threshold in aluminum alloys: (a) comparison of cavitation thresholds in water and Al–17% Si alloys at 20 kHz (adapted from Komarov et al. [40]) and (b) effect of surface tension on cavitation threshold at 260 kHz (adapted from Matsunaga et al. [77]).

demonstrates that the cavitation threshold dramatically increases at higher acoustic frequencies.

Gas dissolved in liquid metals and solid inclusions may significantly affect the cavitation threshold. Therefore, a combined melt–nonmetallic inclusion–gas system should be considered for the analysis of cavitation in real melts. Let us analyze the cavitation strength of aluminum melt within the liquid aluminum–aluminum oxide–hydrogen system.

Aluminum interacts mostly with hydrogen and oxygen. Oxygen is present only in the form of aluminum oxide, Al_2O_3, which is a stable compound resistant to thermal dissociation (though there are some polymorphic modifications of alumina). Alumina is typically present in the melt in the form of suspended micron-size particles, along with alumina films and large particles drawn into the melt from the surface. In commercial aluminum and aluminum alloys, aluminum oxide is contaminated by other oxides formed by additions or impurities (such as Fe, Mg, Cu, Ti, and Si) and also by oxides of transition metals.

Hydrogen forms as a result of the reaction between aluminum and water vapor (more on that in Chapter 3). In molten aluminum, hydrogen exists in atomic form, being dissolved in the liquid phase; there is no experimental evidence for its occurrence in the form of free bubbles. Hydrogen and alumina inclusions have a certain affinity [62, 78]. Aluminum oxide is an active adsorbent. Its surface is poorly wetted by liquid aluminum and adsorbs dissolved hydrogen in the form of ions. According to this model, each alumina particle is surrounded by a hydrogen-rich envelope. Commercial aluminum contains about 0.005% of aluminum oxide, and this small amount controls about 5% of all hydrogen in the melt. Impurities in alumina, e.g., iron, can increase hydrogen adsorption by an order of magnitude [79]. Zirconium, titanium, and other transition metals act in the same direction. Figure 2.23 shows that the cavitation threshold decreases in aluminum melts containing Zr.

The effects of hydrogen and alumina content on cavitation development were specially studied using various grades of aluminum and a commercial Al–6% Mg–0.6% Mn alloy. The composition and characteristics of the starting melts are given in Table 2.6.

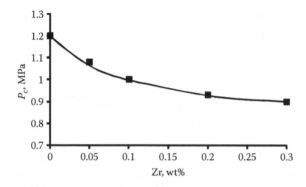

FIGURE 2.23   Effect of Zr on the cavitation threshold of a 1090 aluminum melt at 730°C.

TABLE 2.6

Physical and Chemical Characteristics of Starting Melts

	Alloys and Impurities, wt%			Al_2O_3 Content, wt%	Hydrogen Content, cm³/100 g	Surface Tension, N/m
Melt	Mg	Fe	Si			
A99 (99.9%)	0.003	0.01	0.01	1×10^{-4}	0.2	0.87
A85 (99.85%)	...	0.1	0.1	5×10^{-4}	0.25	0.87
A7 (99.7%)	...	0.15	0.15	5×10^{-4}	0.3	0.87
Al–6% Mg	6.0/0.6 Mn	0.15	0.15	5×10^{-4}	0.65	0.56

The hydrogen content was varied in the melt by adding humidity with a dump refractory plug. The hydrogen concentration was analyzed by the first-bubble method and by vacuum extraction from a solid sample. After the initial melt was prepared, solid synthetic aluminum oxide (α-Al_2O_3) with particle size less than 1 μm was added to the melt using intensive ultrasonication. The bromine–methanol method was used to verify the concentration of these inclusions [80].

Particles of Al_2O_3 are not well wettable in aluminum melts and, under general conditions, do not mix with the melt and remain on the surface. To overcome this difficulty, we injected heated oxide powder into the cavitation region. This procedure is more efficient if a focusing sonotrode is used. The duration of sonication with acoustic power below 0.5 kW was up to 15 min for the melt volume of 1 kg.

It should be noted that prolonged sonication of melts containing suspended Al_2O_3 particles will not only distribute but also disperse and erode oxide particles, thus producing abundant interfaces for the Frenkel–Harvey cavitation nuclei. Indeed, assuming the mass of the melt $m_1 = 0.5$ kg, aluminum oxide content 0.01 wt%, the size of oxide particles $d = 1.0$ μm, and oxide density $\rho = 3.0$ g/cm³, we have:

Mass of a single particle: $m_p = 4/3\ \pi r^3 \rho = 1.57 \times 10^{-12}$ g
Total Al_2O_3 mass in melt: $(0.01\%)m_1 = 0.05$ g
Number of Al_2O_3 particles: $m_1/m_p = 3.2 \times 10^{10}$

When the aluminum oxide concentration increases to 0.1 wt% and 1 wt%, the number of particles in the melt increases to 10^{11} and 10^{12}, respectively, thus considerably increasing the number of cavitation nuclei.

The experimental results on the cavitation threshold are shown in Figure 2.24a. Increased hydrogen content in the melt lowers the cavitation threshold, most significantly at small alumina concentrations, up to 0.005 wt%. The further decrease in the cavitation threshold is associated mainly with a higher content of solid nonmetallic alumina particles: as the Al_2O_3 concentration increases from 0.005 to 0.1 wt%, the cavitation threshold decreases by 47%, from 0.8 MPa to 0.55 MPa. The requirements for the input acoustic power (amplitude) for the developed cavitation

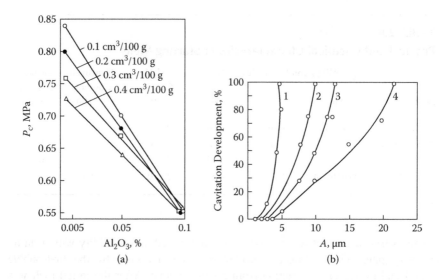

FIGURE 2.24 Effect of hydrogen and alumina concentration on cavitation development in a commercial 1070 aluminum: (a) cavitation threshold at different hydrogen and alumina concentrations and (b) degree of cavitation development versus ultrasonic displacement amplitude A for various oxygen concentrations: (1) 0.03%, (2) 0.008%, (3) 0.005%, and (4) 0.004%.

are less in the case of larger alumina concentrations, as evidenced by the data in Figure 2.24b.

Finally, Figure 2.25 demonstrates the cavitation activity in molten pure aluminum (AA1050) and aluminum alloys AA2324 and AA7475 versus the oxygen content in the melt (recalculated from alumina concentration). Evidently, the ultrasonic cavitation development can be used as an indicator of liquid metal purity with respect to oxide inclusions.

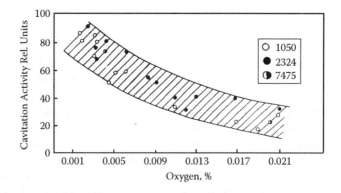

FIGURE 2.25 Effect of oxygen concentration on cavitation activity in melts of 1050, 2324, and 7475 alloys.

2.7 SONOLUMINESCENCE IN MELTS

One of the phenomena that accompany cavitation is sonoluminescence, or emission of light upon collapse of cavitation bubbles. The sonoluminescence was first observed in a sonicated photographic developer liquid by Frenzel and Schultes in 1934 [81] and has been studied extensively ever since. A detailed treatise on this phenomenon can be found in the works of Suslick [82, 83], Crum [84, 85], Margulis [86], and Brenner, Hilgenfeldt, and Lohse [87]. Here we will give some unique results obtained on sonoluminescence of liquid metals.

The development and collapse of a cavitation bubble occurs very quickly, on the microsecond level, and generates extremely high pressure, momentum, and temperature spikes. The pressure was estimated to be in the range of 1 to 10 GPa, temperatures up to 10^5 K, and velocities in the supersonic range [63, 83, 88, 89]. These extreme conditions result in emission of light, either as a result of extreme heat [82, 88] or electrical discharge [50, 84, 86].

Sonoluminescence is a very useful phenomenon for studying the onset of cavitation and its activity; however, the use of this concept in opaque metallic melts presents an obvious challenge. Moreover, it was not clear that this phenomenon actually occurs in the melts. Consequently, research has been performed to verify whether (a) sonoluminescence indeed occurs in metallic melts and (b) it can be used as an indicator of cavitation onset [90]. To facilitate the experiment, only low-melting metals were selected, i.e., Wood's alloy, In, and Sn.

Figure 2.26a gives a schematic diagram of the experimental setup. A corundum crucible (1) with a transparent quartz base (2) was placed in a furnace (3). A sonotrode (4) of a magnetostrictive transducer (5) connected to an ultrasonic generator (6) was immersed in the melt from above. One end of a 250-mm light guide (7) was attached to the crucible bottom, and the other end was connected to a photoelectron multiplier

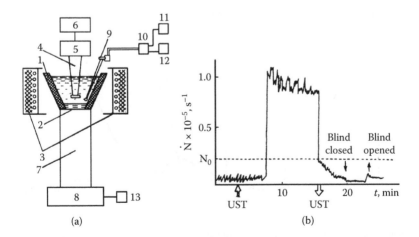

FIGURE 2.26 (a) Setup for studying sonoluminescence in low-melting metallic alloys and (b) effect of ultrasound on sonoluminescence intensity. (After Margulis et al. [90].)

(8). Sonoluminescence was registered with a photon-counting system based on a scintillation spectrometer (13). Simultaneously, cavitation noise was detected with a cavitation probe (9) connected to an amplifier (10), recorder (11), and oscilloscope (12).

Liquid metals are opaque in the visible range; therefore, the photoelectron multiplier receives light only from the cavitation bubbles that are pushed to the transparent crucible bottom by sound pressure and from the emitted light on the internal side of the quartz window (2). Even under these conditions, the scintillation rate $\dot{N}$ obtained at 22 kHz is very responsive to the ultrasound.

Experiments showed that a reasonably strong sonoluminescence occurs in the studied liquid metals. Its intensity was one or two orders of magnitude higher compared to that in concentrated water solutions of KCl. Glow was easily detected, and the scintillation rate was as high as 10^4–10^5 per second. The experimental results showed that sonoluminescence reliably indicates the conditions under which cavitation starts in a Sn melt sonicated at 22 kHz with an ultrasonic intensity of 10^4 W/m².

Brief sonication without cavitation development does not cause any luminescence, while the increase of cavitation intensity to the developed cavitation conditions results in a quick rise of scintillation counts, by one to two orders of magnitude (Figure 2.26b). In the course of developed cavitation, the intensity of sonoluminescence gradually decreases, which may be the result of degassing. The sonoluminescence intensity also decreases with temperature, which is typical of liquids [86], as demonstrated in Figure 2.27a.

The experiments also revealed some effects that were not observed in other liquids. Figure 2.27b gives the dependence of sonoluminescence intensity and integral acoustic pressure of subharmonic frequencies on the ultrasound intensity for

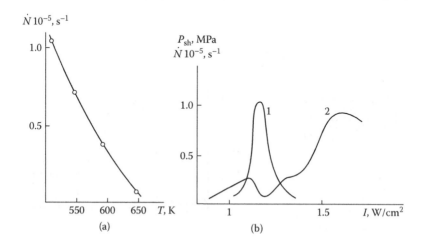

FIGURE 2.27 Sonoluminescence in liquid tin: (a) effect of temperature on sonoluminescence intensity (rate of scintillations) and (b) effect of ultrasonic intensity on sonoluminescence intensity (1) and acoustic pressure of subharmonic components (2) of cavitation noise in liquid tin at 510 K. (Adapted from Margulis et al. [90].)

liquid Sn at 510 K. Each of the curves has a maximum at different intensities. It is worth noting that the position of the sonoluminescence maximum shifts to higher intensities with decreasing distance between the sonotrode and the quartz window. The nonmonotonic change of sonoluminescence intensity may be a result of ultrasound absorption by cavitation bubbles at higher ultrasound intensities. Although the amount of emitted photons increases in the cavitation zone, they cannot reach the transparent quartz window, being scattered on the cavitation bubbles and absorbed by the opaque melt.

The sonoluminescence continues for some time after the ultrasonic processing (cavitation) stops. This is a result of the heating of the liquid phase by the cavitation energy, as illustrated in Figure 2.26b on the right-hand side of the plot with the blind closed (intensity drops) and opened (intensity increases).

The sonoluminescence in metallic melts has a fundamental as well as a practical aspect. The registration of the cavitation threshold and sonoluminescence flux may help in the analysis of gases and solid inclusions in the melt. The time evolution of the sonoluminescence flux may give information about the kinetics of ultrasonic degassing. The application of sonoluminescence can potentially assist in studying the spatial–temporal development of a cavitation zone and individual cavitation bubbles.

2.8 ACCELERATED MIXING AND DISSOLUTION OF COMPONENTS IN THE MELT

The melting of solid metallic alloys as well as the addition of solid metal to the melt requires a certain time to reach the homogeneous distribution of elements throughout the volume of the melt. In addition, a relatively high melt superheat is required to achieve the true single-phase melt condition, i.e., the elimination of short-range crystal-type atomic arrangements [91]. Acoustic cavitation accompanied by active mixing of the melt with acoustic flows accelerates the transition of the melt to a homogeneous state, allowing for lower melt temperatures and shorter holding times [92].

The kinetics of melt homogenization was studied by measuring the density of the melt by γ-ray adsorption. Holding of a Pb–50% Sn melt at 367°C for 5 h (liquidus is 220°C) was not sufficient for the melt to achieve the equilibrium density (Figure 2.28a). At the same time, ultrasonic cavitation at 22 kHz (2 W/cm^2) completely homogenized the melt in only 40 s. Similar results were obtained for an Al–Pb–Na–Ca alloy, which required 3 h at 500°C to achieve the equilibrium melt density without sonication. With cavitation, the melt was homogenized in 1 min. The study of the structure showed that the difference of Ca concentration along the height of the sample decreased from 1.9% to 0.16%, and the particles of Pb$_3$Ca compounds became more rounded.

The dissolution rate was also studied for iron in aluminum. An iron sample was placed on the surface of an aluminum sample with the nominal weight ratio corresponding to 8% Fe. The system was then brought to a temperature of 1130°C, and the density of the aluminum melt was measured. Without ultrasonic treatment, the iron had not completely dissolved in aluminum, even after soaking for 40–60 min. Undissolved particles of Fe were found in the samples. Conversely, when ultrasonic

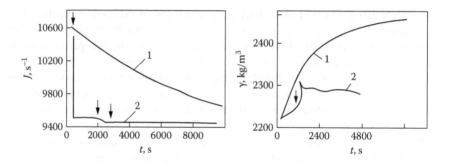

FIGURE 2.28 Effect of ultrasonic cavitation processing on (a) homogenization of a Pb–50% Sn melt and (b) dissolution rate of Fe in liquid aluminum: (1) without sonication and (2) with sonication; arrows point at the times when ultrasound was introduced.

processing was applied for 1 min, complete dissolution was reached after 5 min (Figure 2.28b). Particles of the Al_6Fe compound were evenly distributed throughout the sample after solidification.

The difference in the final equilibrium densities in samples produced with and without sonication indicates that ultrasonic cavitation not only accelerates diffusion processes, but also changes the equilibrium constitution of the melt, similar to the action of high-melt superheat [91].

REFERENCES

1. Blitz, J. 1963. *Fundamentals of ultrasonics*. London: Butterworths.
2. Rozenberg, L.D. (ed.). 1968. *High-intensity ultrasonic fields*. Moscow: Nauka [Translated to English. 1971. New York: Plenum].
3. Flynn, H.G. 1964. Physics of acoustic cavitation in liquids. In *Physical acoustics*, vol. 1, part B, ed. W.P. Mason, 57–172. New York: Academic Press.
4. Suslick, K.S. 1988. *Ultrasound: Its chemical, physical and biological effects*. New York: VCH Publishers.
5. Mason, T.J., and J.P. Lorimer. 1988. *Sonochemistry: Theory, application and uses of ultrasound in chemistry*. Chichester, UK: Ellis Horwood.
6. Brennen, C.E. 1995. *Cavitation and bubble dynamics*. Oxford, UK: Oxford University Press.
7. Ensminger, D., and F.B. Stulen (ed.). 2008. *Ultrasonics: Data, equations and their practical uses*. Boca Raton, FL: CRC Press.
8. Ensminger, D., and L.J. Bond. 2011. *Ultrasonics: Fundamentals, technologies, and applications*. 3rd ed. Boca Raton, FL: CRC Press.
9. Efimov, V.A. 1983. In *Influence of external action on liquid and solidifying metal*, 3–21. Kiev, Ukraine: Institute of Casting Problems, Akad. Nauk UkrSSR.
10. Kapustin, A.P. 1962. *Effect of ultrasound on the kinetics of crystallization*. Moscow: Nauka.
11. Walker, J.L. 1958. In *Proc. ASM Seminar Liquid Metals and Solidification*, 319–36. Metals Park, OH: ASM Publication.
12. Chalmers, B. 1964. *Principles of solidification*. New York: John Wiley & Sons.
13. Hunt, J.D., and K.A. Jackson. 1966. *J. Appl. Phys.* 37:254–57.

14. Frawley, J.J., and W.J. Childs. 1968. *Trans. Met. Soc AIME* 242:256–63.
15. Danilov, V.I. 1956. *Constitution and crystallization of liquids.* Kiev, Ukraine: Izd. Akad. Nauk SSSR.
16. Danilov, V.I., and V.E. Neimark. 1949. *Zh. Eksper. Teor. Fiz.* 19:235–41.
17. Danilov, V.I., and D.E. Ovsienko. 1951. *Zh. Eksper. Teor. Fiz.* 21:879–87.
18. Kazachkovsky, O.D. 1948. In *Collective papers of the laboratory of metals physics,* ed. V.I. Danilov, 103–12. Kiev, Ukraine: Izd. Akad. Nauk UkrSSR.
19. Chvorinov, N.I. 1954. *Crystallization and non-homogeneity of steel.* Prague, Czech Republic: Publishing House of Czechoslovak Academy of Sciences (NČSAV).
20. Balandin, G.F. 1973. *Formation of crystalline structure of castings.* Moscow: Mashinostroenie.
21. Flemings, M.C. 1974. *Solidification processing.* New York: McGraw-Hill.
22. Campbell, J. 1981. *Int. Met. Rev.,* no. 2:71–108.
23. Abramov, O.V. 1972. *Crystallization of metals in an ultrasonic field.* Moscow: Metallurgiya.
24. Vogel, A., R.D. Doherty, and B. Cantor. 1979. In *Solidification and casting of metals,* 518–25. London: TMS.
25. Pilling, J., and A. Hellawell. 1996. *Metall. Mater. Trans.* A 27A:229–32.
26. Jackson, K.A., J.D. Hunt, D.R Uhlmann, and T.P. Seward. 1966. *Trans. Metall. Soc. AIME* 236:149–60.
27. Chernov, A.A. 1956. *Kristallografiya* 1:589–93.
28. Kattamis, T.Z., J.C. Coughlin, and M.C. Flemings. 1967. *Trans. Metall. Soc. AIME* 239:1504–11.
29. Ruvalcaba, D., R.H. Mathiesen, D.G. Eskin, L. Arnberg, and L. Katgerman. 2007. *Acta Mater.* 55:4287–92.
30. Abramov, O.V., and I.I. Teumin. 1970. Crystallization of metals. In *Physical principles of ultrasonic technology,* ed. L.D. Rozenberg, 427–514. Moscow: Nauka [Translated to English. 1973. New York: Plenum].
31. Swallowe, G.M., J.E. Field, C.S. Rees, and A. Duckworth. 1989. *Acta Mater.* 37:961–67.
32. Shu, D., B. Sun, J. Mi, and P.S. Grant. 2012. *Metall. Mater. Trans.* A 43A:3755–66.
33. Kikuchi, Y. 1969. *Ultrasonic transducers.* Tokyo: Corona Publ.
34. Eskin, G.I. 1988. *Ultrasonic treatment of molten aluminum.* Moscow: Metallurgiya.
35. Matauschek, J. 1962. *Einführung in die Ultraschalltechnik.* Berlin: VEB Verlag Technik.
36. Kleppa, O.J. 1950. *J. Chem. Phys.* 18:1331–36.
37. Blairs, S. 2007. *Phys. Chem. Liquids* 45 (4): 399–407.
38. Denisov, V.M., V.V. Pingin, L.T. Antonova, S.A. Istomin, E.A. Pastukhov, and V.I. Ivanov. 2005. *Aluminum and its alloys in the liquid state,* 7–23. Yekaterinburg, Russia: Izd. Ural Otdel. Ross. Akad. Nauk.
39. Battezzati, L., and A.L. Greer. 1989. *Acta Metall.* 37:1791–1802.
40. Komarov, S., K. Oda, Y. Ishiwata, and N. Dezhkunov. 2012. *Ultrason. Sonochem.* 20:754–61.
41. Bergmann, L. 1954. *Der Ultraschall und seine Anwendung in Wissenschaft und Technik.* 6th ed. Zürich: S. Hirzel Verlag.
42. Lauterborn, W., T. Kurz, R. Mettin, and C.D. Ohl. 1999. *Adv. Chem. Phys.* 110:295–380.
43. Zeldovich, Ya.B. 1942. *Zh. Eksper. Teor. Fiz.* 12:525–38.
44. Briggs, H.B., J.B. Johnson, and W.P. Mason. 1947. *J. Acoust. Soc. Am.* 19:664–77.
45. Connolly, W., and F.E. Fox. 1954. *J. Acoust. Soc. Am.* 26:843–48.
46. Plesset, M.S., and A. Prosperetti. 1977. *Ann. Rev. Fluid Mech.* 9:145–85.
47. Khismatullin, D.B. 2004. *J. Acoust. Soc. Am.* 116:1463–73.
48. Il'ichev, V.I. 1969. In *Proc. Akust. Inst. Akad. Nauk SSSR (AKIN)* 6:16–29.
49. Belova, V., D.A. Gorin, D.G. Shchukin, and H. Möhwald. 2010. *Angew. Chem. Inter. Edn.* 49:7129–33.

50. Frenkel, Ya.I. 1945. *Kinetic theory of liquids*. Moscow: Izd. Akad Nauk SSSR [Translated to English. 1955. New York: Dover Publ.].
51. Harvey, E.N., D.K. Barnes, W.D. McElroy, A.H. Whiteley, D.C. Pease, and K.W. Cooper. 1944. *J. Cell. Comp. Physiol.* 24:1–34.
52. Harvey, E.N., W.D. McElroy, and A.H. Whiteley. 1947. *J. Appl. Phys.* 18:162–72.
53. Dobatkin, V.I., R.M. Gabidullin, B.A. Kolachev, and G.S. Makarov. 1976. *Gases and oxides in wrought aluminum alloys*, 65–90. Moscow: Metallurgiya.
54. Nolting, K.B.E., and E.A. Neppiras. 1950. *Proc. Phys. Soc. London. Sect. B* 63:647–85; 64:1032–38.
55. Eskin, G.I., P.N. Shvetsov, and A.I. Ioffe. 1972. *Izv. Akad. Nauk SSSR, Met.*, no. 6:69–74.
56. Minnaert, M. 1933. *Philos. Mag.* 16:235–48.
57. Schmid, J. 1959. *Acustica* 9:321–26.
58. Eskin, G.I., A.I. Ioffe, and P.N. Shvetsov. 1974. *Tekhnol. Legk. Spl.*, no. 1:8–11.
59. Boguslavsky, Yu.Ya. 1967. *Akust. Zh.* 13:23–27.
60. Fyrillas, M.M., and A.J. Szeri. 1994/1995. *J. Fluid. Mech.* 277:381–407; 289:295–314.
61. Crum, L.A. 1980. *J. Acoust. Soc. Am.* 68:203–11.
62. Pimenov, Yu.P., and A.I. Demenkov. 1972. *Tekhnol. Legk. Spl.*, no. 6:33–36.
63. Abramov, O.V., and Yu.S. Astashkin. 1974. In *Application of new physical methods for intensification of metallurgical processes*, 155–67. Moscow: Metallurgiya.
64. Eskin, G.I. 1965. *Ultrasonic treatment of molten aluminum*. Moscow: Metallurgiya.
65. Dubus, B., C. Vanhille, C. Campos-Pozuelo, and C. Granger. 2010. *Ultrason. Sonochem.* 17:810–18.
66. Moussatov, A., C. Granger, and B. Dubus. 2003. *Ultrason. Sonochem.* 10:191–195.
67. Dahlem, O., J. Reisse, and V. Halloin. 1999. *Chem. Eng. Sci.* 54:2829–38.
68. Kumar, A., A.B. Pandit, and J.B. Joshi. 2006. *Chem. Eng. Sci.* 61:7410–20.
69. Yisaf, T.F., and D.R. Buttsworth. 2007. *Ultrason. Sonochem.* 14:266–74.
70. Schenker, M.C., M.J.B.M. Pourquié, D.G. Eskin, and B.J. Boersma. 2013. *Ultason. Sonochem.* 20:502–9.
71. Panov, A.P. 1984. *Ultrasonic cleaning of high-precision parts*. Moscow: Mashinostroeniye.
72. Kanthale, P.M., P.R. Gogatem, A.B. Pandit, and A.M. Wilhelm. 2003. *Ultrason. Sonochem.* 10:331–35.
73. Ishiwata, Y., S. Komarov, and Y. Takeda. 2012. In *Proc. 13th Intern. Conf. Alum. Alloys (ICAA13)*, ed. H. Weiland, A.D. Rollett, and W.A. Cassada, 183–87. Warrendale, PA: TMS.
74. Abramov, O.V., Yu.S. Astashkin, and V.S. Stepanov. 1979. *Akust. Zh.* 35:180–86.
75. Jackson, K.A., and J.D. Hunt. 1965. *Acta Metall.* 13:1212–15.
76. Teumin, I.I. 1962. *Akust. Zh.* 8:372–73.
77. Matsunaga, T., K. Ogata, T. Hatayama, K. Shinozaki, and M. Yoshida. 2007. *Composites: Part A* 38:771–78.
78. Altman, M.B. 1965. *Nonmetallic inclusions in aluminum alloys*. Moscow: Metallurgiya.
79. Arbuzova, L.A., K.I. Slovetskaya, A.M. Rubinshtein, L.L. Kunin, and V.A. Danilkin. 1971. *Izv. Akad. Nauk SSSR, Khimiya*, no. 1:169–70.
80. Dulski, T.R. 1996. *A manual for the chemical analysis of metals*. ASTM Manual Series, vol. 25, 89–90. West Conshohocken, PA: ASTM.
81. Frenzel, H., and H. Schultes. 1934. *Z. Phys. Chem.* B27:421–24.
82. Suslick, K.S., and E.B. Flint. 1987. *Nature* 330:553–55.
83. Flannigan, D.J., and K.S. Suslick. 2005. *Nature* 434:52–55.
84. Crum, L.A. 1995. *Ultrason. Sonochem.* 2:148–52.
85. Matula, T.J., and L.A. Crum. 1998. *Phys. Rev. Lett.* 80:865–68.
86. Margulis, M.A. 1995. *Sonoluminescence and cavitation*. Amsterdam: Gordon and Breach OPA.
87. Brenner, M.P., S. Hilgenfeldt, and D. Lohse. 2002. *Rev. Mod. Phys.* 74:425–84.

88. Suslick, K.S. 1990. *Science* 247:1439–45.
89. Kim, K.Y., K.-T. Byun, and H.-Y. Kwak. 2007. *Chem. Eng. J.* 132:125–35.
90. Margulis, M.A., L.M. Grundel, G.I. Eskin, and P.N. Shvetsov. 1987. *Dokl. Akad. Nauk SSSR* 295:1170–73.
91. Brodova, I.G., P.S. Popel, and G.I. Eskin. 2002. *Liquid metals processing: Application to aluminum alloy production.* London: Taylor and Francis.
92. Popel, P.S., V.A. Pavlov, and G.E. Konovalov. 1985. *Tsvetn. Met.*, no. 7:68–70.

3 Ultrasonic Degassing

It is logical to start the discussion on ultrasonic melt cleaning with degassing, because (a) this is the earliest known application of ultrasonic liquid processing and (b) the removal of gases and solid nonmetallic inclusions assisted by gas bubbles is the basic technology of melt cleaning.

In 1922, in one of the first studies on cavitation in liquids, Boyle [1] reported the possible use of ultrasound for the degassing of liquids. Sörensen [2] quantitatively studied the degassing of water at frequencies from 190 to 950 kHz. Krüger [3] used low-power piezoceramic vibrators for degassing liquid metals, and successfully used ultrasound for degassing molten glass. Bradfield [4] reported degassing of molten aluminum and its alloys with ultrasound at 15 kHz and 26 kHz.

As early as 1950, Eisenreich [5] compared vacuum ultrasonic degassing with vacuum degassing, degassing with chlorine lancing, and sonic and ultrasonic degassing. He pointed out the potential of ultrasonic processing but also mentioned related practical difficulties. Sergeev [6] noted that despite the high potential of ultrasonic degassing, there is a challenge in transferring sufficient ultrasonic power to a large mass of liquid metal.

Indeed, our early investigations [7] demonstrated that the removal of hydrogen from aluminum alloys depends greatly on the acoustic power transferred to the melt and on the development of cavitation. Figure 3.1 compares the degassing kinetics for hydrogen in an A356 melt treated with chlorine salts, ultrasound, vacuum, and ultrasound combined with vacuum. These results were recently reconfirmed, albeit in small-scale laboratory experiments [8–11].

3.1 CAVITATION AND DEGASSING NUCLEI

Any gas (unless intentionally blasted through the melt) is usually completely dissolved in liquid metal, and no free bubbles exist in the liquid volume. Therefore, the theory of ultrasonic degassing—well developed for water [12]—is only applicable to liquid metals after cavitation starts to produce bubbles [13]. In other words, it is only possible to produce the bubbles if the external energy supplied to the melt by ultrasound creates conditions for heterogeneous nucleation of a bubble onto a not-wet solid inclusion.

The nature and origins of the cavitation or degassing nuclei are very important for understanding the mechanisms of ultrasonic degassing in liquid metals. According to modern views, liquid metals and alloys are colloid systems, in which dispersed nonmetallic inclusions, e.g., oxides in liquid Al or Mg, serve as hydrogen concentrators (as well as the cavitation nuclei, see Chapter 2). Experimental results [14] show that pure alumina and even more so alumina contaminated with transition

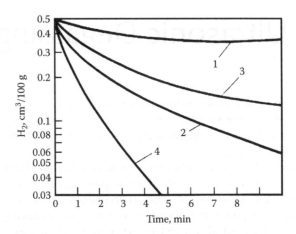

FIGURE 3.1 Kinetics of hydrogen removal from a 10-kg charge of an A356-type melt for different degassing methods: (1) with chlorine salts; (2) with ultrasound; (3) with vacuum; and (4) with ultrasound in vacuum.

metals adsorbs hydrogen in considerable quantities that makes these particles efficient cavitation nuclei and decreases the cavitation threshold (see Figures 2.21c and 2.23–2.25).

We have already discussed in Chapter 2 that free vapor–gas bubbles and not-wettable solid particles can serve as cavitation nuclei in liquids. While both vapor–gas bubbles and solid particles are the appropriate nuclei in water, only solid nonmetallic inclusions may qualify as cavitation nuclei in metallic melts, where free gas bubbles are hardly possible, but gas can exist in capillaries on the surface of the inclusions. This model for cavitation nuclei in liquid metals was described in Section 2.4.1.

Solidifying pure liquid melts (without oxides or other solid impurities) will form supersaturated hydrogen solutions, and also would require large acoustic pressures to initiate cavitation. In real melts containing oxides or other insoluble particles, hydrogen will precipitate on poorly wetted insoluble inclusions, forming configurations as shown in Figure 3.2 [15, 16]. Generally, the precipitation follows the basic rules of heterogeneous nucleation and growth as outlined by Kelton and Greer [17].

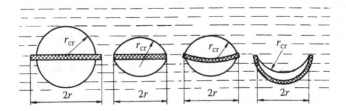

FIGURE 3.2 Possible connections between solid alumina inclusions and hydrogen in an aluminum melt.

Assume that Al_2O_3 particles have the form of flat discs of radius $r_d > r_{cr}$, where r_{cr} is the critical radius of a bubble under equilibrium conditions:

$$\frac{C - C_\infty}{C_\infty} = \frac{\sigma}{(P_{atm} + P_m)r_{cr} + 2\sigma}, \tag{3.1}$$

where C is the hydrogen concentration in the melt, C_∞ is the equilibrium hydrogen solubility at the flat interface, P_{atm} is the atmospheric pressure, P_m is the metallostatic pressure, and σ is the surface tension at the interface.

Hydrogen will form a lens that progressively grows to form a sphere of radius r_{cr}, and then this bubble will grow without further limitations, as shown in Figure 3.2a. In this process, hydrogen lenses will remain on flat particles at almost any hydrogen concentration above the equilibrium solubility. If $r_d < r_{cr}$, the situation changes, and the bubble grows only until it is bound by a spherical segment of radius r_d (Figure 3.2b).

The situation when the surface of the substrate particle is curved (Figure 3.2c) or has some concave segments such as fractures and slots is of special interest. It is also most common in reality. In this case, hydrogen forms lenses with negative curvature at hydrogen concentrations significantly below the equilibrium concentration. The calculations show that r_{cr} ranges from 15 to 30 μm in aluminum melts with typical hydrogen concentrations (i.e., <0.5 cm³/100 g) [16]. As a result, the crevices and cracks that are smaller than 30 μm will be filled with hydrogen under normal pressure and concentration conditions.

From the condition of stable suspension of alumina particles in the aluminum melt, Makarov [15] estimated that the volume of molecular hydrogen adsorbed at the surface of alumina inclusions is approximately half of the total volume of the inclusions. This amount will also depend on the morphology of the oxide and other nonmetallic inclusions: the more tortuous the surface, the stronger the hydrogen adsorption is.

It is very difficult to study scattered fine aluminum oxide particles in liquid metal directly. Their morphology can be investigated only indirectly, by studying the structure of sediments on a filter. Figure 3.3 presents the fracture of a five-layer filter made of 0.6 × 0.6-mm mesh glass cloth. These photos were obtained with a scanning electron microscope. They show an agglomerate of oxide particles (Figures 3.3a,b) and individual micron-size oxide particles (Figures 3.3c,d) on the surface of the filter. The rough oxide surface is clearly visible, demonstrating fractures and hollows.

It is noted that the transformation of γ-Al_2O_3 films to α-Al_2O_3 compact particles decreases the amount of adsorbed hydrogen [16]. In any case, there is a strong relationship between the amount of nonmetallic inclusions and the amount of molecular hydrogen in the aluminum melt. At the same time, the presence and concentration of alumina in the melt does not affect the solubility of hydrogen in the liquid aluminum [15], only influencing the fraction of molecular hydrogen adsorbed on the surface of the inclusions.

Table 3.1 illustrates the relationship between alumina and molecular hydrogen in liquid aluminum at 700°C with a hydrogen concentration of 0.25 cm³/100 g [18]. These data clearly show that the amount and the fraction of molecular hydrogen in

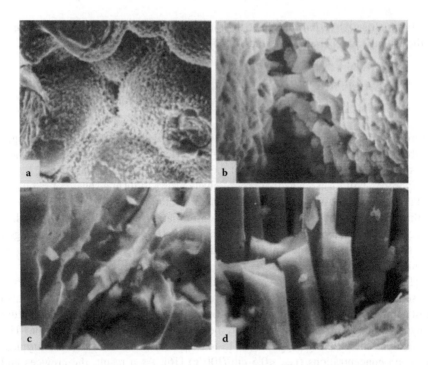

FIGURE 3.3 Sediment of nonmetallic inclusions collected on a five-layer filter (0.6 × 0.6-mm mesh glass cloth) after fine filtration of 6 tons of molten AA2324 alloy: (a) agglomerate of 5–100-µm Al_2O_3 particles at 0.3 mm from the filter surface; (b) the same, ×1000; (c) 1–2-µm Al_2O_3 particles at the surface of the glass cloth between the second and the third layers of the filter; and (d) the same, ×1000.

liquid aluminum are rather small, with the volume of free hydrogen at typical alumina concentrations varying between 0.01 and 0.2 vol%. However, free hydrogen plays an important role in liquid aluminum, as its amount is sufficient to maintain the suspension of small alumina inclusions in the melt volume and because it provides cavitation and degassing nuclei.

According to a theory of acoustic degassing that Kapustina [12] suggested for liquids with existing vapor/gas bubbles, the degassing is controlled by the pulsating

TABLE 3.1

Estimated Concentration of Molecular Hydrogen on Alumina Inclusions

Alumina content in the melt, wt%	0.03	0.01	0.005	0.001
Alumina volume, cm³/100 g Al	0.0084	0.0028	0.0014	0.00028
Hydrogen adsorbed on alumina particles at 700°C, cm³	0.00437	0.00143	0.00072	0.00014
Molecular hydrogen fraction, %	0.52	0.16	0.08	0.016

bubbles that accumulate dissolved gas due to its diffusion from the liquid in the rarefaction stage of bubble oscillation and recombination to the molecular form inside the bubble. The bubbles then grow, coalesce, and eventually float to the surface. According to Kapustina, the role of cavitation is in acceleration of the process due to multiplication of bubbles and more active diffusion of the dissolved gas into the small bubbles oscillating in a nonlinear manner. In addition, intense cavitation produces acoustic flows and secondary convective flows that contribute to bubble distribution and flotation. Water is an example of such a liquid, with oxygen bubbles readily present in the liquid volume. As a result, the degassing threshold for water (i.e., the sound intensity that leads to gas liberation from the liquid phase) is always lower than the cavitation threshold.

The situation is quite different for liquid metals, where vapor–gas bubbles do not usually exist and where their formation requires cavitation of the liquid. In this case, the degassing and cavitation thresholds must coincide. The cavitation nuclei are of the same origin as the degassing nuclei and, as we have discussed previously, are represented by gas adsorbed on the surface of poorly wetted inclusions. While the cavitation threshold shows the starting point of degassing, the degree of cavitation development determines the degassing in melts. In this process, the disruption of the dynamic equilibrium in the melt–oxide–hydrogen system by cavitation is strongly controlled by the concentration of solid oxide inclusions.

The results shown in Figure 2.24 demonstrate that the cavitation threshold depends on the content of solid aluminum oxide inclusions in the liquid aluminum, so that this measurement can even be used in rapid tests of the amount of solid inclusions before continuous casting. Figure 3.4 further illustrates the effect of solid inclusions by the degassing kinetics of 99.99% pure Al contaminated with α-Al$_2$O$_3$ and γ-Al$_2$O$_3$, which are known to have different adsorption ability with respect to

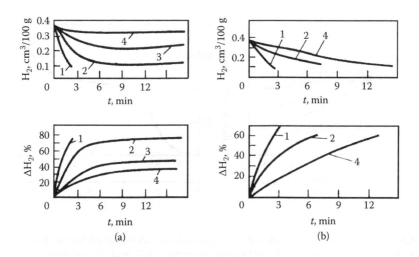

FIGURE 3.4 Kinetics of the ultrasonic degassing of Al99.99 (AA1090) aluminum melt contaminated with (a) γ-Al$_2$O$_3$ and (b) α-Al$_2$O$_3$ to the concentration (1) 0.05 wt%, (2) 0.01 wt%, (3) 0.01 wt% (calcinated γ-Al$_2$O$_3$), and (4) 0.1 wt%.

hydrogen [16]. Degassing curves 2–4 show that for oxide concentrations exceeding typical concentration (0.005 wt%) in the initial pure metal, the degassing becomes slower, the efficiency of the ultrasonic degassing decreases, and the residual hydrogen concentration in the melt increases. There is a seemingly contradictory situation: contamination with solid inclusions lowers the cavitation threshold and, hence, accelerates the onset of the degassing, but it also slows down the process proper and decreases its efficiency. To understand this paradox, we need to recall that the presence of small nonmetallic particles in the melt slows down the dissociation of supersaturated hydrogen solutions in aluminum due to hydrogen adsorption by the particles [19, 20]. On the other hand, the same very adsorption facilitates the formation of cavitation nuclei (see Section 2.4.1 and Figures 2.23–2.25). The nonuniform hydrogen distribution near nonmetallic inclusions assists the transformation of cavities generated around particles into gas bubbles. This is due to the fact that near these particles, the melt is supersaturated with hydrogen, regardless of the type of hydrogen–particle bonds.

These bubbles, in turn, are able to take part in the degassing process; they pulsate, collapse, generate new small bubbles, interact with adjacent bubbles, and grow to a size determined by the Stocks law, forcing them to float to the surface of the pool. To put it another way, the cavitation threshold and formation of single hydrogen bubbles near nonmetallic inclusions determine the start of degassing, i.e., in liquid metals the cavitation threshold coincides with the degassing threshold. Indeed, if we consider the efficiency of ultrasonic degassing as a function of ultrasonic intensity transferred to the melt, we obtain curves similar to those shown in Figure 3.5 for commercially pure aluminum (1 kg melt, 40-mm cylindrical sonotrode). These curves can be easily related to the formation and development of acoustic cavitation in melts (see Figure 2.24).

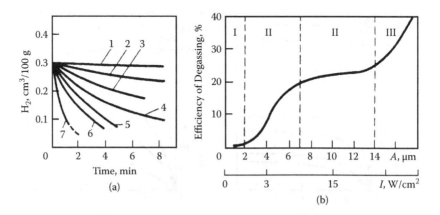

FIGURE 3.5 Effect of cavitation development on the kinetics of ultrasonic degassing of commercially pure aluminum AA1070: (a) different horn null–peak displacement amplitudes A equal to: (1) 2 μm, (2) 3 μm, (3) 5 μm, (4) 10 μm, (5) 15 μm, (6) 18 μm, and (7) 22 μm; and (b) effect of ultrasonic intensity on ultrasonic degassing: (I) precavitation, (II) cavitation threshold, and (III) developed cavitation.

Analyzing the family of curves in Figure 3.5a, one can conclude that (1) degassing is absent at ultrasonic intensities below 1.0 W/cm^2 and (2) degassing rate and intensity increase in proportion to the intensity at ultrasonic intensities above 2.5 W/cm^2. Presenting the efficiency of ultrasonic degassing of commercial aluminum as an isochrone (Figure 3.5b), one can see three specific regions, illustrating the generation and development of ultrasonic cavitation in liquid melts.

Indeed, region I, where ultrasonic degassing is almost absent, may be called the region of precavitation treatment. Region II, where the efficiency of degassing first increases rapidly and then gradually stabilizes, refers to the ultrasonic treatment regimes around the cavitation threshold. The generation and development of cavitation changes the ratio between the ultrasonic power transferred to the melt and the ultrasonic power consumed in the melt [21]. With the development of cavitation, the wave impedance (the product of density ρ and sound velocity c) of the melt decreases. As a result, the ratio between the sound intensity transferred to the melt and the degassing efficiency changes.

With a further increase in ultrasonic intensity, cavitation reaches a steady level, and we can define an additional region of regimes for ultrasonic treatment—region III, where the efficiency of ultrasonic degassing increases in a linear manner with the ultrasonic intensity. In this region, metals are treated in the regime of developed cavitation, where the wave impedance $\rho_c c_c$ is smaller than the wave impedance of the noncavitating melt $\rho_0 c_0$. Our data show that the formation and development of cavitation decreases the acoustic impedance of the aluminum melt by approximately a factor of 7—from 17 to 2.4 kg/(m^{2}s).

For even greater intensities of ultrasound introduced into the melt, one may propose the existence of an additional region, where the efficiency of degassing is drastically reduced. Physically, this state of melts is characterized by an increase in acoustic impedance under well-developed cavitation. This increase is so high that bubbles dispel all droplets of liquid from the cavitation zone, and oscillations are entirely absorbed near the radiating face and do not penetrate the treated volume. However, such a regime is hardly achievable because it requires very high magnitudes of acoustic energy that are possible only within very small volumes, whereas the practice of cavitation treatment of liquids (melts) tends to maximally expand the treated volumes and minimize the ultrasonic power.

3.2 MECHANISM OF ULTRASONIC DEGASSING

The degassing, irrespective of physical and technical means, is dependent on the concentration of dissolved gas in the liquid. This concentration is not a constant, but depends on several factors, most important of which are temperature, vapor pressure, and limit solubility.

Liquid aluminum and its alloys react actively with gases, thus forming nonmetallic impurities. One of the most important gases is hydrogen that finds the way to the liquid metal through the interface between the melt and the atmosphere. The main sources of hydrogen are the molecular hydrogen in air and water moisture or vapor in the atmosphere. The latter reacts with liquid aluminum at the surface of the melt and produces alumina and hydrogen through Reaction (3.2). The resultant atomic hydrogen is dissolved in the aluminum, and Al_2O_3 is deposited at the surface or dispersed in the liquid. Hydrogen that is not dissolved, or hydrogen that precipitates during degassing or solidification, forms molecular hydrogen (Reaction 3.3). Water vapor can react with liquid Al, producing molecular hydrogen as well; this will mostly dissolve back in the air (Reaction 3.4) [16].

$$3 H_2O + 2 Al \rightarrow 6 H + Al_2O_3 \tag{3.2}$$

$$H + H \rightarrow H_2 \tag{3.3}$$

$$3 H_2O + 2 Al \rightarrow Al_2O_3 + 3 H_2 \tag{3.4}$$

It is important to understand that the solubility of hydrogen in liquid aluminum is not a constant or a fixed number. The solubility depends on the conditions at the interface between the hydrogen-containing medium (atmosphere or bubble) and the liquid metal (surface or bulk). A quasi-equilibrium solubility exists for each combination of the hydrogen concentration in the atmosphere (humidity), in the melt (dissolved hydrogen) and the pressure (air pressure and partial pressure of hydrogen).

The practical importance of dissolved hydrogen comes from the sharp decrease of its solubility with aluminum solidification: Dissolved hydrogen can be measured up to 0.65 cm^3/100 g in liquid aluminum just above the melting temperature, while just below this temperature the solubility drops down to 0.034 cm^3/100 g [22]. During solidification, this difference causes the excess hydrogen to precipitate and, being trapped between the solid dendrites, form porosity. Gas porosity combined with shrinkage porosity is detrimental to the mechanical properties of the final products, especially to their fracture toughness, fatigue endurance, and ductility. Moreover, hydrogen that has not had time to precipitate and formed supersaturated solid solution with aluminum will precipitate during downstream processing, e.g., homogenization, extrusion, or hot rolling, thereby forming delaminations and secondary porosity, especially harmful in thin-gauge products or surface-critical applications.

Conversely, the interaction of molten aluminum with oxygen directly produces solid oxides. The partial vapor pressure of oxygen at 727°C is very low (4.406 × 10^{-42} kPa), so the solubility of oxygen in liquid aluminum is not an issue [16]. Aluminum oxides occur in two basic modifications: α-Al_2O_3 and γ-Al_2O_3. The α-modification is the equilibrium form, which is resistant to temperature and occurs in the form of corundum. The γ-modification and other metastable versions, such as δ and κ, naturally form in liquid aluminum as films and particles, and are eventually transformed into the α-modification upon melt heating and holding. As we discussed in Section 3.1, alumina plays a significant role in the kinetics and efficiency of degassing.

The hydrogen content of primary aluminum poured from the cell is under 0.15 cm^3/100 g, and the oxide concentration is about 0.006–0.007 wt%. However, during remelting, alloying, and casting, the metal becomes heavily contaminated with nonmetallic impurities and saturated with hydrogen.

In the preparation of melts of aluminum and its alloys, the main sources of hydrogen are water vapor in the furnace atmosphere and water adsorbed by charge components, fluxes, lining, and so on. Carbohydrates present in the burning gas may also contribute to the hydrogen enrichment of melt in gas-heated furnaces. In liquid metals, the hydrogen solubility is proportional to the concentration of atomic hydrogen in the atmosphere; however, its content in air is negligible (about 5×10^{-5} vol%), so the main source of hydrogen is the reaction (Reaction 3.2) between liquid aluminum and water vapor.

Information about relative humidity (RH) on the day of degassing can be converted to hydrogen concentration (H) in the atmosphere using the following formula deduced from data provided by Waite [23]:

$$H \text{ (cm}^3/100 \text{ g)} = 0.1772 \text{RH} \text{ (\%)} + 0.0394.$$ (3.5)

The limit solubility of hydrogen (C_{lim}, cm^3/100 g) in a liquid aluminum alloy can be calculated as a function of temperature (K) as follows [24]:

$$\log C_{lim} = -3050/T + 2.94.$$ (3.6)

Thermodynamic analysis of Reaction (3.2) [15, 16] shows that the partial pressure of hydrogen is extremely high even at low pressures of water vapor. At 727°C and a water vapor pressure of 1.33 kPa (typical atmospheric value), the equilibrium partial pressure of hydrogen at the liquid–gas interface reaches a huge value of 8.87×10^6 GPa, so the hydrogen content of the melt might be as high as 3.24×10^5 cm^3/100 g. This means that all available hydrogen can be dissolved in liquid aluminum, and that relatively small atmospheric humidity may lead to high hydrogen concentration in the melt. One cubic meter of air contains about 10 g of water, which is equivalent to 1 g of hydrogen. This one gram of hydrogen being dissolved in a metric ton of liquid aluminum may produce about 3% porosity [16].

The hydrogen produced from water vapor is dissolved in liquid aluminum. When its concentration reaches the concentration reflecting the equilibrium between liquid aluminum and molecular hydrogen at the current ambient pressure conditions, the dissolution stops, and atomic hydrogen will have a driving force to recombine into molecules and leave the melt. As a result of these two processes, there will be a dynamic equilibrium between atomic hydrogen intake (regassing) and molecular hydrogen expel from the melt (degassing). This equilibrium can be shifted if the pressure, temperature, humidity, or interface conditions change. The general possibilities for the variation of hydrogen content in liquid aluminum after ultrasonic degassing are illustrated in Figure 3.6a [25]. It is important to note that the degassing process is usually faster than that of regassing [12].

Of course, not all hydrogen is adsorbed in molten metal; part of it recombines into molecules and remains in the atmosphere. The oxide film formed at the metal–gas interface also hampers the penetration of hydrogen into the molten volume. This film can be as thick as 200 nm, and its chemical composition depends on the composition of the particular alloy; in pure aluminum, it usually consists of γ-Al$_2$O$_3$.

Alloying elements may influence of hydrogen concentration in the aluminum melt in three ways [16]. Firstly, some elements change the solubility of hydrogen in the liquid aluminum, e.g., Mg and Li increase solubility, while Cu, Si, and Fe decrease

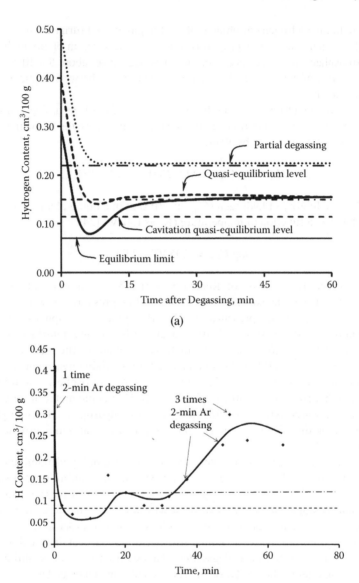

FIGURE 3.6 Kinetics of degassing and regassing: (a) different possible scenarios of variation in hydrogen content in liquid aluminum after degassing and (b) regassing after Ar rotary degassing of an aluminum alloy. (Courtesy of T. Pabel.)

it [26]. Also, the solubility of hydrogen in the solid Al can be affected. Secondly, some elements change the nature of the oxide layer on top of the melt, affecting the hydrogen permeability. For example, Mg changes the composition of the surface layer from alumina to spinel, which increases its permeability, as spinel film is much less continuous and much weaker than alumina. As a result, the processes of degassing and regassing are accelerated. Beryllium, on the contrary, strengthens the oxide film and prevents regassing of the melt. Thirdly, an alloying element can act as a surfactant and change the interfacial energy at the surface of a bubble forming in the melt during degassing. Magnesium thus assists in forming larger bubbles that more easily float to the surface, thereby accelerating degassing.

Aluminum alloys would typically have different levels of hydrogen content: Commercially pure Al will have between 0.2 and 0.3 cm^3/100 g, while Al–Si and Al–Cu alloys will retain and pick up more hydrogen than pure aluminum, from 0.4 to 0.5 cm^3/100 g. In the case of Al–Mg alloys, the typical hydrogen levels are 0.4 to 0.6 cm^3/100 g; these alloys may pick up more hydrogen, but at the same time release more hydrogen. For a given charge of liquid aluminum, hydrogen content can be naturally reduced to 0.1–0.2 cm^3/100 g (degassing) given sufficient time (up to 1 h) and typical conditions (750°C, 30% humidity) [16, 22].

Natural degassing takes a long time and is impractical for industrial applications, so different methods have been proposed for accelerating this process. Two types of degassing methods are currently used for aluminum alloys: gas purging (rotary and lance systems) and vacuum degassing. Chlorine-containing gases, however efficient they are, have been replaced with inert gases, mostly Ar, due to environmental considerations. Bubbles formed by purged gas create numerous interfaces that promote recombination of hydrogen to molecular form and evacuate this gaseous hydrogen from the melt. The number and size of the bubbles along with the forced convection seem to be the main parameters of the process [16, 23]. Vacuum degassing is based on the decreased pressure above the melt surface that should result in a decrease in the quasi-equilibrium hydrogen solubility and facilitate degassing [27]. Additionally, the decreased pressure helps in evacuating the bubbles from the melt, accelerating the process of natural degassing.

In most experiments and in industrial practice, the degassing process is performed until the desirable concentration of hydrogen (usually about 0.1 cm^3/100 g) is achieved. After that, the melt is cast. It is known that in degassing large volumes, some time is required to finalize the process of degassing by allowing the bubbles to float to the surface. What is much less studied is the process of regassing, or what could happen to the degassed melt after the end of the degassing process. Regassing is seldom reported, but there are some data showing that it is not an unusual phenomenon. General considerations of gas/liquid equilibrium attest for both processes, i.e., degassing and regassing, occurring simultaneously, will eventually reestablish a dynamic equilibrium that reflects the environmental and process conditions. Regassing was observed experimentally after the end of rotary Ar-assisted degassing (see Figure 3.6b), with ambient humidity reported to be responsible for that [23, 28].

Ultrasonic degassing had been suggested quite some time ago as an environment friendly, robust, and efficient means of melt degassing [13, 29]. So let us now consider the mechanisms of ultrasonic degassing of molten metal in detail.

The gas solubility in the liquid phase depends on the pressure and, in the case of sonic processing, on the pressure variation in the sound wave. Chernyshev [30] theoretically considered the solubility of gases in melts subjected to low-frequency vibrations. Using Sieverts's equation for gas solubility and the variation of pressure

due to acceleration in the acoustic wave, he arrived at the following expression for
the vibration-dependent solubility:

$$C = k\frac{\omega}{2\pi}\sqrt{P_0 + \rho h(1 + A\omega^2\sin(\omega t))}, \qquad (3.7)$$

where P_0 is the atmospheric or hydrostatic pressure, h is the liquid (metal) head, and
t is time. Integrating this equation, one may obtain a relation for the solubility averaged over the sound-wave period, which results in estimates close to the limit gas
solubility at stationary conditions. Such an estimate would result in the conclusion
that vibrations are not practical for degassing. However, Sieverts's law is only valid
for the constant interface between gas and liquid, which is not the case for oscillating bubbles.

The oscillation of a bubble in the acoustic field brings about a special type of
convection that is called *rectified diffusion*. As a result, the gas transfer from the
liquid phase into the bubble becomes possible even when the difference between the
average gas concentration in the liquid C_0 and the gas concentration at the bubble/
liquid interface C_g is not large. The gas concentration at the bubble interface can be
written as [12]:

$$C_g = C_p[1 + 2\sigma/(R_0 P_0)], \qquad (3.8)$$

where R_0 is the equilibrium bubble radius, P_0 is the hydrostatic pressure, and C_p is the
equilibrium gas concentration in the liquid phase.

When the bubble compresses, the gas concentration inside increases, and the gas
diffuses to the liquid. Upon bubble expansion, the opposite process takes place. As
the bubble surface (hence, interface available for diffusion) becomes larger upon
expansion than that upon compression, the diffusion rate is higher in the rarefaction
stage than in the compression stage of the oscillation. This results in gradual filling
of the bubble with the gas dissolved in the liquid. In other words, the oscillating
bubble acts as a pump, extracting gas from the liquid phase. In addition to the rectified diffusion, microscopic acoustic streams generated in the viscous boundary layer
around the bubble take their part in the mass transfer, bringing fresh liquid phase to
the surface of the pulsating bubble.

The amount of gas M that is transferred to the bubble during several cycles N of
ultrasonic wave oscillation can be estimated as [31]:

$$M = \frac{16}{15}\sqrt{5\pi}\frac{z_0}{\rho_1}t^2 C_0\sqrt{Dt}\sum_{n=1}^{N}\frac{1}{\sqrt{n}}, \qquad (3.9)$$

where z_0 is the tensile stress acting on the bubble, C_0 is the gas concentration in the
liquid (g/cm³), t is the time of bubble expansion during one period of sound, D is the
gas diffusion coefficient, and ρ_1 is the liquid density.

The actual gas solubility in the liquid phase under conditions of cavitation will
be lower than the quasi-equilibrium solubility. There exists a limit until which the

gas can be extracted from the liquid phase by cavitating bubbles. This limit was estimated to be about 50% of the quasi-equilibrium gas solubility under given environmental conditions. This was first established for degassing water from oxygen [32] and then confirmed for degassing aluminum from hydrogen [13, 33]. The actual value can be even smaller due to the hysteresis of gas diffusion [32]. Figure 3.6a shows this as the cavitation quasi-equilibrium level. Under conditions of cavitation, the instantaneous solubility can be described as [32]:

$$C = \frac{C_0}{2} + \frac{1}{\pi} \left[\frac{C_0}{\sin\left(\frac{C_0}{C_A}\right)} - C_A + \sqrt{C_A^2 - C_0^2} \right], \tag{3.10}$$

where C_0 and C_A are the gas solubilities at atmospheric and acoustic pressure, respectively.

Kapustina [12] reports that the quasi-equilibrium concentration achieved under ultrasonic degassing depends on $P_A^2 t_{qe}$ (here t_{qe} is the time required to reach the quasi-equilibrium concentration), decreasing with this product. The apparent independence of the quasi-equilibrium concentration of the sound pressure is because the increased sound pressure (amplitude) results in shortening the time t_{qe}, so the product of those does not change much. The quasi-equilibrium concentration depends also on the frequency of oscillations in a form of $fe^{-\beta f}$ [12]. This function changes with maximum.

As we discussed previously in Section 3.1, liquid metal typically does not contain free gas bubbles unless they are intentionally introduced by lancing or blasting of gas. A sound wave with pressure above the cavitation threshold propagating through a liquid metal generates cavitation bubbles in the melt, which are then filled with gas through rectified diffusion, after which they oscillate, grow, coalesce, and float.

The ultrasonic degassing of liquid metal is a process of three simultaneous stages [5, 12, 29, 34]: (1) gas bubbles form on cavitation nuclei and grow in the ultrasonic field, accumulating hydrogen through rectified diffusion (if the liquid contains small bubbles, this stage consists only of their diffusion growth); (2) separate bubbles coalesce under the action of the Bjerknes and Bernoulli forces; and (3) bubbles float to the surface of the molten metal.

The first stage was considered in Section 3.1 (degassing nuclei) and in Chapter 2, where we discussed the diffusion growth of a bubble under cavitation conditions (see Figures 2.9–2.13). The rectified diffusion occurs in the precavitation regime, as considered by Kapustina [12], and under cavitation, as discussed by Boguslavsky [31].

It should be noted that contradictory opinions exist on the connection between cavitation and degassing. For example, Kapustina [12] believes that cavitation is not necessary for degassing. And this might be true for water, with readily available air bubbles, which were the main object of Kapustina's study. In this case, micro- and macrobubbles pulsate, grow due to rectified diffusion and coalescence, and eventually escape from the liquid.

Conversely, our investigations with light alloys demonstrated that the efficiency and the very occurrence of ultrasonic degassing is a function of cavitation development (see Figure 3.5). In our opinion, the effect of cavitation is in the formation and multiplication of bubbles, and the enhanced diffusion flow of gas to the bubbles is

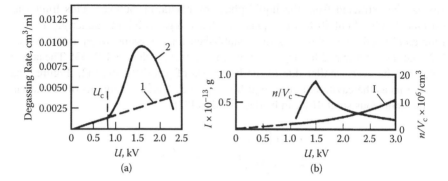

FIGURE 3.7   Effect of ultrasonic intensity on (a) the degassing of water and (b) the concentration of cavitation bubbles n/V_c (n is the number of bubbles, and V_c is the volume of cavitation region) and the diffusion flow I to the bubble: (1) precavitation mode, (2) developed cavitation. (Adapted from Kapustina [12].)

due to a significantly increased area of the gas–liquid interface for bubbles oscillating in the cavitation mode.

Figure 3.7 illustrates the effect of ultrasonic cavitation on the degassing kinetics in water [12]. The experiments were done with water irradiated by a focused sonotrode operating at 500 kHz. Figure 3.7a shows that the amount of oxygen extracted from the water increases dramatically when the ultrasonic intensity overcomes the cavitation threshold. As it follows, the rate of mass transfer under cavitation exceeds the rate in the precavitation mode by a factor of 3. However, with further increase of ultrasonic intensity, the efficiency of ultrasonic degassing decreases, falling to theoretical (dashed) curve 1, which presents the degassing rate under assumption that no cavitation occurs.

The reason for such a degassing behavior is primarily in bubble multiplication upon cavitation. This increase in the number of bubbles is very rapid because it follows chain-reaction dynamics. The process is as follows: A collapsing bubble loses its stability and breaks into a multitude of small bubbles that find themselves in the area of elevated temperature and pressure. These bubbles expand more easily during the rarefaction phase than stationary cavitation nuclei and, during the following compression phase, they collapse more actively, generating new bubbles. However, the increase in the sound pressure (ultrasound intensity) results in the increase of the stable bubble radius; so at a certain bubble size, the time required for its collapse may become equal or larger than the half-period of oscillation, as illustrated in Figure 3.7b (here the voltage 1.7 kV corresponds to the time of bubble collapse equal to the half-period of the acoustic wave [35]). In this case, the pressure at the final stage of collapse decreases, and the multiplication of bubbles slows down, which is reflected in the descending branch of curve 2 in Figure 3.7a and in curve n/V_c in Figure 3.7b.

Recognizing the value of the data given in Figure 3.7a, we should however note that these results were obtained in a unique experiment that used a very powerful ultrasonic concentrator capable of producing ultrasonic intensities above 2 kW/cm^2

at 500 kHz in a focal volume of 1 cm^3. In practice, ultrasonic intensities applied to liquid metals do not exceed 100 W/cm^2 even at lower frequencies (10–20 kHz), so these intensities correspond to the left-hand branch of the degassing curve in the cavitation mode (see Figure 3.5). In addition, the cavitation threshold is known to depend significantly on sound frequency, so the cavitation processing may be more appreciable at lower frequencies because of the higher number of cavitation bubbles and intense acoustic flows.

The dynamics of size and pressure in the cavitation bubbles are given in Section 2.4.2 for 1.5–2.5 periods of the sound wave. At a sound pressure of 0.2 MPa (the precavitation mode of ultrasonic degassing), bubbles grow only slightly and hardly pulsate. In the bubbles of all simulated sizes, the pressure also varies only slightly. As the sound pressure increases up to levels exceeding the cavitation threshold (P_A = 1 MPa), the bubbles start to grow more actively. Bubbles of smaller initial radii begin to pulsate, and their internal pressure decreases appreciably.

A further increase in the sound pressure P_A up to a level of 5.0 MPa results in a considerable increase of bubble size for all observed initial radii, and the second maximum appears on the bubble growth curves. The most appreciable growth is seen in the bubbles whose initial radii are comparable with the size of solid non-metallic inclusions suspended in the melt (about 1 µm). In the developed cavitation mode, the pressure in bubbles decreases greatly: when a cavity with a 1-µm initial radius expands, the pressure falls below 1 kPa. Analysis of numerical solutions of the dynamic cavity equations (see Figures 2.9–2.13) shows that cavitation is a necessary condition for liquid-metal degassing.

If we take a bubble with the initial radius $R_0 = 1$ µm, which is approximately equal to the size of solid inclusions in liquid metal, we notice that in the precavitation and threshold modes, such a bubble pulsates only slightly around the equilibrium position, and the bubble pressure almost coincides with the atmospheric pressure. An increase of the sound pressure to above the cavitation threshold of 1.0 MPa and the transition to the developed cavitation mode considerably changes the pattern. In cavitation, the motion of bubbles with $R_0 = 1$ µm develops a characteristic cavitation pattern: Bubbles actively expand during 1–2 periods and then collapse. Because the bubbles expand greatly in this mode (by more than three orders of magnitude), relatively large bubbles (up to 1 mm in size) are formed. It is fair to suggest that the collapse of such large bubbles will create a multitude of bubbles in the size range from 1 to 100 µm. While the growth and collapse of small bubbles leads to the multiplication of gas nuclei, medium and large bubbles can coalesce and float to the surface of the liquid-metal pool.

If we take into account the diffusion of hydrogen into the bubbles, the behavior will be altered. The collapse of small bubbles will be hindered. The amplitude of vibration for larger bubbles will decrease until they become very large. However, the simulations show that hydrogen diffusion into the bubble will noticeably change the mass of hydrogen inside the bubbles only in the developed cavitation mode (see Table 2.3).

Thus, the redistribution of dissolved hydrogen over pulsating cavitation bubbles is characteristic of the actual ultrasonic treatment of melts in the developed cavitation mode. In this process, the initial content of solid nonmetallic inclusions and the initial ambient and processing conditions play a considerable role in degassing.

The earlier that cavitation starts, the lower is the sound intensity that is required for the beginning of hydrogen redistribution to the bubbles. In other words, the new metastable dynamic equilibrium is established sooner in the melt–aluminum oxide–hydrogen system. The higher the initial concentration of cavitation nuclei (or, for an aluminum melt, the higher the concentration of solid nonmetallic inclusions of aluminum oxide), the sooner the reproduction of bubbles stabilizes and the efficiency of the ultrasonic degassing ceases to increase with the sound intensity. This can be seen as the shift of the dynamic equilibrium by the solid inclusions (alumina) under conditions of developed cavitation.

After the bubbles are created by cavitation and start to absorb gas dissolved in the liquid phase, they also start to interact with each other. Pulsating bubbles are either attracted (when oscillating in phase) or repulsed (when oscillating in antiphase). The forces of their interaction are called *Bjerknes forces*. It is important to note that the bubbles with similar dimensions are usually attracted to each other, whereas the bubbles with different sizes are repulsed. The calculations [12] show that the distance at which the bubbles are attracted to each other and then coalesce increases with the bubble diameter, sound pressure amplitude, and oscillation frequency. The time required for coalescence should be larger than the time within which the bubbles are passing each other at a distance when the attraction is possible. The maximum possibility of coalescence will be for bubbles of the resonance size. The coalescence of bubbles is assisted by acoustic streaming and convective flows.

The bubbles that grow to a substantial size due to gas intake and coalescence will float to the surface of the melt. The velocity of their rise to the surface depends on the size of the bubble, the convective regime in the liquid volume, the viscosity of the melt, and the surface tension. For liquid aluminum at 700°C, this velocity changes from 0.1 to 0.82 m/s for the bubble diameters from 0.4 to 3 mm and does not change much with further increase in the bubble size [18]. The real residence time of such bubbles in the melt volumes used for degassing (depth up to 2 m) is therefore limited to a couple of seconds, which is not sufficient for reaching the full absorption capacity. Ultrasonic cavitation, on the one hand, intensifies the diffusion of hydrogen to the bubbles and, on the other hand, induces stirring of the melt volume that increases the residence time of smaller bubbles and accelerates the evacuation of larger bubbles.

3.3 ULTRASONIC DEGASSING IN A STATIONARY VOLUME

Historically, degassing was initially studied and implemented in stationary volumes, i.e., in furnaces, ladles, crucibles. This is also a typical setting for laboratory experiments. Let us first look at the main parameters of the degassing process as studied in the laboratory. A series of such experiments have been performed recently in the United States [9, 36], in Switzerland and Portugal [37], in China and Japan [38], and in the U.K. [25]. Figure 3.8 illustrates the influence of various process parameters on the degassing kinetics and efficiency.

It is clear that there are important process parameters, i.e., the extent of cavitation (both in terms of energy and dimension of the cavitation region), the chemical composition of alloy (alloying elements and inclusions), and the treated volume and processing time.

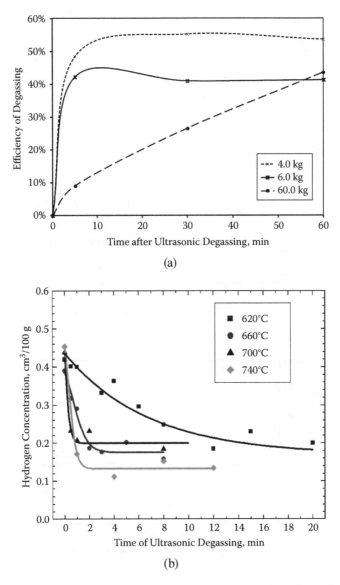

(a)

(b)

FIGURE 3.8 Effects of various process parameters on degassing of aluminum alloys: (a) effect of treated volume (A380, 2-min degassing); (b) effect of processing time and temperature (0.2 kg, A356) (adapted from Xu, Han, and Meek [9]). (*continued*)

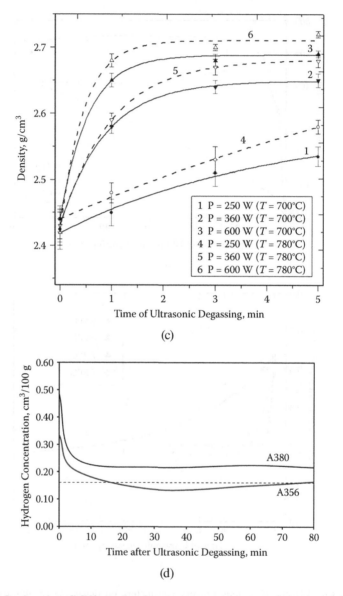

(c)

(d)

FIGURE 3.8 (*continued*) Effects of various process parameters on degassing of aluminum alloys: (c) effect of ultrasonic energy and melt temperature (4 kg, A380 type alloy) (adapted from Puga et al. [39]); (d) effect of alloy type (4 kg, 2-min degassing) (adapted from Alba-Baena and Eskin [25]).

Ultrasonic energy: Ultrasonic energy or the processing power can be estimated in different ways, e.g., input power of the generator (kW), amplitude of the sonotrode (μm) at a given frequency (kHz), and input acoustic power to the liquid (W or W/cm²). In any case, it is shown that the increase in the ultrasonic power results in the increased efficiency of degassing. The time of degassing for 10-kg A356 melt decreases from 25 to 10 min when the ultrasonic power increases from 6 to 10 W/cm² [29]. A rule of thumb for the input ultrasonic power is about 6–8 W/kg [29]. Figures 3.5 and 3.8c illustrate the importance of ultrasonic power for the efficiency of degassing [39].

Melt temperature: Melt temperature affects the efficiency of degassing though hydrogen solubility, diffusivity, and the melt viscosity. The recommended range of Al melt temperatures is from 700°C to 760°C, where the efficiency of the degassing increases with the melt temperature as the viscosity of the melt decreases and the diffusivity of hydrogen increases [9, 29, 37]. At lower temperatures, the viscosity increases and prevents efficient bubble removal, and the diffusion of hydrogen from the melt to the bubbles slows down. At higher temperatures, the intake of hydrogen due to its increased solubility in the liquid phase becomes comparable with the amount of hydrogen removed by degassing. Table 3.2 shows the duration of ultrasonic degassing of an A356 alloy to the quasi-equilibrium concentration versus temperature and viscosity. At temperatures below 700°C, the efficiency of degassing decreases, not only because of an increased viscosity that hampers pulsations of cavitation bubbles, their coagulation, and floating, but also because the diffusion coefficient of hydrogen decreases with temperature in liquid metals, thus decreasing the rate of one-way diffusion of hydrogen from the liquid solution to bubbles under the action of alternating sound pressure. The optimum temperature for A356 alloy would be around 730°C. Figures 3.8b,c further illustrate the importance of melt temperature for degassing of A356 and A380 alloys. Our results also show that optimum melt temperature for an AMg6 (Al–6% Mg–0.6% Mn) alloy is around 710°C.

Inclusions: The presence of insoluble inclusions (as we have discussed previously) provides interfaces for cavitation and gas bubble nucleation. As

---

TABLE 3.2
Effect of Temperature on the Duration of Ultrasonic Degassing of an A356 Alloy

Temperature, °C	Viscosity (μ), Pa·s	Hydrogen Content, cm³/100 g		Duration, min
		Initial	Final	
710	0.0089	0.36	0.1	10
730	0.0089	0.37	0.1	8
760	0.0082	0.34	0.1	10
790	0.0079	0.32	0.14	11

a result, the cavitation threshold is decreased and the degassing starts at a lower ultrasonic intensity. At the same time, the absorption of hydrogen at the inclusions retards the degassing, and the earlier cavitation onset results in earlier reached dynamic equilibrium in the Al–oxide–hydrogen system. Therefore, the presence of inclusions may decrease the overall efficiency of degassing. On the other hand, the absorption of hydrogen to the inclusions and the formation of an attached bubble may trigger flotation of small inclusions, effectively cleaning the melt of both inclusions and hydrogen.

Alloy composition: The alloy composition may affect the efficiency of degassing by the change of hydrogen solubility (increases with Mg, Zn, Li), change of viscosity (decreases with the amount of alloying elements), and change of surface tension (decreases with Li, Bi, Pb, Mg, Sb) [40]. However, there is not enough experimental data to make quantitative conclusions about the effects of specific alloying elements. Figure 3.8d shows that the presence of Mg in an A356 alloy makes a difference in the kinetics of melt degassing as compared to a Mg-free A380 alloy. Our data on wrought 5XXX (Al–Mg), 2XXX (Al–Cu–Mg), and 7XXX (Al–Zn–Mg–Cu) alloys given in Figure 3.9 attest that (at least for these alloy types) the efficiency of ultrasonic degassing depends much more strongly on the intensity of ultrasound than on the alloy composition.

Treatment time and volume: Figures 3.8a–c and 3.9 illustrate the importance of the time and volume as important process parameters. In all cases, the degassing curve reaches the limit (plateau) when further treatment does not result

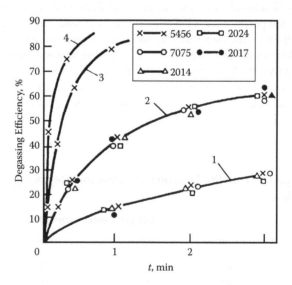

FIGURE 3.9 Efficiency of ultrasonic degassing of various industrial aluminum alloys versus treatment duration for several sound intensities: (1) 36 W/cm^2, (2) 45 W/cm^2, (3) 68 W/cm^2, and (4) 95 W/cm^2.

in additional effect, which reflects the quasi-equilibrium solubility of hydrogen under given processing conditions (see Figure 3.6a). The initial degassing efficiency is very high, but on increasing the time, the efficiency decreases. There is no readily available recipe to calculate the time and ultrasonic intensity required to degas a given volume. It is reported that the degassing of up to 10 kg of melt takes up to 10 min, with most of the degassing occurring in the first 2–5 min, and can be done with a single sonotrode. Larger volumes will require much more time for the release of gas bubbles, as shown in Figure 3.8a for a 60-kg melt. When it comes to larger volumes, several sonotrodes need to be used, or a motion of sonotrodes in the volume should be effected. An empirical formula has been developed for the time required to efficiently degas a liquid aluminum bath 20 to 250 kg in volume [29]:

$$t = 2 \times 1.2^{(0.02m-1)} \times \frac{m^{0.5}}{n}\left(1 + \frac{740}{T}\right), \tag{3.11}$$

where t (in min) is the treatment time sufficient to efficiently degas the liquid volume with mass m (in kg); n is the number of sonotrodes, and T is the melt temperature (in °C). The coefficient 2 implies that two sessions of degassing are made, with an interval between to allow the bubbles to float. The time will be longer if a lower input energy is used.

The degassing of a stationary volume by short sonication sessions with idle intervals between them proves to be a good option. Figure 3.10 gives an example of ultrasonic degassing of a 2-kg charge of an A380 alloy by 2-min sessions. It is evident that each of the degassing sessions led to an additional decrease in hydrogen concentration, while the idle periods allowed for gas release from the melt through bubble flotation.

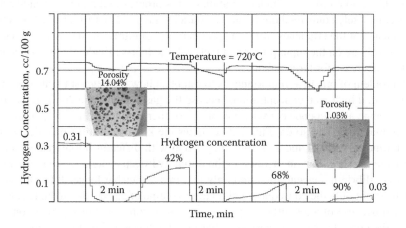

FIGURE 3.10 Kinetics of hydrogen removal from A380 melt (2 kg) as measured by Foseco Alpek-H detector.

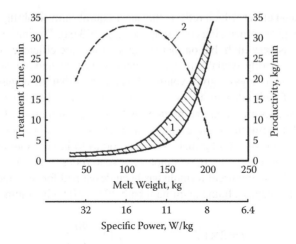

FIGURE 3.11 Effect of ultrasonic degassing efficiency on the specific power and melt weight: (1) the treatment time and (2) the productivity. (Adapted from Eskin [29].)

The following grid illustrates the duration (min) of ultrasonic degassing for various masses of melts to achieve 50% degassing:

Mass, kg	50	80	100	120	150	180	200	250
US degassing, (min)								
First	6	4	5	6	8	10	10	14
Pause		5	5	6	5	5	5	5
Second		5	5	6	8	10	20	20

It should be noted that the implementation of this process in a number of industrial trials completely confirmed our recommendations. Figure 3.11 further illustrates the dependence of the efficiency (time and productivity) of ultrasonic degassing in a stationary volume on the input power and melt volume (mass). Evidently the efficiency starts to decrease for large volumes.

A problem of practical importance is to establish how the degassing rate depends on the initial hydrogen concentration, because the gas content in melts varies across a wide range and depends considerably on atmospheric humidity, the preparation of charge materials, and so on. Figure 3.12 shows the kinetics of ultrasonic degassing of aluminum AA1070 and a casting alloy A356 with different initial gas contents. These curves show that an increase in the initial hydrogen content does not alter the quasi-equilibrium concentration and only requires somewhat longer times to reach it. This is not only the consequence of a larger volume of gas needed to be removed, but also because of a low efficiency of gas removal from a stationary melt volume, where the upper layers are degassed earlier than the lower and bottom layers.

The next section will show that for melts treated in a flow with a shallow ($H \leq 100$ mm) liquid pool, the degassing rate increases considerably. Conversely, when the

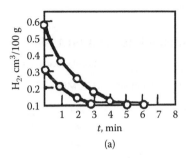 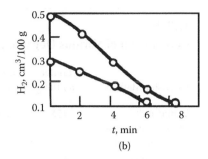

(a) (b)

FIGURE 3.12 Effect of the initial hydrogen content on the rate of ultrasonic degassing of (a) commercial aluminum (AA1070) and (b) a casting alloy (A356).

mold height is increased to $H \geq 500$ mm, as is the case in commercial degassing in holding furnaces, the duration of ultrasonic degassing increases.

The ultrasonic degassing of aluminum was implemented in foundries for precision investment casting, sand casting, die and high-pressure die casting, and injection casting. Let us look at the example of sand casting. A special ultrasonic degassing system UZD-200 had been developed in 1959 for degassing up to 250 kg of melt in stationary volume (see Figure 1.1b). The installation (in stationary UZD-200 and mobile UZD-100 versions) consisted of a 10-kW generator that fed four magnetostrictive transducers that worked in a sequence with a time gap of 15–20 s. The frequency was 19.5 kHz and the total acoustic power was 1.6 kW. The system was equipped with a time relay that allowed for a programmed degassing schedule. The waveguiding system was at this time made of a steel extension and a Ti sonotrode. However, later Ti was substituted by a Nb alloy. A356 and A361.1 alloys were melted in a gas furnace of 3-tonne capacity, and then 200-kg portions were taken by a ladle, where the melt was modified and cleaned using various degassing techniques, including ultrasonic degassing. The refined metal was then cast in sand molds to obtain 30-kg shape castings. At all stages of cleaning, samples were taken and the hydrogen concentration was measured. Then the castings were X-rayed and their properties were measured. Table 3.3 summarizes these results for castings of an A361.1 alloy. It can be easily seen that ultrasonic degassing significantly increases the density of cast metal and makes it possible to obtain almost pore-free castings (rank 1 in the porosity scale).

One of the important advantages of ultrasonic degassing as compared to the widely used Ar-rotary degassing is a drastically reduced amount of dross formed on the surface of the liquid metal during degassing processing. Intense rotation in an Ar-blasting degasser and vigorous Ar bubbling result in a highly disturbed surface of the liquid bath, frequently accompanied with vortex formation. As a result, a thick layer of alumina dross containing a considerable amount of metallic aluminum is formed at the surface. This dross should be removed and recycled; otherwise, the loss of aluminum becomes economically significant. In contrast, ultrasonic degassing occurs with a very quiet melt surface and much less turbulence in the melt bulk. The amount of dross decreases by a factor 5 to 10 [9, 41]. For example, after 15-min

TABLE 3.3

Comparison of Various Degassing Methods for an A361.1 Alloy

Degassing Method	H$_2$ Content, cm^3/100 g	Density, g/cm^3	Porosity Number	Tensile Properties	
				UTS, MPa	El, %
Starting melt	0.35	2.660	4	200	3.8
Ultrasonic degassing	0.17	2.706	1–2	245	5.1
Vacuum treatment	0.2	2.681	1–2	228	4.2
Argon blasting	0.26	2.667	2–3	233	4.0
Hexachloroethane	0.3	2.665	2–3	212	4.5
Flux	0.26	2.663	3–4	225	4.0

Note: UTS = ultimate tensile strength; El = elongation.

degassing of 150 kg of an A356 alloy in a stationary ladle, the amount of dross was 1800 g after Ar-rotary degassing and just 340 g after ultrasonic degassing, the degassing efficiency being the same [41]. This clearly gives an edge to the ultrasonic degassing.

3.4 ULTRASONIC DEGASSING IN THE FLOW

The requirement for processing of large, industrial-scale volumes of melt, especially in large foundries and continuous-casting plants, shows a limit for batch degassing operations. Another approach needs to be used, and the processing of the melt flow seems like a logical and viable possibility. In large melting/casting operations, it is more appropriate to relocate the cleaning of melts from the melting or holding furnace to the zone of metal transfer, somewhere en route from the furnace to the mold. One of the examples of degassing in the melt flow is the combination with vacuum degassing shown in Figure 1.3b. Recently there was an idea to combine the ultrasonic degassing with Ar-lancing in a vessel through which the melt is constantly flowing [42]. Despite some suggested technological schemes, ultrasonic degassing in melt flow has not yet found the way to foundries, but it was successfully used in direct-chill (DC) casting.

In DC casting of light alloys, ultrasonic processing can be implemented in three basic variants (Figure 3.13): (1) in the liquid pool of a solidifying ingot (the sonotrode penetrates the melt surface), (2) in the mold (oscillations are transferred to the melt through mold walls), and (3) in an intermediate volume between the furnace and the mold (the oscillations are transferred to the melt from the top or the bottom).

Earlier investigations showed that the treatment of the melt in the liquid pool of an ingot is less promising than in an intermediate volume between the furnace and the mold. This is because of the liquid pool temperatures around the liquidus when the viscosity of the melt is fairly high, and hydrogen bubbles need to move in the direction opposite to the acoustic flow [13]. Therefore, the ultrasonic treatment in a DC

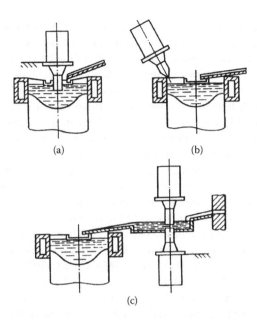

FIGURE 3.13 Possible variants of the ultrasonic degassing of melts in DC casting: (a) in the liquid bath, (b) in the mold with excitation of the mold walls, and (c) in the flow between the furnace and the mold.

casting mold is mostly applicable to grain refinement of the ingot structure, as will be shown in Chapters 5 and 7. As for the treatment with ultrasound transferred through the mold walls, this method requires such high acoustic energy levels that oscillations can produce stresses exceeding the fatigue strength of the mold, rapidly destroying it. Lower, subsonic frequencies can be efficiently used, and such schemes are available elsewhere [43]. There was one interesting scheme suggested by Seemann and Staats where the melt was treated in a launder by consumable sonotrodes, but it has not been commercially implemented (see Figure 1.3a)

First industrial trials on ultrasonic degassing in melt flow were performed in the USSR in the early 1960s during DC casting of aluminum alloys using a setup similar to that described previously for the batch ultrasonic degassing (UZD-200). The difference was in the arrangement of sonotrodes in line, as shown in Figure 3.14. The steel waveguides were about 2 m long, and the sonotrodes with "mushroom"-shaped tips were made from Ti, which assured (at the available generator power) an amplitude of 10–15 μm [44]. By taking into account that DC casting involves high flow rates and relatively low melt temperatures, a principle of multiple ultrasonic processing of melt flow was used.

The launder was constructed in such a way that it contained a section of ultrasonic processing and a section of gas release. In the former, hydrogen bubbles were formed under cavitation conditions; and in the latter, the bubbles grew and floated to the melt surface. Released hydrogen manifested itself by burning sparks above the melt surface.

The industrial trials involved simultaneous casting of two flat ingots, with and without ultrasonic processing, so that a proper comparison could be made (Figure 3.14b). In the case of round billets (Figure 3.14c), one half of the billet was cast with and the other one without ultrasonic processing. The flow rate was about 70 kg/min and the ultrasonic intensity was about 5 W/cm². The results demonstrated that the ultrasonic degassing in the melt flow allowed for a 1.5–2 times decrease in hydrogen concentration in the melt, as illustrated in Table 3.4. One can notice that the efficiency of degassing commercially pure aluminum is less than for more-concentrated alloys. This might be a consequence of its higher purity in solid inclusions with corresponding lesser cavitation development. The density measurement of the ingots and billets showed that the density was increased by 15%, which corresponded to a 1.5–2 times decrease in porosity. The amount of defects (porosity, nonmetallic inclusions) decreased by a factor of 5–8, e.g., from 0.82 to 0.1 mm/cm² in a 460-mm billet of a AA2038-type. The mechanical properties were also improved (see Chapter 7).

This experience was later extended to DC casting of various aluminum alloys, including Al–Mg (2% to 6% Mg), Al–Zn–Mg–Cu (AA7055-type), and Al–Cu–Mg (AA2038 and AA2214 types). In most cases, 40–50-mm conical or cylindrical tips (sometimes with mushroom-shaped endings) were used to transfer ultrasonic oscillations to the melt. Commercially produced generators and magnetostrictive

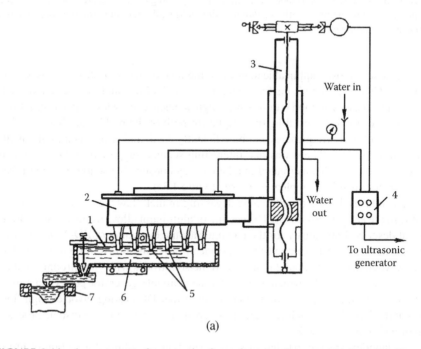

(a)

FIGURE 3.14 Arrangement of sonotrodes for melt processing en route from the furnace to the mold in DC casting. (a) Schematic: (1) degassing section, (2) assembly of ultrasonic transducers, (3) positioning arm, (4) terminal box, (5) sonotrodes, (6) melt; and (7) DC casting mold. (*continued*)

(b)   (c)

FIGURE 3.14 *(continued)* Arrangement of sonotrodes for melt processing en route from the furnace to the mold in DC casting. (b) DC casting of flat ingots; (c) DC casting of round billets.

TABLE 3.4

Concentration of Hydrogen Before and After Ultrasonic Degassing Upon DC Casting of Aluminum Alloys

			Hydrogen Concentration, cm³/100 g	
Alloy	Ingot/Billet Size, mm	Casting Speed, mm/min	No Degassing	Ultrasonic Degassing
AD (AA1030)	1040 × 300	123	0.28	0.18
AD (AA1030)	1040 × 300	70	0.25	0.19
AA2024	1480 × 210	123	0.41	0.24
AK6 (AA2038)	250 diam.	70	0.33	0.18
AK6 (AA2117)	460 diam.	35	0.4	0.21
AD1 (AA1070)	350 diam.	44	0.2	0.10
AMg2 (AA5017)	350 diam.	65	0.42	0.3

transducers with automatic frequency control and tuning were employed. The number of ultrasonic sources was varied, depending on the ingot diameter, melt flow rate, and the desired degree of degassing.

Ultrasonic degassing is most valuable for weldable wrought Al–Mg alloys, especially with high Mg concentration. An industrial-scale degassing plant was designed and manufactured for casting large flat ingots (1700 × 300 mm) from an AMg6 Russian Grade (6% Mg, 0.6% Mn). The degassing was performed in a specially designed section of a launder at 20 m downstream from a 40-tonne holding furnace. The melt flow rate was up to 100 kg/min. From 4 to 12 thyristor generators, each with 4.5 kW in power, excited matching magnetostrictive transducers. Each of the transducers was capable of delivering to the melt a maximum of 1 kW of acoustic power. Two schemes of ultrasound input were tried: from the bottom of the launder and from the top of the melt [45]. The latter version proved to be more reliable and efficient.

The efficiency of ultrasonic degassing with regard to the acoustic power introduced to the melt is given in Table 3.5 and in Figure 3.15. Similar results were obtained when casting round billets from the same alloy. Table 3.6 summarizes the results in relation to the ingot diameter and the number and intensity of ultrasonic sources. The efficiency of this process shows distinct dependence on the metal flow rate (billet diameter) and acoustic power (or the number of sources) conveyed to the melt [13].

Our experience with in-flow ultrasonic degassing of different alloys at different scales allows us to conclude that the 50% degassing efficiency is possible under the following conditions:

Number of ultrasonic sources	2–3	5–7	9–10	11–12
Melt flow rate, kg/min	10–15	30–50	75–85	90–100

The study of fracture surfaces and direct analysis of inclusions in aluminum and magnesium alloys demonstrates an additional effect of ultrasonic degassing: cleaning of the melt from solid inclusions. The amount of inclusions can be decreased by 30–50% as a result of ultrasonic degassing in the melt flow, as described previously.

Degassing of magnesium alloys is not a standard procedure. However, ultrasound degassing is shown to have an exceptional effect on the hydrogen content in magnesium alloys, where the hydrogen solubility is one order of magnitude greater than in aluminum melts. Figure 3.16 shows the effect of ultrasonic degassing on the hydrogen content in a large (165 × 550 mm) flat ingot of a MA2-1 (AZ31B) magnesium alloy [46].

Despite these successful industrial applications, the further development and spreading of this experience was hindered by the bulkiness of the equipment and lack of optimization of melt flow. Current efforts are concentrated on understanding the interaction between the melt flow, cavitation field, and acoustic streaming via physical and numerical modeling. Figure 3.17 demonstrates that the flow management via dams and baffles in a launder, as well as the orientation of the sonotrode, may considerably affect the flow pattern and the residence time of the melt in the acoustic field [47].

TABLE 3.5

Hydrogen Content (H$_2$), Relative Porosity (P), and Attenuation of Ultrasound Upon Nondestructive Testing (δ_{us}) in Flat 1700 × 300-mm Ingots of an AMg6 Alloy for Various Acoustic Powers Used in Ultrasonic Degassing in the Melt Flow During DC Casting

| Acoustic Power, kW | H$_2$ in Melt, cm^3/100 g | | Ingot Properties | | | | | |
| | No Degassing | US Degassing | No Degassing | | | Ultrasonic Degassing | | |
			H$_2$ in Solid, cm^3/100 g	P, %	δ_{us}, dB/cm	H$_2$ in Solid, cm^3/100 g	P, %	δ_{us}, dB/cm
4	0.60	0.45	0.41	n/a	n/a	0.36	n/a	n/a
5	0.60	0.42	0.44	0.6	1.93	0.35	0.51	1.75
7	0.58	0.39	0.41	0.66	1.83	0.33	0.55	1.67
11	0.56	0.29	0.38	0.56	1.80	0.20	0.40	1.50

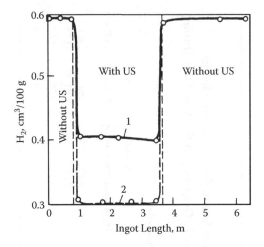

FIGURE 3.15 Hydrogen content variation in dependence on ultrasonic degassing upon DC casting of an AMg6 alloy (1700 × 300-mm ingot; 80 kg/min flow rate): (1) 9 kW and (2) 11 kW acoustic power.

TABLE 3.6

Efficiency of the Ultrasonic In-Flow Degassing of an AMg6 Alloy in Relation to the Billet Diameter, and the Number and Intensity of Ultrasonic Sources with 40-mm Diameter Radiating Face

Billet Diam., mm	Number of Sources	Ultrasonic Parameters A, μm	I, W/cm²	Hydrogen Content,[a] cm³/100 g Initial	Final	Degassing Efficiency, %
127	1	15	30	0.4	0.25	37
127	1	15	30	0.39	0.25	25
204	1	5	3	0.67	0.46	25
204	1	10	15	0.67	0.39	40
204	1	15	30	0.67	0.35	43
204	1	20	60	0.67	0.28	58
370	1	12	40	0.31	0.26	13
370	2	20	60	0.48	0.24	50
370	3	20	50	0.51	0.19	60

[a] Data obtained by vacuum extraction from the billet.

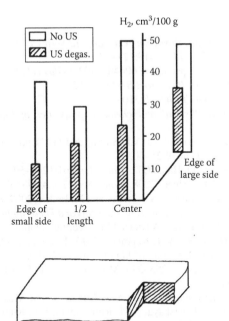

FIGURE 3.16 Effect of ultrasonic degassing on the hydrogen content in a large (165 × 550 mm) flat ingot of a MA2-1 (AZ31B) magnesium alloy.

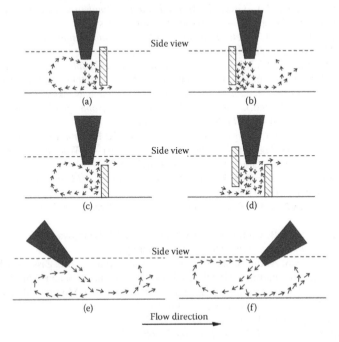

FIGURE 3.17 Flow patterns in the liquid flow under the action of ultrasound with different flow management. The results were obtained by water modeling. (Adapted from Zhang [47].)

REFERENCES

1. Boyle, R. W. 1922. *Trans. Royal Soc. Canada, 3rd Series* 16:157–62.
2. Sörensen, Ch. 1936. *Ann. Phys.* 418:121–37.
3. Krüger, F. 1938. German patent no. 604486, 1931; *Glastech. Ber.*, no. 7:233–36.
4. Bradfield, G. 1950. *Proc. Phys. Soc. B* 63 (5): 305–21.
5. Eisenreich, H. 1950. *Die Technik* 5:310–15.
6. Sergeev, S.V. 1952. *Physical and chemical properties of liquid metals.* Moscow: Oborongiz.
7. Altman, M.B., D.V. Vinogradova, V.I. Slotin, and G.I. Eskin. 1958. *Izv. Akad. Nauk SSSR, Otd. Tekhn. Nauk,* no. 9:25–30.
8. Xu, H., T.T. Meek, and Q. Han. 2007. *Mater. Lett.* 61:1246–50.
9. Xu, H., Q. Han, and T.T. Meek. 2008. *Mater. Sci. Eng. A* 473:96–104.
10. Puga, H., J. Barbosa, E. Seabra, S. Ribeiro, and M. Prokic. 2009. In *Fourth Intern. Conf. on Advances and Trends in Engineering Materials and Their Applications, AES—ATEMA '2009.* Hamburg, Ottawa: AES-Advanced Engineering Solutions (ISBN 0-9780479).
11. Haghayeghi, R., H. Bahai, and P. Kapranos. 2012. *Mater. Lett.* 82:230–32.
12. Kapustina, O.A. 1970. Degassing of liquids. In *Physical principles of ultrasonic technology,* ed. L.D. Rozenberg, 253–336. Moscow: Nauka [Translated to English. 1973. New York: Plenum].
13. Eskin, G.I. 1988. *Ultrasonic treatment of molten aluminum.* Moscow: Metallurgiya.
14. Arbuzova, L.A., K.I. Slovetskaya, A.M. Rubinshtein, L.L. Kunin, and V.A. Danilkin. 1971. *Izv. Akad. Nauk SSSR, Khimiya,* no. 1:169–70.
15. Makarov, G.S. 1983. *Cleaning of aluminum alloys by gases.* Moscow: Metallurgiya.
16. Dobatkin, V.I., R.M. Gabidullin, B.A. Kolachev, and G.S. Makarov. 1976. *Gases and oxides in wrought aluminum alloys,* 65–90. Moscow: Metallurgiya.
17. Kelton, K.F., and A.L. Greer. 2010. *Nucleation in condensed matter: Application in materials and biology,* 165–97. Oxford: Pergamon Press.
18. Makarov, G.S. 2011. *Billets of aluminum alloys with magnesium and silicon for extrusion.* Moscow: Intermet Engineering.
19. Pimenov, Yu.P., and A.I. Demenkov. 1972. *Tekhnol. Legk. Spl.,* no. 6:33–36.
20. Kunin, L.L., Yu.P. Pimenov, L.A. Arbuzova, and V.A. Danilkin. 1971. *Physical chemistry of surface phenomena in melts,* 99–103. Kiev, Ukraine: Naukova Dumka.
21. Rozenberg, L.D. (ed.). 1968. *High-intensity ultrasonic fields.* Moscow: Nauka [Translated to English. 1971. New York: Plenum].
22. Campbell, J. 1993. *Castings.* 2nd rev. ed. Oxford, UK: Butterworth-Heinemann.
23. Waite, P.D. 1998. *Light metals 1998,* ed. B. Welch, 791–96. Warrendale, PA: TMS.
24. Opie, W.R., and N.J. Grant. 1950. *Trans. AIME* 188:1237–41.
25. Alba-Baena, N., and D.G. Eskin. 2013. *Light metals 2013,* ed. B. Sadler, 958–62. Warrendale, PA: TMS/Wiley.
26. Anyalebechi, P.N. 2003. *Light metals 2003,* ed. P.N. Crepeau, 857–72. Warrendale, PA: TMS.
27. Altman, M.B., E.B. Glotov, R.M. Ryabinina, and T.I. Smirnova. 1970. *Purification of aluminum alloys in vacuum.* Moscow: Metallurgiya.
28. Zhang, L. 1996. A kinetic study of hydrogen absorption and degassing behaviour of DURALCAN composites. MSc thesis, University of Quebec at Chicoutimi.
29. Eskin, G.I. 1965. *Ultrasonic treatment of molten aluminum.* Moscow: Metallurgiya.
30. Chernyshev, I.A. 1953. *Lit. Proizvod.,* no. 10:13–18.
31. Boguslavsky, Yu.Ya. 1967. *Akust. Zh.* 13:23–27.
32. Lindström, O. 1955. *J. Acoust. Soc. Am.* 27:654–71.
33. Eskin, G.I. 1998. *Ultrasonic treatment of light alloy melts.* Amsterdam: Gordon and Breach OPA.

34. Eskin, G.I. 1968. *Ultrasonic degassing of molten metal*. Moscow: Mashinostronie.
35. Sirotyuk, M.G. 1967. *Akust. Zh.* 13:265–69.
36. Meek, T.T., Q. Han, and H. Xu. 2006. *Degassing of aluminum alloys using ultrasonic vibrations*. Report ORNL/TM-2006/61. U.S. Department of Energy.
37. Puga, H., J. Barbosa, E. Seabra, S. Ribeiro, and M. Prokic. 2009. *Mater. Lett.* 63:806–8.
38. Li, J., T. Momono, Y. Tayu, and Y. Fu. 2008. *Mater. Lett.* 62:4152–54.
39. Puga, H., J. Barbosa, J. Gabriel, E. Seabra, S. Ribeiro, and M. Prokic. 2011. *J. Mater. Process. Technol.* 211:1026–33.
40. Korolkov, A.M. 1960. *Casting properties of metals and alloys*. Moscow: Izd. Akad. Nauk SSSR.
41. Pabel, T., and H.N. Villa Sierra. 2013. Report D6.2, FP7 Project Ultragassing, Grant Agreement No. 286344.
42. Han, Q., H. Xu, and T.T. Meek. 2010. Degassing of molten alloys with the assistance of ultrasonic vibration. U.S. patent 7682556 B2, 23.03.2010.
43. Herrmann, E., and D. Hoffmann. 1980. *Handbook on continuous casting*. Dusseldorf, Germany: Aluminium Verlag.
44. Livanov, V.A., G.I. Eskin, and N.A. Genisaretsky. 1968. *Tsvetn. Met.*, no. 6:82–84.
45. Eskin, G.I., and P.N. Shvetsov. 1977. In *Metals science and casting of light alloys*, 8–17. Moscow: Metallurgiya.
46. Gur'ev, I.I., G.I. Eskin, and A.E. Ansyutina. 1978. In *Magnesium alloys*, 136–40. Moscow: Nauka.
47. Zhang, L. 2013. Ultrasonic processing of aluminum alloys. PhD thesis, Delft, Delft University of Technology.

4 Ultrasonic Filtration

4.1 EFFECT OF ULTRASOUND ON SURFACE TENSION AND WETTING

The control of surface tension is paramount for any filtration technique, as the surface tension determines the capillary pressure that is needed to be overcome for the liquid to penetrate through filter channels. Wetting (characterized by the wetting angle) is the other parameter determining the ease of liquid penetration through the filter. Equation (4.1) illustrates this by the metallostatic head required for the melt passage through the filter:

$$H = \frac{4\sigma \cos(\theta - 90°)}{\rho g a}, \tag{4.1}$$

where σ is the surface tension, ρ is the melt density, a is the mesh size, θ is the contact angle, and g is the gravity acceleration. Similar properties are important for introducing nonmetallic particles into the metal-matrix composites, as will be discussed in Chapter 9.

The characterization of wetting and surface tension under dynamic conditions—such as under the action of ultrasonic waves and cavitation—is not well developed, and indirect methods are most commonly used instead. However, there have been attempts to quantify the effect of ultrasonic vibrations on the wetting and surface tension using various experimental techniques adapted to dynamic conditions.

The most accepted experimental method of measuring wettability is the sessile-drop technique. In this method, the shape and contact angle of a drop of the liquid metal on the flat substrate are measured using optical techniques. Both the contact angle and the surface tension can be measured. The method requires a very stable substrate to make the observation of the wetting possible and inert atmosphere to prevent oxidation of the liquid/solid and liquid/gas interfaces. To reach steady-state conditions, the experiment can be continued for hours. The dependence of wetting on the substrate roughness and contact time makes the results less reproducible. This method is not directly applicable to studying the wettability under dynamic conditions. For example, application of ultrasonic vibrations in the cavitation mode to the substrate or the liquid will result in immediate dispersion of the liquid droplet, preventing any measurements from being taken.

Sessile-drop-type experiment can be performed in a precavitation mode. It was shown that 50-kHz vibrations at 50 W decrease somewhat the wetting angle of Hg on alumina and water on silica and alumina [1]. This effect becomes more pronounced with the time of vibrations up to 5 min, and this was explained by improved penetration of the liquid phase into crevices of the substrate surface. Recently, a version of this technique was applied to studying the wettability of graphite by Al and Al–Ti melts [2]. The droplet of melt was placed in a closed metallic heated chamber with the graphite substrate at the bottom, and ultrasonic vibrations (20 kHz, 50 W) were applied from the top of the chamber by short 25-µs pulses. The chamber was evacuated and, at the end of experiment, filled with Ar to cool the droplet. The geometrical changes in the droplet shape were monitored in situ by a high-definition camera. The results showed that the wetting angle decreases significantly when the ultrasonic vibrations are applied, from 150°–160° before processing to 45°–50° after 10 min of holding after the ultrasonication. The reasons behind the improved wetting are: (a) the destruction of alumina film surrounding the droplet in the case of wetting of graphite with liquid aluminum and (b) enhanced reactive wetting by forming TiC in the case of wetting graphite with Al–Ti alloys.

Despite limited success, the sessile-drop technique does not seem to be suitable for assessing wettability under dynamic conditions. As a result, more techniques have been developed and used, most specifically for the composite materials. The schematic illustration of various options and settings is given in Figure 4.1. Dipping experiments wherein a ceramic plate is dipped in the liquid and the surface contact angle (meniscus) is observed and measured allow for somewhat dynamic experimental conditions [3, 4]. The contact angle can be measured in situ (especially under wetting conditions, when the liquid phase "climbs" the dipped plate) or in the solid state after sectioning the sample. This technique has been

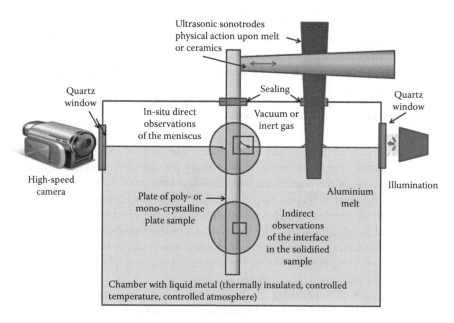

FIGURE 4.1   Setup for experimental assessment of wetting characteristics under dynamic conditions.

tried with some success for studying the effect of ultrasound on the wetting of graphite by liquid aluminum [5]. In this case, a graphite plate was excited by direct contact with a sonotrode. After 1 min of ultrasonic processing at 400 W, 21 kHz, and 800°C, the graphite plate was wetted with liquid aluminum, as was evidenced by strong adhesion of solid aluminum to the graphite plate. The explanation of this phenomenon was linked to the removal of absorbed gas and alumina oxide film by intensive streaming and cavitation. Similar experiments were performed by us with an alumina plate, only in this case the ultrasound was transmitted to the melt. One side of the plate was exposed to high-intensity ultrasonic processing under cavitation mode (18.5 kHz, 1 kW input power, 720°C) and the other side not. Figure 4.2 demonstrates obvious improvement in wettability of the alumina plate by ultrasonic processing, with an even nucleation of aluminum grains happening on the (upper) wet alumina surface, while on the other (lower) side there is no good adhesion.

Wettability can be assessed by studying the spreading of liquid metal on a substrate. The experiments with various low-melting alloys and metallic or ceramic substrates showed that the application of ultrasonic vibrations to the substrate results in almost immediate wetting. This effect is enhanced by increasing temperature and ultrasound amplitude. A summary of experimental results can be found elsewhere [6].

Yet another set of techniques used for wetting assessment involves infiltration of a ceramic perform (filter). The pressure required for the melt penetration through the filter or the penetration depth at a known constant pressure can be recalculated to obtain the surface tension [4].

$$P = S_f \left(\sigma_{sl} - \sigma_{sg} \right) = S_f \sigma_{lg} \cos \theta, \tag{4.2}$$

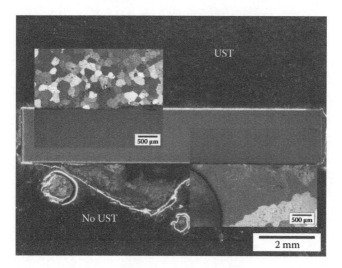

FIGURE 4.2 Wetting of alumina plate with liquid aluminum: Upper part was subjected to ultrasonic cavitation, whereas the lower part was not. (Courtesy T.V. Atamanenko.)

where S_f is the surface area of the interface per unit volume of the filter; σ is the surface energy or surface tension; subscripts s, l, and g denote solid, liquid, and gas; and θ is the contact angle ($\theta < 90°$ means wetting).

This method can be used for dynamic wetting conditions through application of melt processing on the top of the filter. The measured pressure is close to the capillary pressure but is influenced by the wetting conditions, percolation limits of the filter, and the infiltration length. Therefore, the results obtained will not represent the local change in surface tension under the dynamic conditions, but rather an average of different values changing along the filter depth. Cavitation, in addition, may obscure results by changing the capillary pressure through pressure pulses emitted by collapsed bubbles.

It is possible to suggest a technique that can assess wetting at the solid–liquid interface not affected by the changes in the liquid–gas surface tension. For that, one can use the scheme shown in Figure 4.3, which is a variation of the dipping scheme. The meniscus height Z_{max} or the rise of the capillary column Z_c can be measured, representing the contact angle θ at the solid–liquid interface according to the following relationships [7]:

$$Z_{max} = l_c(1 - \sin\theta)^{1/2} \quad \text{and} \quad Z_e = \frac{l_c \cos\theta}{r}, \tag{4.3}$$

where l_c is the capillary length: $l_c = 2\sigma_{lg}/\rho_l g$ (σ_{lg} is the surface tension between liquid and gas phases, ρ_l is the liquid density, and g is the gravity constant), and r is the capillary radius. Under similar experimental conditions (temperature, pressure, atmosphere), the capillary length is constant, and hence the measured meniscus height or capillary column can be taken as a quantitative characteristic of wettability that

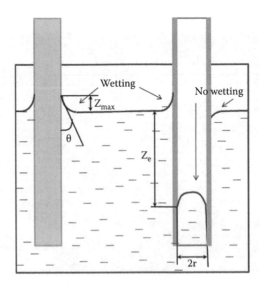

FIGURE 4.3 Scheme illustrating the wetting parameters that can be measured using dipping of a plate or a capillary in a sonicated liquid.

will be affected only by the properties of the liquid/solid interface and the dynamic conditions in the melt.

Despite all the efforts, an experimental method that allows the reliable assessment of wettability in the ceramic–liquid-metal systems is currently not available. Most of research on infiltration and assessment of wettability and surface tension is done by qualitative or semiquantitative methods. The improvement of infiltration and wetting by ultrasonic cavitation is, however, beyond doubt and has been extensively used in manufacturing of composite materials, as we will discuss in Chapter 9. In this chapter, we will focus on melt filtration with the aim of cleaning.

4.2 SONOCAPILLARY EFFECT

The ability of sonicated liquids to penetrate narrow capillary channels with a higher velocity and to a greater depth than can be expected from the overall pressure and surface tension was reported as early as in the 1960s [8]. This effect was first studied in metal cutting, where cooling oil penetrates into a region near the cutting tool. This phenomenon was called the *ultrasonic capillary effect* or the *sonocapillary effect*.

This effect was extensively studied in the following years, and the results were summarized in a monograph [9]. The crucial role of cavitation in the sonocapillary effect was proved both theoretically and experimentally [10, 11]. The liquid rise in a capillary increases by an order of magnitude under developed cavitation conditions. Other ultrasonic phenomena, such as variations in menisci and the vibration of capillary walls, may enhance the motion of liquids in capillaries, but they are far from being crucial factors.

Although the fundamental studies of this phenomenon are ongoing, the importance of its role in many processes and mechanisms is beyond any doubt. For a number of metallurgical procedures—melt degassing, filtering, wetting of solid inclusions, forming of cavitation and solidification nuclei, manufacturing of composite materials, insert casing, and precision casting—the sonocapillary phenomenon is essential.

The fundamental description of the sonocapillary effect is based on the following assumptions [10]: (1) the liquid volume is infinite compared to the capillary volume; (2) a cavity forms, grows, and collapses at the capillary entry repeatedly in each oscillation period T; (3) the capillary walls do not affect the cavity collapse, so that bubbles remain spherical until the end of the collapse phase; (4) a shock wave generated by the collapse acts on the liquid column in the capillary with the force

$$F_1 = PS, \qquad (4.4)$$

where P is the pressure in the shock wave and S is the cross-section area of the capillary; (5) this force F_1 acts for a very short time ($t = 10^{-8}$ s) during the collapse phase, and the liquid then moves by inertia throughout the rest of the oscillation period $(T_v - t)$; and (6) the liquid in the capillary is treated as a concentrated mass m.

This last assumption allows one to use the momentum conservation law in the simplest form:

$$Fdt = d(mv), \tag{4.5}$$

where F is the vector sum of two opposite forces: F_1, which forces the liquid to penetrate the capillary after cavity collapse, and F_f, the viscous friction force that opposes penetration and is given by Poiseuille's formula

$$F_f = v\frac{4\eta_f H}{r^2}S, \tag{4.6}$$

where η_f is the friction coefficient, H is the liquid column in the capillary, r is the capillary radius, and v is the velocity of the liquid.

Assuming that the liquid mass intake in the capillary over 1 period of oscillation T_v is several times smaller than the initial liquid mass in the capillary, the mass $m = \rho_1 SH$ can be taken as constant (ρ_1 is the density of the liquid). Then, Equations (4.5) and (4.6) can be rewritten as:

$$P - vK = \frac{m}{S}\frac{dv}{dt}, \tag{4.7}$$

where $K = \frac{4\eta_f H}{r^2}$. Solving Equation (4.7) for the time range within one pressure pulse $0 \leq t \leq \tau$, one can find the velocity of liquid rise after a single pressure pulse:

$$v_i = \frac{P}{K}\left(1 - \exp\left(\frac{KSt}{m_{i-1}}\right)\right), \tag{4.8}$$

where t is time, and m_{i-1} is the liquid mass in the capillary at the moment before the ith pulse is applied.

In most cases, e.g., at typical values of $P = 0.5$ MPa, $\tau = 10^{-8}$ s, and $r = 200$ μm, the first term of solution expansion will suffice. Then

$$v_i = \frac{P\tau}{\rho}\frac{1}{H_{i-1}}. \tag{4.9}$$

Therefore, the liquid inside the capillary acquires velocity v_i during pulse τ and then continues moving with this velocity until the next bubble collapses. In this case, the liquid rise during one oscillation period T_v will be

$$\Delta H_i = v_i(T_v - \tau) = \frac{P\tau(T_v - \tau)}{\rho H_{i-1}} \tag{4.10}$$

so that the resulting level due to the sonocapillary effect will be

$$H = H_0 + \sum_{i=1}^{n}\Delta H_i, \tag{4.11}$$

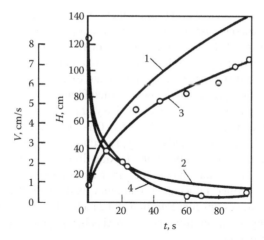

FIGURE 4.4 Comparison of calculated (1, 2) and experimentally measured (3, 4) rising height H (1, 3) and rising velocity V (2, 4) for a glass capillary submerged into water sonicated at 18 kHz. (After Kitaigorodsky and Drozhalova [10].)

where H_0 is the column height determined by the surface tension and generic capillary forces, and n is the number of periods (oscillations) at the moment of reading.

In derivation of these relationships, it was assumed that during the bubble collapse at the entry to the capillary, the pressure P acting on the liquid in the capillary is equal to the pressure generated by the bubble collapse P_{max} [10]. The results obtained with this simplified analysis agree well with the experimental data, as demonstrated in Figure 4.4. In these experiments, however, only a single bubble collapse affects the rise of water column (due to the sizes of the bubble and the capillary diameter), and in addition, the shock wave spreads in the capillary cross section, effectively decreasing the pressure.

These and other effects were taken into account in a sonocapillary theory [9]. This theory includes asymmetry in the boundary conditions for a collapsing cavity, when it loses its spherical shape and implodes, emitting a cumulative jet of liquid. This cumulative jet is assumed to be responsible for the elevation of liquid level in the capillary. Repeated with a frequency determined by the probability of bubble occurrence and collapse near the capillary entry, the cavity collapse and jets produce accumulated liquid rise ΔH, resulting in the sonocapillary effect. In the interval between two successive jets reaching inside the capillary, the liquid can escape from the capillary, and the liquid column can decrease.

4.3 BASICS OF FILTRATION

An aluminum melt is a heterogeneous system containing coarse (up to 5–10 μm) and fine (up to 1 μm) nonmetallic inclusions, either indigenous to the melt processing (oxides, carbides) or intentionally put in the melt, e.g., for grain refinement (borides, carbides) [12]. The filtration process as applied to aluminum melts deals mostly with the former, and mainly with oxides. Molten aluminum interacts with oxygen to form aluminum oxide, which covers the liquid metal with a strong dense film and also exists as a homogeneous suspension of fine 0.1–0.3-μm particles in the liquid bulk. The content of suspended aluminum oxide is small, varying from 0.002% to 0.2% [13]. The surface oxide layer can be easily entrained in the bulk as a result of melt transfer and surface disturbance, creating thin and extended oxide films and folded bi-films [14]. The most probable thickness of such films is 0.1–1 μm, and their extension maybe as long as several millimeters. They are nonuniformly distributed over the melt, and the contamination coefficient is 0.1–1.0 mm^2/cm^2, depending on the purity of the metal, the melting and casting procedure, the refinement procedure, and the casting conditions. The composition of oxide film is almost identical for aluminum alloys that do not contain magnesium. In magnesium-alloyed aluminum, spinel and magnesia are formed rather than aluminum oxide, with the content and ratio varying with the magnesium concentration in the melt. Melts may also contain pieces of refractory lining, slag, and flux, these particles being quite coarse and measuring up to several mm in size.

Fine aluminum oxide particles, as a result of their association with hydrogen (see Sections 3.1 and 3.2), noticeably reduce the diffusion mobility of hydrogen in the melt, thus preserving a high hydrogen content in the liquid phase and increasing the probability of gaseous porosity in castings and other gas-induced defects in semifinished products. Large inclusions result in weaker cross sections of the products. They act as crack initiators, for example, for cold cracks in DC-cast billets and ingots [15]. Oxide inclusions may also cause deformation slivers appearing in forged, stamped, and extruded products. These defects can lead to shape distortion, corrugation, and cracking of deformed products. Nonmetallic inclusions may reduce the fatigue strength, corrosion resistance, anodizing response, and other such properties of semifinished products.

The refinement of aluminum alloys from nonmetallic inclusions is, therefore, a major technological operation in the production of extruded and rolled items. Industrial refinement procedures are based on flotation (inert- or active-gas bubbling and vacuum extraction of bubbles), adhesion (fluxing and filtration through active filters), and mechanical separation of inclusions by filtration in screen, foam, or deep-bed filters [16].

Filtration through glass-fiber cloth is attractive because it is inexpensive, easy to implement, and compatible with the existing temperature–flow-rate casting regimes. At the same time, the efficiency of one-layer filters—even with 0.4 × 0.4-mm mesh—is low because they can capture only large inclusions. Fine inclusions with sizes much smaller than the mesh size cannot be retained and thus pass through the filter. Better cleaning can be achieved with multilayer filters or with one-layer glass-cloth filters of smaller mesh. But then, as a result of poor wettability and the tortuousness of the channels, these filters cannot be penetrated by the melt under normal casting conditions, as this requires an impractical metal head, as illustrated in Figure 4.5. It is necessary to find a way to overcome the capillary pressure in filter channels and to allow liquid metal to pass through easily, leaving the main portion of impurities in the filter. The next section shows how this can be done using ultrasonic cavitation.

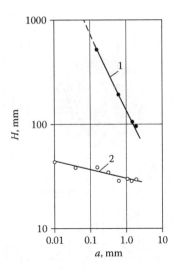

FIGURE 4.5 Effect of the mesh size of a glass-fiber multilayer filter on the height H of the liquid-metal column required for the metal to pass through the filter: (1) without sonication and (2) with sonication.

The filtration process involves several mechanisms: (1) mechanical arrest of larger particles with a size larger or close to the cross section of a filter channel; (2) inertia-driven collision of medium-size particles with the filter fabric; and (3) peripheral interception and/or adhesion of small particles to the filter fabric controlled by the decrease of the free energy of the filter–inclusion–melt system. Figure 4.6 gives a schematic of all three of these main mechanisms, with the particle r_0 representing the first mechanism, r_3 representing the second, and r_2 being the third. In the case of r_1, the particle goes through the filter carried by the flow.

The first mechanism is obvious. Accumulation of large particles, mainly on the top surface of the filter, forms sediment and clogs the filter, preventing further penetration of the melt. Therefore, multistage filtering systems with initial separation of large particles are used in industry. For example, a first-stage filter can be made of glass-fiber cloth with mesh sizes from 0.4 × 0.4 mm to 1.3 × 1.3 mm formed by intertwined threads of 0.6–0.7-mm diameter composed of 9–11-μm fibers. Such filters are capable of mechanically retaining solid nonmetallic inclusions with a size of $2r_0$ (Figure 4.6), which is larger than the size of the mesh [17].

For smaller particles, the retaining mechanism of the filter will be different. For a certain size r_3, which is determined by the Stokes number (Stk, see Equation 4.13), particles no longer move with streamlines. Inertia (viscosity, gravity) tends to straighten the trajectories of these particles in those points where the stream changes its direction (see Figure 4.6). As a result, the probability of a particle being retained by the filter increases. This mechanism is called *inertial impaction* [16, 18].

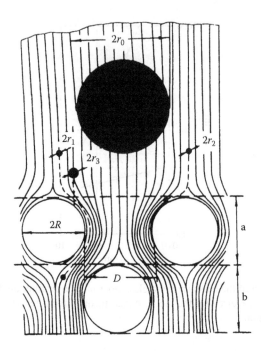

FIGURE 4.6 Motion of solid nonmetallic inclusions in aluminum melt relative to the multilayered screen filter: (a) first layer and (b) second layer.

Under conditions of potential melt flow (Re = 40–700), particles will precipitate on the filter if [18]

$$\text{Stk}_{cr} = \frac{2\text{Re}\,\rho_m r_3^2}{9\rho_i R^2} \ge 0.08, \tag{4.12}$$

where ρ_m and ρ_i are the densities of melt and nonmetallic inclusion, respectively.

In screen filters for aluminum melts, metal stream velocities are usually around 0.3–0.4 m/s, so the smallest size of particles retained by the screen filter by inertial capture can be estimated as $2r_3 \approx 60$ μm [18]. In the case of potential flow, the probability of a particle being precipitated in inertial impact is determined as

$$E_i = \frac{\text{Stk}^2}{\left(\text{Stk} + 0.06\right)^2}. \tag{4.13}$$

For the critical value of Stk = 0.08 (Equation 4.12), the probability of precipitation $E_i = 0.33$. Hence, only one-third of inclusions with critical and larger sizes can be extracted from the flow by inertial impact at a turning point near filter elements. Consequently, a proper filter design must include multiple changes of flow direction, which can be achieved, for example, by increasing the number of layers in a screen filter.

Even smaller particles are carried out with the flow, clear of filter threads. As can be seen in Figure 4.6, streamlines may deflect small particles of size $2r_1$ from the frontal collision with glass threads of radius R. They will touch the filter fabric only if they approach its surface at a distance that is equal or less than their radius. In this case, they can be captured by a peripheral interception mechanism, and the precipitation coefficient E_{capt} can be assumed to be [17]

$$E_{capt} \approx \frac{3r_2}{R}. \qquad (4.14)$$

It is not difficult to estimate that the probability of a fine particle meeting the filter body is very small, and does not exceed 0.005 for 1-µm particles, i.e., only 5 particles of 1000 have a chance to touch the filter fabric. This probability increases with the size of inclusion and equals 0.05 for 10-µm particles.

Adhesion of solid nonmetallic inclusions from the aluminum melt to the filter fabric (for small particles with r_1 and r_2) is a thermodynamic process powered by the decrease in the specific surface free energy of the inclusion–melt–filter system. Before the inclusion contacts the filter, the specific free energy is $\sigma_{im} + \sigma_{mf}$, and after the contact it is σ_{if} if there is no melt layer between the inclusion and the filter (the subscripts i, m, and f stand for inclusion, melt, and filter, respectively). Adhesion occurs only if the variation of free energy ΔW is positive, i.e.,

$$\Delta W = (\sigma_{im} + \sigma_{mf}) - \sigma_{if} > 0. \qquad (4.15)$$

Taking into account the Young equation,

$$\sigma_{mf} \cos\theta = \sigma_{if} - \sigma_{im}, \qquad (4.16)$$

we obtain

$$\Delta W = \sigma_{mf}(1 - \cos\theta), \qquad (4.17)$$

where θ is the contact angle.

In aluminum melts, where solid nonmetallic inclusions appear predominantly in the form of oxides that are not well wet by the melt, the decrement of specific surface free energy is always positive. Consequently, upon contact with the filter surface, solid nonmetallic oxide inclusions tend to adhere to the filter. The adhesion works well when the surface-tension values between inclusions and the melt (hydrophobic inclusions) and between the melt and the filter (nonwettable filter) are higher than those between the inclusions and the filter material. In this case, the inclusion is rejected by the melt and pushed toward the filter surface. In the same way, fluxes are used to improve the filtration efficiency, as the flux wets inclusions better than the melt and, therefore, "extracts" particles from the melt. At higher temperatures, especially above 800°C, the wetting of inclusions and filter material by the aluminum melt improves (θ decreases), and, thermodynamically, the cleaning of aluminum melt by filtering becomes less efficient.

4.4   MECHANISM OF ULTRASONIC MELT FILTRATION

The cleaning of melts from solid nonmetallic impurities is an important process in the modern production of castings and ingots of aluminum alloys. One of the efficient means of doing this is filtration of the melt. Combined with degassing (see Chapter 3), this process refines liquid metals and enhances the purity of aluminum alloys required for sensitive application such as memory disks, foils, and large extruded panels.

Let us consider filtration through a cloth of caustic-free alumino-borosilicate glass fiber. Single-layer filters that are widely used in industry to capture coarse particles cannot retain fine inclusions. Clearly, the smaller the mesh, the finer are the inclusions retained by the filter and, hence, the better is the filtration. In principle, the efficiency can be enhanced by using multilayer filters with 0.4–0.6-mm mesh. However, the penetration of liquid metal through a two-layered filter with 0.6×0.6-mm mesh requires 500 mm of the liquid metal head. A smaller mesh (0.4×0.4 mm) or three layers with 0.6×0.6-mm mesh makes filtration impossible, even for metal heads larger than 500 mm [19]. The reason is that the capillary pressure required becomes too great. This capillary pressure is affected not only by narrow filter passages, but also by poor wettability of filter materials and high surface tension of liquid metals (see Equation 4.1).

Acoustic cavitation has the potential to solve the problem due to the sonocapillary effect (see Section 4.2). To check whether the melt can pass through the capillaries of multilayer glass filters, we carried out experiments with aluminum alloys AA7075, AA2024, and AA7055 using filters with 1 to 9 layers of glass cloth with 0.6×0.6-mm mesh. These filters can be characterized as follows:

Number of layers in contact	1	3	5	9
Effective capillary size, mm	0.6	0.15	0.04	0.01

Figure 4.5 demonstrates that at least a 1000-mm column of liquid metal is required for melt penetration through fine-screen filters under normal conditions (curve 1). This makes filtration impracticable. On the other hand, the melt sonicated to create developed cavitation conditions ($P_{max} \approx 100$ MPa) near the surface of a multilayer filter can overcome the capillary pressure and friction forces. As a result, the melt easily passes through the filter capillary channels, and the required metal-lostatic head H_{us} does not exceed 30–40 mm (curve 2). At each oscillation period, the height can be increased by the value determined by Equation (4.10). The pulse duration is much smaller than the period T_v of ultrasound: $\tau \ll T_v$ (at a frequency of 18 kHz, $T_v = 56 \times 10^{-6}$ s and $\tau = 10^{-8}$ s). Because cavitation pressure pulses are short, fine filtration in an ultrasonic field starts after 10–60 s of melt sonication.

In a fine filter, the porous body (several layers of glass cloth with 0.6×0.6-mm or 0.4×0.4-mm mesh) is characterized by relatively short capillary channels no longer than a few millimeters, with a very tortuous profile. A flow of melt in such a channel will be governed by a mechanism somewhat different from that for a straight capillary.

The effective cross section of a capillary will vary during filtration because small particles of solid nonmetallic inclusions will stick to the walls. In addition, this filtration will be governed by the regularities of a flow through porous material with a multitude of channels, rather than through a single channel. Assuming that the flow is laminar, the rate of metal flow through a multilayer filter can be described by the Hagen–Poiseuille equation and should be independent of time:

$$Q = \frac{n\pi\Delta P R^4}{8\mu L},$$
(4.18)

where n is the number of capillary channels per unit surface, ΔP is the pressure drop at the entry to the capillary channel, R and L are the radius and length of the channel, and μ is the viscosity of liquid metal.

However, if molten metal contains fine particles smaller than the channel cross section (typically under 5 μm), the metal flow rate becomes time dependent, mainly because these particles adhere to the channel walls. In addition, nonmetallic particles larger than the channel cross section will be retained by the filter as sediment on the surface. As a result, the apparent length of the channels will increase, and the effective surface of the filter will decrease.

Melt sonication induces developed cavitation accompanied by acoustic streaming at the filter surface, which disperses the sediment. In the ideal case, the metal flow through a filter can be maintained at a constant rate during relatively long intervals of filtration. The sonocapillary effect is induced by collapsing cavitation bubbles located at the entry to the capillary (filter channel) [10]. Excessively powerful ultrasonic processing may create such strong acoustic streams that the cavitation cloud at the surface of the filter is "washed away." As a result, the influx of liquid into capillaries ceases instantaneously. This effect, however, is much less pronounced in the case of a large upper surface area of the filter. In this case, collapsing bubbles will always be available for individual capillaries. Another feature of filtration based on the sonocapillary effect is that the acoustic energy sufficient to maintain the flow through capillaries is 5–10 times smaller than the energy required to initiate the process, because viscous friction caused by the capillary walls needs to be overcome only in the initiation stage of filtration.

Our experiments with filtering melts through multilayered filters showed the following main features.

1. Since filtration proceeds through a multitude of capillary channels in a filter, we can disregard the dependence of the filtering process on the intensity of ultrasound associated with development of intense acoustic streaming. However, the intensity of ultrasound must not exceed certain maximum levels that can produce erosion of the filter. To avoid this, the face of a sonotrode must be kept at a certain distance from the filter.

2. Once the filtration process has begun and the filter capillaries have become wetted, the process can continue even after the ultrasonic source is switched off. According to our data, the period of such postcavitation filtration may

be as long as 3–6 min after an initiation stage of 20–60 s. This inertial feature allowed us to develop a process of dynamic filtration in which the operating ultrasonic sonotrode reciprocates or performs a rotational motion over the filter surface.

3. Since ultrasonic filtration is conducted with a shallow layer of melt above the filter (at most 30–60 mm), the melt is also degassed efficiently; thus one may speak of the combined refining action of ultrasonic treatment.

When ultrasonic treatment is combined with the addition of special inoculants, efficient purification is accompanied by grain refining.

4.5   FILTERING THROUGH MULTILAYER SCREEN FILTERS: USFIRALS PROCESS

We have already considered general issues of filtering without ultrasonic processing (Section 4.3) and the mechanism of ultrasonic-aided filtration (Section 4.4). Let us now look at the technology that takes advantage of ultrasonic cavitation and the sonocapillary effect and dramatically increases the efficiency of filtration (see Figure 4.5). This procedure, developed in the 1970s, is called Usfirals (ultrasonic filtration and refining of aluminum) and has been patented in major industrial countries [20]. The sonicated melt, driven by the pulses caused by collapsed cavitation bubbles, eases the penetration of the melt through the filter [21], thereby allowing the use of several layers of glass fiber and providing the flow necessary for the Usfirals filtering process. The technology was successfully tested under industrial conditions upon DC casting of various types of commercial aluminum alloys.

Our experiments with a filter of 10–15 layers of glass cloth with 1.0 × 1.0-mm mesh, conducted under the conditions of the Usfirals process, proved that such a filter could combine efficient cleaning with high rates of melt flow through the filter. The performance of this multilayer filter system was tested by casting 270-mm-diam. billets of alloys AA5086, AA7075, and AA2024. We found that the process continues steadily only under conditions of active cavitation.

We also analyzed the performance of a glass-fiber filter with 0.4 × 0.4-mm mesh with the acoustic cavitation formed in the melt close to the filter surface. Using Equation (4.18), we determined R, L, and n and came to the conclusion that increasing the number of layers above five noticeably reduces the melt penetration through the filter, decreases the melt throughput, and limits the possibility of casting large ingots.

A solution may be found in a filter design that would combine layers with small (0.4 × 0.4 mm) and large (1.0 × 1.0 mm) meshes. In this filter, the filtering of fine inclusions from the melt would be governed by the path of the melt along the filter fibers rather than by pore size (R, n). Acoustic cavitation is essential here to ensure that the melt passes through the filter, thereby overcoming the forces of capillary pressure and viscous friction.

Further industrial trials were continued with multilayered filters with the mesh size 0.4 × 0.4 mm. In this case, the penetration of the melt through the filter becomes possible only with the aid of cavitation, as illustrated in Table 4.1.

TABLE 4.1

Penetration Time for a Melt Flow Through a 0.4 × 0.4-mm Mesh Glass-Fiber Filter as a Function of the Number of Layers and Ultrasound Intensity

Number of Layers	Melt Penetration Time, s		
	Without Cavitation	Incipient Cavitation	Developed Cavitation
1	10	10	5
2	no flow	15	10
3	no flow	no flow	18
5	no flow	no flow	45

Metallographic analysis of the filter sediment conducted after filtering 5–6 tons of liquid metal (alloy AA7075) showed that coarse films of aluminum oxide were retained before the first layer (at the filter surface). To examine the distribution of small inclusions, we had to use scanning electron fractography. Fractograms in Figure 4.7 revealed that the filter sediment fracture retained oxide particles from 8 to 100 µm in size. These oxides stuck not only as individual particles, but also as conglomerates of particles measuring up to 500 µm. After the second layer, and especially after the third, the retained volume of aluminum oxide particles decreases, and the size of these particles is also reduced to 5–20 µm. As might be expected, a considerable proportion of the smallest particles (5–10 µm) is retained at the filter fibers. These results are also summarized in Table 4.2. Evidently, the main load in the filtering operation is on the first three layers.

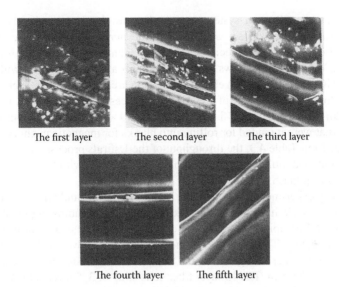

The first layer   The second layer   The third layer

The fourth layer   The fifth layer

FIGURE 4.7 Fractograms of a five-layered screen filter with 0.4 × 0.4-mm mesh (×2500).

TABLE 4.2

Characterization of Fine Filtration Through a Five-Layered Glass-Fiber Filter with 0.4 × 0.4-mm Mesh

Layer Number	Quantity of Oxide Inclusions Retained [a]	Size of Retained Inclusions, µm
Filter surface	Agglomeration	N/A
1	Very large	10
2	Large	5–10
3	Small	2–5
4	Minimal	2.0
5	Negligible	1.0

[a] Scanning fractography data (Figure 4.7).

In parallel with fractography, we conducted a layer-by-layer neutron analysis of the sediment for the content of oxygen. This analysis corroborated with the fractography data and showed that the main proportion of aluminum oxide was retained near the filter surface at a distance of 1.0–1.5 mm, as can be seen in the following grid:

Depth into filter, µm	10–20	200	400	600	1000	1500
Oxygen content, wt%	0.08–0.12	0.07	0.04	0.02	0.01	<0.01[a]

[a] Concentrations below instrumental sensitivity.

The ultrasonic filtration system was further developed to make the best use of the multilayer filter surface and improve the performance of the casting system. A UZF-1 system with a reciprocating movement of the sonotrode is shown in Figure 4.8a. This system allows filtration through a multilayer filter with a cross section of 110×330 mm and an effective filtration area of 108.5 cm^2. Table 4.3 gives the technological parameters of the Usfirals process conducted with the UZF-1 machine during DC casting of flat ingots ($90 \times 550 \times 1500$ mm) from an AA5186-type alloy that were intended for rolling into blanks for magnetic memory disks. As can be seen from Table 4.3, the throughput of the Usfirals process can be improved by using combined multilayered filters consisting of layers with 0.4×0.4-mm mesh and 1.0×1.0-mm mesh.

Laboratory experiments with different numbers of layers and different mesh sizes of the glass-fiber cloth allowed us to find an optimum combination of filter properties. Table 4.4 compiles the results of over twenty tests carried out using 0.7 kg of an AA6063 alloy. The filtration was performed under developed cavitation. The increase in the number of layers with 0.4×0.4-mm mesh reduces the melt flow rate from 0.85 to 0.3 kg/min. The 10-layer filter consisting of 3–5 layers of 0.4×0.4-mm mesh glass-fiber cloth alternating with 7–5 layers of 1.0×1.0-mm mesh glass cloth does not reduce the filtering capacity of the system.

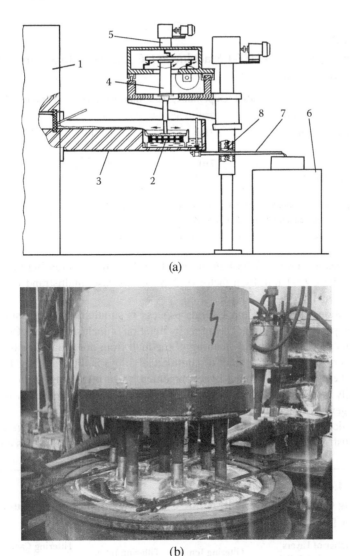

(a)

(b)

FIGURE 4.8 Ultrasonic filtration systems: (a) schematic with a sonotrode reciprocating movement over the multilayered glass-fiber filter with (1) holding furnace, (2) multilayered filter, (3) launder, (4) ultrasonic transducer with sonotrode, (5) reciprocal movement mechanisms, (6) casting machine with a mold, (7) casting feeder (trough), and (8) ascending (descending) mechanism; and (b) practical implementation of combined ultrasonic filtering and grain refinement (Usfirals).

Table 4.5 shows the specific filtering capacity in laboratory experiments recalculated to the size of a real filter in an industrial-scale UZF-1 system. A 10-layered combined filter as compared with a 10-layered 0.4×0.4 mesh filter improves the filtering capacity from 8.7 to 14–26 kg/min with the same filter area (108 cm²), which is sufficient for casting of large ingots up to 650 mm in diameter. Further increased

TABLE 4.3

Filtering Capacity of a UZF-1 Machine in DC Casting of Ingots from an AA5186-Type Alloy

Filter Design	Melt Temperature, °C	Starting Time, s	Filtration Time, min
5 layers of 0.4 × 0.4 mm	720	45	15
5 layers of 0.4 × 0.4 mm + 3 layers of 1.0 × 1.0 mm	710	60	12
5 layers of 0.4×0.4 mm + 5 layers of 1.0 × 1.0 mm [a]	710	75	11

Note: Ingot cross section 90 × 550 mm; casting speed 105 mm/min.

[a] Alternating layers in a multilayered filter.

filtering capacity (up to 50–75 kg/min) can be achieved by increasing the filtering surface in the UZF-1 unit by a factor of 2–3. In this case, multistrand DC casting becomes possible.

An additional benefit of the Usfirals process is simultaneous degassing of the melt accompanied by additional cleaning by flotation of inclusions. The efficiency of the Usfirals process in general cleaning of the melt from nonmetallic (gaseous and solid) impurities upon DC casting of commercial alloys is illustrated by Tables 4.6 and 4.7. The purification of melt is accompanied by an increasing grain size. This additionally proves that the amount of solid inclusions, which may act as substrates for heterogeneous nucleation of aluminum, decreases.

In practice, the ultrasonic filtration can be combined with ultrasonic grain refinement. Industrial experiments showed the feasibility and advantages of this approach.

TABLE 4.4

Filtering Capacity Under Ultrasonic Cavitation Conditions as a Function of Filter Design

Number of Layers		Filtering Temp.,	Filtering Time,	Filtering Capacity	
0.4 × 0.4 mm	1 × 1 mm	°C	min	kg/min	kg/cm²·min
5	...	710–740	1	0.85	0.15
10	...	700–730	2.0–2.5	0.3	0.08
15	...	710	2.5	0.3	0.08
3	7 [a]	750	0.8	0.9	0.24
5	5 [a]	710–740	1.5	0.52	0.13
5	5 [a]	730	1.5	0.69	0.18
7	3	780	3	0.25	0.07
9	1	770	5	0.15	0.04

[a] Alternating layers.

TABLE 4.5
Effect of Filter Design on the Capacity of a UZF-1 Unit

Number of Layers		Specific Capacity,	Capacity of UZF-1,
0.4 × 0.4 mm	1 × 1 mm	kg/cm²·min	kg/min
5	...	0.15	16.28
10	...	0.08	8.68
15	...	0.08	8.68
3	7	0.24	26.04
5	5	0.13	14.1
5	5 [a]	0.18	14.53

[a] Alternating layers.

TABLE 4.6
Effect of Usfirals Process on Metal Quality Achieved in Commercial Aluminum Alloys AA2324 and AA7475

Alloy	Ingot Diam., mm	Filter Mesh, mm	No. of Layers	H_2, cm³/100 g	O_2 in Filter Sediment, %[a]	Inclusions, mm²/cm²
AA2324	650	0.6	5	0.13	0.08–0.2	0.005
		1.3	1	0.16		0.025
AA7475	650	0.6	2	0.14	0.6–0.1	0.008
		1.3	1	0.16		0.021
AA7475	830	0.6	3	0.18	0.12–0.4	0.005
		1.3	1	0.25		0.035

[a] Oxygen concentration in the solid metal is 0.01%.

TABLE 4.7
Effect of Usfirals Process on Metal Quality Achieved in a Commercial AA6063 Alloy

	H_2, cm³/100 g	O_2, %	Grain Size, μm
Before filtration	0.5	0.03	140–150
Usfirals, 5 layers, 0.4 × 0.4 mm	0.3	0.017	210–220

Figure 4.8b demonstrates a casting process where the filter box is placed between the holding furnace and the DC casting mold. Two ultrasonic transducers with sonotrodes are submerged in the melt and slowly rotate above the filter. The filtered melt is then supplied to the large round mold, where ultrasonic processing continues in the billet sump, being administered by an array of transducers and sonotrodes.

We can conclude that cavitation-aided filtration can be successfully applied under industrial conditions, improving the performance of existing filtering systems. Although it has been applied for glass-fiber screen filters, the general features and attained benefits should be the same in other filtration systems. In particular, this method can be very efficient in foundries where filtration is performed through rigid ceramic filters.

REFERENCES

1. Hitchcock, S.J., N.T. Carroll, and M.G. Nicholas. 1981. *J. Mater. Sci.* 16:714–32.
2. Li, Y.L., and T.G. Zhou. 2013. *Metall. Mater. Trans.* A 44:3337–43.
3. Naidich, Ju.V. 1981. *Progr. Surf. Membr. Sci.* 14:353–484.
4. Mortensen, A., and I. Jin. 1992. *Int. Mater. Rev.* 37:101–28.
5. Polacovic, A., P. Sebo, J. Ivan, and Z. Augustinicova. 1978. *Ultrasonics* 16:210–12.
6. Abramov, O.V. 1998. *High-intensity ultrasonics. Theory and industrial applications*, 291–313. Amsterdam: OPA Gordon and Breach.
7. Eustathopoulos, N., M.G. Nicholas, and B. Drevet. 1999. *Wettability at high temperatures*, 48–51. Amsterdam: Pergamon/Elsevier.
8. Konovalov, E.G., and I.K. Germanovich. 1962. *Dokl. Akad. Nauk Belorus. SSR* 6 (8): 492–93.
9. Prokhorenko, P.P., N.V. Dezhkunov, and G.E. Konovalov. 1981. *Ultrasonic capillary effect*. Minsk: Nauka i Tekhnika.
10. Kitaigorodsky, Yu.I., and V.I. Drozhalova. 1977. *Application of ultrasound in metallurgy, Trans. Moscow Inst. Steel and Alloys*, no. 90:12–16.
11. Malykh, N.V., and V.M. Petrov. 2003. In *Proc. XIII Session Rus. Acoust. Soc.*, 41–44. Moscow: Russian Acoustical Society.
12. Jaradeh, M.M., and T. Carlberg. 2012. *Metall. Mater. Trans.* B 43:82–91.
13. Makarov, G.S. 1983. *Cleaning of aluminum alloys by gases*. Moscow: Metallurgiya.
14. Campbell, J. 2006. *Metall. Mater. Trans.* B 37:857–63.
15. Lalpoor, M., D.G. Eskin, D. Ruvalcaba, H.G. Fjær, A. Ten Cate, N. Ontijt, and L. Katgerman. 2011. *Mater. Sci. Eng.* A 528:2831–42.
16. Davis, J.R. (ed.). 1993. *Aluminum and aluminum alloys: ASM specialty handbook*, 213–18. Materials Park, OH: ASM International.
17. Makarov, G.S., Yu.P. Pimenov, and G.I. Eskin. 1995. In: *Metals science, casting and processing or alloys*, 187–201. Moscow: VILS.
18. Apelian, D., and R. Mutharasan. 1980. *JOM* 32 (9): 14–20.
19. Makarov, G.S., P.E. Khodakov, G.A. Valova et al. 1977. *Tekhnol. Legk. Spl.*, no. 1:53–56.
20. Dobatkin, V.I., G.I. Eskin., S.I. Borovikova, et al. 1986. Method for continuous casting of light-alloy ingots. U.S. patent 4564059, filed 13.06.1981, issued 14.01.1986 (also U.K. patent 2100635, Swiss patent 651253, French patent 2507933, Canadian patent 1194675, German patent 3126590).
21. Eskin, G.I. 1988. *Ultrasonic treatment of molten aluminum*. Moscow: Metallurgiya.

5 Ultrasonic Grain Refinement

5.1 MECHANISMS OF ULTRASONIC GRAIN REFINEMENT

The dynamic means of affecting solidification were briefly discussed in Section 2.1. The main contributions to this topic are summarized in the literature [1–7]. One may recall that two main mechanisms were considered: (a) undercooling of the cavitation bubble surface during the expansion phase of oscillations and (b) undercooling of the liquid phase resulting from the instantaneous increase of pressure during cavitation bubble collapse (according to the Clapeyron equation). The latter mechanism seems most probable, as the decrease of bubble surface temperature does not exceed 1 K, while the change of the melting point as a result of bubble collapse can reach tens of degrees and approach $0.2T_m$ [4]. For example, for 99.99% pure aluminum, the increase in melting temperature changes with the pressure as shown in the following grid:

P, MPa	0.1	500	1000	2000	4000
t_0,°C	660.5	690	720	780	830

In addition to dynamic nucleation, multiplication of solidification nuclei by activation of heterogeneous substrates was suggested in the 1930s–1950s [8–11], involving improved wetting, decreased surface tension, and enhanced heterogeneous nucleation in the available insoluble substrates such as oxides, carbides, etc.

Direct observations of transparent analogues seem to confirm that nucleation is indeed facilitated by ultrasonic cavitation. Figure 5.1 shows early experiments performed in the 1960s with crystallization of supersaturated ammonium chloride (NH_4Cl) [7]. Sonication was performed by a conical sonotrode with a 0.1-mm thin plate attached and put in direct contact with the solution superheated to 80°C. Ultrasonic frequency was 20 kHz at an intensity of 2–3 W/cm^2 (an amplitude of 10 µm). Ammonium chloride solidifies in dendritic morphology under normal conditions, as can be seen in Figure 5.1 (lower row). Under ultrasonic conditions, the entire observed volume (about 15 mm^3) becomes opaque due to the almost instantaneous formation of fine hexagonal particles with uniform size distribution. Upon further ultrasonic processing, a fine equiaxed structure is formed. In the following years, advances in high-speed imaging allowed for more specific observations of the interaction between cavitation and solidifying material. Swallowe et al. [12] studied the solidification of camphene. Ultrasound was transmitted via a glass strip attached to an ultrasonic soldering iron working at 36 kHz. In this work, both nucleation of

Ultrasonic Cavitation

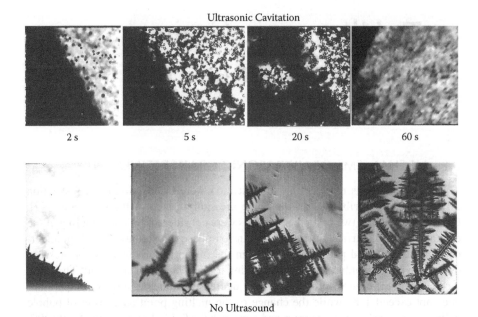

No Ultrasound

FIGURE 5.1 Solidification of a drop of supersaturated ammonium chloride solution (×28): upper row, with ultrasound; lower row, without ultrasound.

the solid phase in the ultrasonic field and fragmentation of growing dendrites by oscillating cavitation bubbles were observed, as illustrated in Figure 5.2. Interesting evidence of dynamic nucleation of 15 wt% water solution of sucrose was reported by Chow et al. [13]. A commercial 20-kHz transducer was used in a setting resembling ultrasonic melt treatment. Ice nuclei were formed at a distance from the sonotrode almost immediately after an ultrasonic pulse, and they grew to equiaxed crystals, as shown in Figure 5.3.

The fragmentation of the solid phase under dynamic action was suggested by many as the main mechanism of structure refinement [12, 14–18]. Mechanical fracture and liquid-metal embrittlement [17, 19, 20] and solute redistribution and local dendrite branch remelting [21–24] were considered as mechanisms (see also Section 2.1). Acoustic and secondary flows assist in transporting the formed fragments to the solidification front and in the melt bulk.

Destruction of dendrites by cavitating bubbles has been demonstrated on transparent analogues [12, 25, 26]. Studies on a transparent succinonitrile–1 wt% camphor solution showed that the fragmentation of dendrites occurs by either explosion of a cavitation bubble or by its oscillation in the vicinity of a dendrite [26]. A cold ultrasonic probe working at 20 kHz with an amplitude of 4 µm was used to induce individual cavitation bubbles. Figure 5.4 gives a sequence of frames capturing the fragmentation event by an exploding bubble.

In the following sections, different mechanisms of grain refinement by ultrasonic cavitation are considered in more detail.

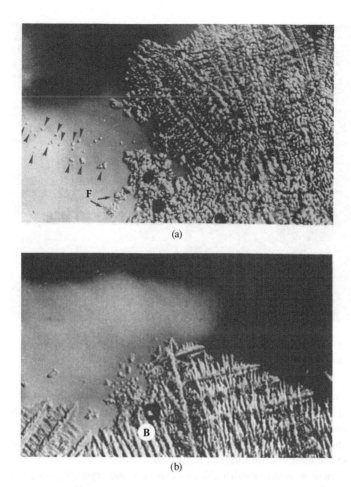

(a)

(b)

FIGURE 5.2   Effect of ultrasonic processing in a camphene solution on (a) dynamic nucleation, where the smaller particles on the left are those nucleated in the liquid, while larger particles shown as F are fragments; and (b) fragmentation, where B is a cavitation bubble inside the dendrite that causes fragmentation by oscillations. (Adapted from Swallowe et al. [12].)

5.2   ACTIVATION OF SUBSTRATES

Solidification of real melt always occurs heterogeneously and on available substrates that are either naturally present (indigenous impurities) or deliberately added (exogenous grain refiners) to the melt. In aluminum and magnesium alloys, the former are represented by oxides and carbides and the latter by borides, carbides, and primary phases. The term *activation* is usually applied to indigenous particles and includes the phenomena of wetting, formation of stable or metastable surface layers, deagglomeration, and nonequilibrium solidification.

Let us take a closer look at the particles that can be activated and involved in solidification in light alloys. In aluminum, the most likely indigenous inclusions are represented by alumina (aluminum oxide). Alumina may exist in various crystal

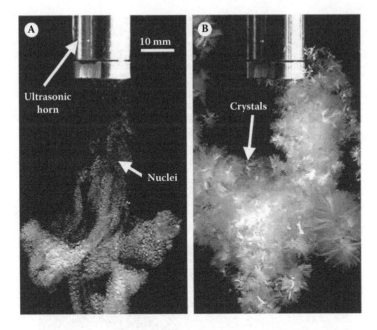

FIGURE 5.3 Dynamic nucleation of ice ahead of a sonotrode: (a) immediately after ultrasonic pulse and (b) after 5 s. (Adapted from Chow et al. [13].)

modifications, depending on the temperature range as well as the origin and presence of hydrogen and other alloying elements. There is uncertainty about the crystal structure and the temperature range for the existence of "intermediate" forms of alumina, and Table 5.1 gives crystallographic data of some alumina modifications that exist in the temperature range of liquid aluminum.

Each of these structure modifications exhibits some crystallographic planes that match the lattice parameter of the aluminum solid solution. The transition between

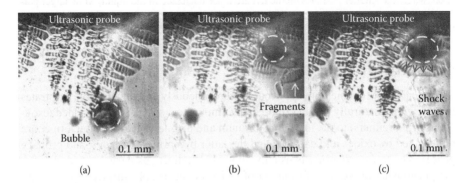

FIGURE 5.4 Fragmentation of a dendrite by an exploding cavitation bubble: (a) bubble contacts the dendrite; (b) oscillating bubble breaks dendrite branches; and (c) exploded bubble emits shock wave that further fragments and disperses dendrite branches. (Adapted from Shu et al. [26].)

TABLE 5.1
Crystal Structure and Temperature Range of Existence of Different Crystal Forms of Alumina

Alumina Type	Crystal Structure	Lattice Parameters, nm	Temperature Range of Existence, °C	Crystal Plane Spacing Closest to (Al): $(100)_{Al} - 0.404$ nm $(200)_{Al} - 0.202$ nm $(111)_{Al} - 0.2337$ nm
γ	Cubic, $Fd\,3m$ or amorphous	$a = 0.7911$	500–850	$(400) - 0.1997$ nm
δ	Cubic	N/A	850–1050	$(114) - 0.407$ nm
χ	Cubic or hexagonal	$a = 0.557$ $c = 0.864$	200–850	$(400); (104) - 0.198$ nm $(321); (202) - 0.211$ nm
θ	Monoclinic, $C2/m$	$a = 1.1854$ $b = 0.2904$ $c = 0.5622$ $\beta = 103.83°$	870–1150	$(112) - 0.2019$ nm
κ	Orthorhombic, $Pna2_1$	$a = 0.4834$ $b = 0.8310$ $c = 0.8937$	900–1000	Close-packed planes 0.2254 nm
α (stable)	Trigonal, $R\,3c$	$a = 0.47589$ $c = 1.2991$	Above 1000	$(113) - 0.2085$ nm

Source: Grandfield, Eskin, and Bainbridge [27].

the lower-temperature modifications of alumina upon heating is, in reality, continuous and related to the dehydration (removal of hydrogen from the crystal lattice). Upon increasing the melt temperature, δ-alumina (or γ-alumina, which is sometimes considered to be the same phase) transforms to χ-alumina (700°C–850°C), and then the stable α-alumina is formed at above 1000°C [28]. During cooling from 1000°C, the stable modification α-alumina is either retained (at high cooling rates) or is transformed to χ-alumina (at low cooling rates) without δ-alumina forming [28]. It is important to note that the oxide film formed on the surface of molten aluminum under typical melting and casting conditions consists of γ-alumina.

The transformation of one type of alumina into another may improve wettability [29]. As mentioned in Chapter 3, hydrogen tends to dissolve in alumina and absorb at the surface of oxide particles, preventing the wetting. This interaction of hydrogen and oxides is one of the reasons why the contamination of aluminum melts with (as well as cleaning of) oxides and hydrogen goes hand in hand. Melt superheat results in the formation of stable α-alumina, which is inert to hydrogen [28]. As a result, the layer of absorbed hydrogen is removed, and the oxide inclusion comes into direct contact with the aluminum melt. It has been shown that very high melt superheating (above 1000°C) of commercially pure aluminum reduces the grain size and promotes the columnar–equiaxed transition [30]. This behavior can be explained by the

TABLE 5.2

Interatomic Disregistry in Selected Oxide–Aluminum Systems at 660°C

Interface	Orientation Relationship $\{uvw\}<uvw>_{oxide} \parallel \{hkl\}<hkl>_{Al}$	Interatomic Distance in Direction [uvw]	Interatomic Distance in Direction [hkl]	Disregistry, %
$\alpha Al_2O_3/Al$	$\{0001\}<10\bar{1}0> \parallel \{100\}<001>$	0.82818	2×0.41212	–0.48
$\gamma Al_2O_3/Al$	$\{111\}<110> \parallel \{111\}<110>$	0.56310	2×0.29141	3.38
Al_2MgO_4/Al	$\{111\}<110> \parallel \{111\}<110>$	0.57462	2×0.29141	1.41
MgO/Al	$\{111\}<110> \parallel \{111\}<110>$	0.30036	0.29141	3.07

Source: Fan and Wang [31].

fact that alumina fine particles, which are inert with respect to liquid aluminum at temperatures below 900°C–950°C, start being wetted by liquid aluminum and act as solidification sites at higher temperatures.

Table 5.2 shows the lattice disregistry calculated for interatomic distances in the most close-packed directions of various oxides, and the data are extrapolated to 660°C using the thermal expansion coefficient. It is evident that oxides have good potency for being efficient substrates for heterogeneous nucleation of aluminum.

The presence of magnesium in aluminum alloys changes the nature of indigenous oxides, as magnesium has greater affinity for oxygen than aluminum [28]. Even small concentrations of magnesium, 0.01–1 wt%, result in the formation of mixed oxides $MgO \cdot Al_2O_3 + MgO$. At magnesium contents above 1%, the magnesium oxide MgO almost completely substitutes alumina. Eventually, spinel Al_2MgO_4 may be formed, which has very good potential to be an efficient substrate for aluminum (see Table 5.2).

Magnesium oxide has an FCC crystal structure with $a = 0.4211$ nm for the pure compound and $a = 0.4234$ nm for the oxide extracted from the magnesium alloy melt [32]. The role of magnesium oxide in grain refining of aluminum is not reported, though it may have a good potency (see Table 5.2). In magnesium alloys, there is enough evidence to suggest that it may play a significant part in grain refinement [33]. An orientation relationship may exist between crystal lattices of Mg and MgO along the magnesium crystal plane (0002) ($d = 0.2617$ nm) and magnesia crystal plane (111) ($d = 0.2445$ nm): $[1\bar{2}10]_{Mg} \parallel [01\bar{1}]_{MgO}$; $(0002)_{Mg} \parallel (111)_{MgO}$, giving the crystallographic disregistry of about 5.46% [32].

If melt superheating is not a practical way to activate inclusions in aluminum alloys, the situation is different in magnesium alloys, where melt superheating is one of the first patented techniques of grain refinement in these alloys, more specifically in Mg–Al alloys [34]. Depending on the alloy composition, the melt superheating is 150°C–260°C above the liquidus, followed by relatively rapid cooling to the pouring temperature. The most accepted mechanism of such grain refinement in magnesium alloys involves nucleation of magnesium on Al_4C_3 particles that are formed by the reaction between carbon and solute aluminum. In the case of adding carbon particles, especially of small size, to the magnesium melt, the wetting becomes an issue.

Without wetting, the reaction of carbide formation cannot proceed. Improved wetting "activates" carbon particles and facilitates their reaction with solute Al to form the carbide, and grain refinement follows [35]. It is interesting that the superheating seems to work even if no special additions of carbon or carbide have been made. This phenomenon is called *native grain refinement* and occurs only in Mg–Al alloys. It is attributed to the natural uptake of carbon from the crucible, tools, or protective atmosphere at high temperatures. In any case, this is a typical example of activation, when carbon inclusions are transformed to the aluminum carbide by the reaction activated by wetting and increased melt temperature.

Ultrasonic cavitation proved to be an efficient means for activation of inclusions. The activation of inclusions by ultrasonic cavitation has been demonstrated recently for pure aluminum with mixed-in oxide surface film, as shown in Figure 5.5 [36], and for Mg–Al alloys with added carbon black nanoparticles [35]. An even more striking effect can be achieved by adding small amounts of alumina to the melt of a concentrated alloy such as AA7055. The results given in Figure 5.5e demonstrate the possibility of getting the nondendritic structure at a lower ultrasound intensity by adding more alumina substrates.

Cavitation treatment turns "hydrophobic" particles into active solidification sites by the following mechanism [37]. Any microscopic solid particle that has affinity to the solidifying phase has a potential to become an active solidification site. This affinity can be due to the match of crystal structures or due to the presence of a special adsorbed layer, or even the matrix solid phase, on its surface. In the latter case, the stability of such a solidification site can be assured only when the adsorbed layer or the solid phase is thermally stable within some temperature range above the liquidus of the alloy. Such conditions can be met in discontinuities like microcracks where, due to the capillary effect, the melting temperature of the alloy is much higher than the equilibrium liquidus. The increase in the melting point under conditions of negative curvature (concave particle) is described by the Gibbs–Thompson equation [38]:

$$T_m^r = T_m^\infty - \frac{2\Gamma}{r}, \tag{5.1}$$

where T_m^r is the melting point of a concave particle inside a crevice (see Figure 5.6), T_m^∞ is the melting point of a particle with flat interface, r is the curvature (negative in the case of the concave particle), and Γ is the Gibbs–Thompson coefficient depending on the surface tension, density, and latent heat.

However, the presence of a gaseous phase at the surface and in the surface imperfections of nonmetallic particles hinders the access of the liquid phase to the inclusion, thus preventing the wetting and filling of the imperfections with the melt. Therefore, the majority of the inclusions remain inert with regard to the solidification.

During ultrasonic treatment with the intensity higher than the cavitation threshold, a cavitation bubble is formed close to the capillary opening filled with gas. In this place, the cavitation strength of the melt is weakened by the presence of a gaseous phase (see Chapter 2). The formed cavitation bubble starts to grow and pulsate; after several periods of oscillation it collapses, producing a high-energy pulse or a cumulative jet. Both phenomena result in the sonocapillary effect—filling

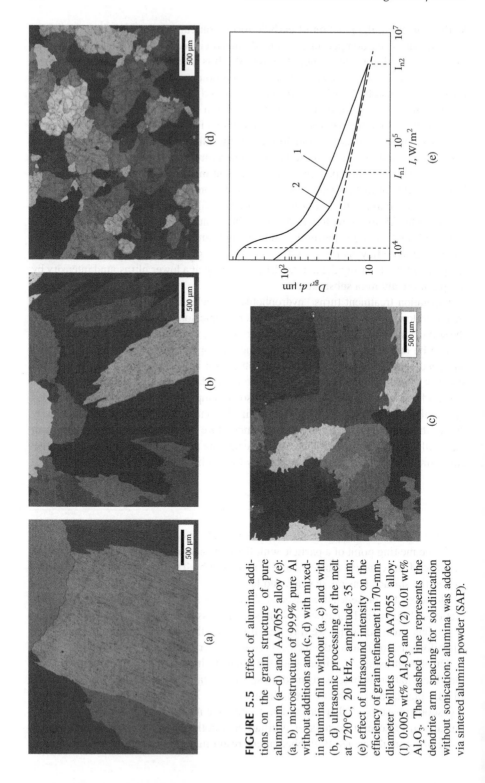

FIGURE 5.5 Effect of alumina additions on the grain structure of pure aluminum (a–d) and AA7055 alloy (e): (a, b) microstructure of 99.9% pure Al without additions and (c, d) with mixed-in alumina film without (a, c) and with (b, d) ultrasonic processing of the melt at 720°C, 20 kHz, amplitude 35 μm; (e) effect of ultrasound intensity on the efficiency of grain refinement in 70-mm-diameter billets from AA7055 alloy: (1) 0.005 wt% Al_2O_3 and (2) 0.01 wt% Al_2O_3. The dashed line represents the dendrite arm spacing for solidification without sonication; alumina was added via sintered alumina powder (SAP).

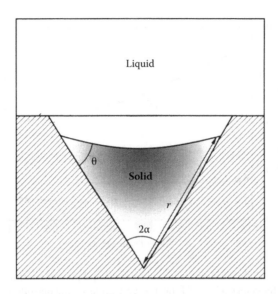

FIGURE 5.6 Scheme showing the solid concave particle with high melting point inside a crack or crevice at the surface of a solid inclusion.

of the capillary opening with the melt (see Section 4.2), followed by the subsequent solidification of this melt. The dynamics of a cavitation bubble with its collapse and the generation of a pressure pulse at a level well exceeding 100 MPa provides enhanced filling of capillaries of almost submicroscopic sizes, i.e., smaller than a micrometer. At the same time, the particle is stripped of absorbed gas and becomes accessible by the surrounding melt. As a result of this activation and locally changed equilibrium conditions (Equation 5.1), the solidified alloy inside capillary openings (cracks) of the particle stays solid at a temperature of the surrounding melt and acts as a perfect solidification site for the matrix melt. The same mechanism should be valid for nucleation of any primary solidifying phase: solid solution, intermetallic, or silicon. Actually, the activation of nonmetallic impurities may facilitate nucleation and refine any primary phase that solidifies in the alloys, as we will discuss further in this chapter and in Chapter 6.

The analysis of structure of ingots from a high-strength aluminum alloy produced with and without ultrasonic cavitation treatment during DC casting shows that the number of active nuclei increases by several orders of magnitude after the cavitation treatment (Figure 5.7). For example, the cavitation treatment in the case of small-sized ingots (65–74 mm) enables the activation of nucleation substrates with density up to 10^9 per cm^3 as compared to 10^3 per cm^3 without sonication. In the case of middle-sized ingots (270–285 mm), this difference reduces to four orders of magnitude; for large-sized ingots (830–845 mm), this difference reaches three orders of magnitude.

Current views on grain refinement are based on the concept of growth restriction, when the final grain size is a function of two contributions. The first is the amount of potent heterogeneous nuclei or substrates that can be activated by various means, e.g., constitutional undercooling formed by solute accumulation in the liquid phase.

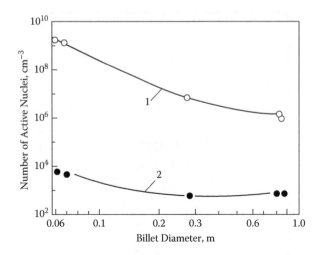

FIGURE 5.7 Effect of cavitation treatment in the liquid pool of continuously cast ingots (65–830 mm in diameter) with UST on the activation of uncontrolled solid inclusions and modifying additions in Al–Zn–Mg–Cu–Zr alloys: (1) casting with UST, (2) casting without UST.

The second is the restriction of growth of new grains due to the solute accumulation at the solid/liquid interface and due to the release of latent heat by neighboring new grains. There is also a way to separate the influence of potency and the impact of the number of the substrates on the final grain size. The reader is referred to StJohn et al. [39] for details.

With respect to ultrasonic cavitation treatment, it is suggested to write the dependence of grain size D_{gr} on the parameters of the substrates as

$$D_{gr} = 5.6 \left(\frac{Dz\Delta T_n}{vQ} \right) + \frac{1}{\sqrt[3]{f(A)N_v}} \tag{5.2}$$

where D is the diffusion coefficient of solute in the melt; $z\Delta T_n$ is the incremental amount of undercooling required to activate the next nucleation even ahead of the solidification front; v is the growth rate of the solid–liquid interface; Q is the growth restriction factor determined as $Q = mC_0(k - 1)$ (where m is the slope of the liquidus, C_0 is the alloy composition, and k is the partition coefficient); N_v is the number density of the substrate; and A is the ultrasonic amplitude [40].

Application of this approach to cavitation-aided grain refinement of magnesium alloys showed that the ultrasonic processing results in a sharp increase of the amount of solidification nuclei, which is consistent with the mechanism of activation.

Figure 5.8a shows the experimental data on the influence of the ultrasonic cavitation development (quantified as the squared amplitude of vibrations) on grain refinement of several binary Mg–Al alloys. There are clearly three ranges in this dependence, i.e., before the cavitation starts, the cavitation threshold, and the developed cavitation. The increasing concentration of solute aluminum decreases the grain size, which is especially significant for solidification of the melt not subjected to cavitation or treated at the cavitation threshold. In the developed cavitation conditions, the grain size drastically decreases, but the solute effect can still be traced. The data for the grain size obtained under cavitation conditions are replotted in Figure 5.8b for different oscillation amplitudes and with regard to the inverse growth restriction factor ($1/Q$) [40]. The slopes of the lines for amplitudes 10 μm through 25 μm are the same, while their intercept with the Y-axis decreases. This is interpreted as the increased amount of active substrates (activation) with the same potency or of the same nature. The slope of the line reflecting 6.5-μm amplitude is greater and signifies the transition from underdeveloped to developed cavitation. In this case, the potency of the particles may also change.

Interesting results were obtained with intentional additions of 0.25 wt% of 42-nm carbon particles to Mg–Al alloys [35]. The alloys were then treated with ultrasound at 700°C using a 20-kHz transducer with 4.3 kW/cm^2 acoustic power (48-μm amplitude). Figure 5.8c gives the results replotted against the inverse growth restriction factor. It is evident that the addition of carbon particles introduces substrates for heterogeneous nucleation (decreasing slope), thereby increasing the amount of active nuclei (decreasing intercept). Cavitation melt treatment additionally increases the amount and potency of the nucleating particles as it improves wetting and dispersion of the carbon particles and facilitates their reaction with solute aluminum to form aluminum carbide.

5.3 DEAGGLOMERATION AND DISPERSION OF PARTICLES

One of the mechanisms of ultrasonic grain refinement is the improvement of distribution of substrates, which includes their deagglomeration and dispersion. The optimum size of a substrate to be active at the melt undercooling that occurs during solidification is 1 to 5 μm. The particles that are introduced in the melt via grain-refining master alloys, e.g., TiB$_2$ or TiC particles in standard Al–Ti–B or Al–Ti–C grain refiners, are of the right size but are mostly agglomerated in clusters up to 30–50 μm in size. During dissolution of a master alloy, only a small fraction of particles becomes active as the substrates for grain nucleation, due to both the presence of larger particles (3–5 μm) that nucleate first and due to the fact that large agglomerates are arrested by filters or settle to the bottom of the melt. In the worst-case scenario, large agglomerates find their way to the final casting and become a casting defect. Therefore, the deagglomeration and dispersion of potential substrates is a means for improving efficiency of the grain refiners. A similar approach is important for the technology of metal-matrix composite materials.

The particles are held together in agglomerates by adhesion, capillary forces, and van der Waals forces. The strength of the agglomerate depends on the sizes of constituent particles. Based on the assumption of agglomerates as a collection of spherical particles, Rumpf [41] calculated the tensile strength of an agglomerate and suggested that

$$\frac{P}{Ar} = \frac{f_v}{1 - f_v} \frac{F}{d^2} \tag{5.3}$$

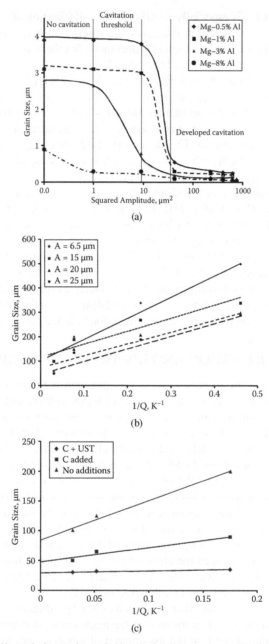

FIGURE 5.8 Effect of ultrasonic cavitation and solute concentration in Mg–Al alloys on the efficiency of grain refinement through activation of substrates: (a) grain refinement versus aluminum concentration and intensity of ultrasonic cavitation; (b) dependence of the grain size in (a) on the inverse growth-restriction factor with respect to the ultrasonic amplitude (adapted from StJohn et al. [40].); and (c) dependence of the grain size on the inverse growth-restriction factor with respect to additions of carbon particles and ultrasonic cavitation (adapted from Bhingole and Chaudhari [35]).

where P is the total force of binding, Ar is the cross-sectional area of the agglomerate, f_v is the volume fraction of the particles in the agglomerate, F is the interparticle binding force, and d is the average diameter of the particle. The left-hand side term represents the tensile strength of the agglomerate. The smaller the particle, the higher the tensile strength and thus the higher the shear stress that has to be applied to break the clusters. Figure 5.9 represents the relation between the size of the particles and the tensile strength of the agglomerate and also shows the contribution of different binding forces.

Cavitation is known for its erosion effect on solid surfaces. The erosion can even be used as a measure for the cavitation intensity. A criterion for erosion activity may look like [42]

$$Z = \frac{V_{max}}{V_{min}\ f\Delta t},\qquad(5.4)$$

where V_{max} and V_{min} are the cavitation bubble volumes at the rarefaction and collapse stages, f is the ultrasonic frequency and Δt is the portion of the oscillation period when the collapse occurs. For spherical bubbles with radius r, V can be substituted for r^3. This is a dimensionless criterion that characterizes the energy transformation during bubble oscillation and collapse, and it can be calculated analytically or numerically. It can be shown that the erosion activity weakly depends on surface tension, density, and viscosity of the liquid, but strongly depends on the gas pressure inside the bubble. The increase of pressure from 1.32 to 13.3 kPa decreases Z from 2.78×10^6 to 7.6×10^2 [43]. As we have shown in Chapter 2, the hydrogen pressure inside the cavitation bubble depends on the acoustic pressure or the ultrasound intensity. The gas pressure decreases from 10^{-3} MPa at the cavitation threshold (acoustic pressure 0.2 MPa) to 10^{-9} MPa under developed cavitation (acoustic pressure 10 MPa). As a result, the criterion of erosion activity in liquid aluminum is very high, as is the efficiency of cavitation action on various interfaces in the melt.

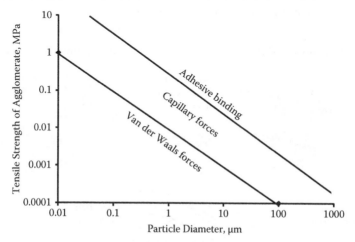

FIGURE 5.9 Tensile strength of an agglomerate vs. the size of constituent particles. (Adapted from Rumpf [41].)

High-intensity ultrasonic oscillations create vast number of microscopic bubbles that are distributed within the volume by acoustic and secondary flows. The bubbles preferentially form at the interfaces and gas pockets. Therefore, the agglomerates of the particles and particles themselves are ideal nuclei for cavitation. The mechanisms of deagglomeration can be represented as follows: The cavitation bubbles are formed at the interfaces of the particle/gas pocket/liquid. These bubbles pulsate intensely, loosening the agglomerates, and then implode. The resultant pressure and momentum pulses literally rip the agglomerates apart. The local pressure generated (up to 500 MPa) far exceeds the forces that hold together the particles in agglomerates, i.e., up to 1 MPa (capillary and adhesive forces) [41, 43]. The flows that are generated by the cavitation zone distribute the particles further in the volume. The process is schematically depicted in Figure 5.10.

This mechanism was experimentally verified by remelting a commercial Al–3% Ti–1% B master alloy rod at 780°C–820°C and ultrasonic processing of the melt at 720°C–800°C with subsequent casting in a metallic mold. In some experiments, the master alloy was diluted with pure aluminum in 1:2 ratio. A 5-kW transducer working at a frequency of 18 kHz and producing an amplitude of 40 μm was used. The ultrasonic processing resulted in partial deagglomeration of clusters containing small TiB_2 particles and the refining of large Al_3Ti primary intermetallics (from 20–25 μm to 7–8 μm), as illustrated in Figure 5.11. In other words, the cavitation treatment of the master alloy releases individual TiB_2 particles 1–3 μm in size from agglomerates and makes them available as substrates for heterogeneous nucleation of aluminum grains.

The efficiency of deagglomeration upon ultrasonic melt processing is illustrated in Figure 5.12 for a small-scale DC-casting simulation experiment [44, 45] and in Figure 5.13 for a larger-scale DC casting [46, 47]. The addition of an AlTiB master alloy rod was done in a launder where the ultrasonic processing of the melt flow was also performed. A significant additional grain refinement can be achieved when combining traditional grain-refining rod addition with the cavitation treatment. The effect depends also on the cooling rate during solidification, as this determines the critical size of a substrate that can be activated [47].

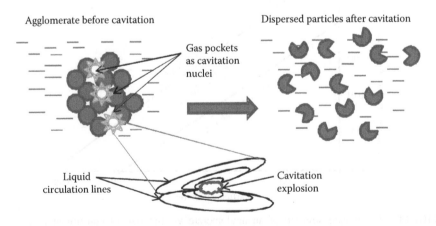

FIGURE 5.10 Diagram showing the mechanism of cavitation deagglomeration.

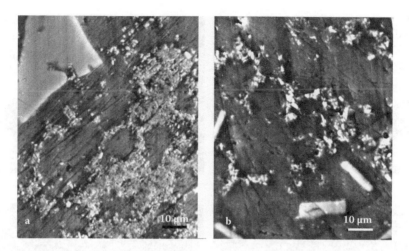

FIGURE 5.11 Microstructure of a commercial Al–3% Ti–1% B master alloy (a) remelted and cast without ultrasonic processing and (b) remelted and subjected to ultrasonic cavitation prior to casting.

The ultrasonic cavitation can also be applied to the process of manufacturing grain-refining master alloys, which are produced by reaction of molten aluminum with a mixture of KBF_4 and K_2TiF_6 salts at 800°C–850°C. The best result is obtained when the mixture is processed twice, first in the stage of the reaction and afterwards during solidification of the master alloy [48]. As a result, an Al–5% Ti–1% B master alloy contains uniformly distributed small 20-μm blocky Al_3Ti particles and loose clusters of 0.65–1.4-μm TiB_2 particles, which are easily dispersed in aluminum melt upon adding the master alloy. The efficiency of the grain

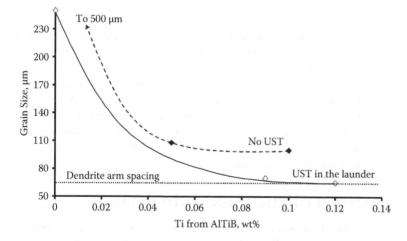

FIGURE 5.12 Effect of an Al–5% Ti–1% B master rod addition on the grain size of a 2017-type alloy. Rod addition and ultrasonic processing are performed in the melt flow upon casting a 40-mm-diameter billet. (Courtesy of S.G. Bochvar and G.I. Eskin [43].)

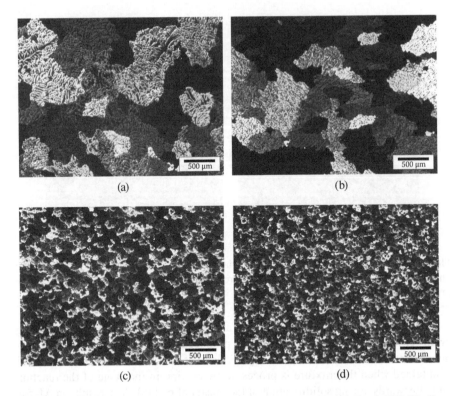

FIGURE 5.13 Microstructure of a 200-mm billet from an Al–4% Cu–0.03% Ti alloy produced by DC casting: (a) not grain refined; (b) not grain refined, ultrasonic melt treatment in the launder; (c) addition of 0.04% Ti via Al–3% Ti–1% B rod in the launder; and (d) same as (c) with ultrasonic melt treatment in the launder. (Courtesy of T.V. Atamanenko [46].)

refiner is improved, with the attainable grain sizes of commercially pure aluminum below 50 μm as compared to 130 μm with a commercial master alloy (on adding 1 wt% of the grain refiner) [49].

In magnesium alloys, the use of grain-refining master alloys is limited. Recently, a product (Nucleant 5000) containing fine carbon particles was developed by Foseco for Mg–Al alloys, and Mg–Zr master alloys are used to grain-refine Al-free magnesium alloys. In both cases, it has been shown that the ultrasonic cavitation processing of the melt increases the amount of active nucleating particles via the mechanisms of deagglomeration and dispersion [50].

5.4 REFINEMENT OF NUCLEATING INTERMETALLICS

Grain refinement can be achieved by additions of elements that form primary intermetallics with good crystallographic match with the matrix solid solution, i.e., aluminum or magnesium. In aluminum alloys, titanium aluminide and scandium aluminide are well known to possess structural features required for powerful grain-refinement effect; in magnesium alloys, zirconium forms a primary phase that is used in Al-free alloys for grain refinement.

The role of zirconium in aluminum alloys is controversial. On one hand, Zr has a peritectic phase diagram with Al and forms primary aluminide Al_3Zr at hyper-peritectic concentrations. On the other hand, this aluminide has average grain-refining ability and, in addition, tends to form large particles that can be harmful for mechanical properties. It has been known since the 1960s that the addition of Zr in combination with ultrasonic treatment results in considerable grain refinement of aluminum alloys [7]. Later on, the essential role of small Ti additions was demonstrated [37], and the combined effect of Zr, Ti, and ultrasonic processing has been explained [51, 52].

Figures 5.14–5.16 show that efficient grain refinement requires a certain combination of concentrations of Zr and Ti in addition to the ultrasonic cavitation treatment. If the concentration of Zr is sufficient to produce primary intermetallics (>0.12% Zr) and a small amount of Ti (0.05–0.08%) is present in the melt, then ultrasonic processing results in very strong grain refinement, even with the formation of nondendritic structure (Figures 5.14b, 5.15a, and 5.16c,d). When the concentration of Zr is much lower, then it is required to have the hyper-peritectic concentration of Ti to achieve such a degree of grain refinement [37] (Figure 5.14a). Further research demonstrated that the temperature range of ultrasonic processing is very important and should be within the temperature range of primary solidification of the Al_3Zr phase (below 725°C for 0.2% Zr), whereas the concentration of Zr should not exceed the peritectic concentration by much [51] (Figure 5.15b). It is important to note that the presence of titanium is essential (compare Figures 5.16a,c) and that ultrasonic processing plays a vital role (compare Figures 5.16b,c). The effect of grain refinement becomes more vivid in concentrated alloys, e.g., of 2XXX, 6XXX, and 7XXX series (see Figures 5.14b and 5.16c).

The analysis of mechanisms of grain refinement in Al–Zr–Ti alloys reveals that the main reason for the observed strong effects of ultrasonic melt treatment is the increased potency of the primary particles. This is evidenced by the decreased slope and the same intercept of the dependencies in Figure 5.15c. A thorough investigation shows that the ultrasonic treatment performed in the range of primary solidification of the Al_3Zr phase causes significant refinement of the primary particles, as demonstrated in Figures 5.17a,b. In addition, the refined particles appear in the center of grains, which hints that these particles may indeed act as nucleants. Titanium at the given low concentrations does not form its own phases but, rather, dissolves in Al_3Zr and in aluminum. Titanium being dissolved in Al_3Zr changes slightly (and unfavorably for nucleation) the lattice parameters of the tetragonal equilibrium DO_{23} type phase, increases internal stresses, and promotes cracking of the primary particles that facilitates their further fragmentation by cavitation [52]. As a result of these processes, the nucleation of aluminum on Al_3Zr particles requires larger undercooling, both due to their changed lattice parameters and the decreased size, i.e., the nucleation undercooling increases from 0.02 to 0.2 K. This allows for a delayed nucleation when numerous small Al_3Zr particles 1–5 μm in size formed by cavitation become active and more potent than large primary intermetallics formed in the alloys without ultrasonic processing.

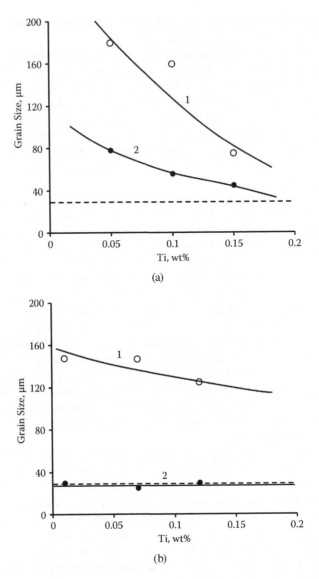

FIGURE 5.14 Effect of ultrasonic processing (80 W/cm²) and the concentration of Zr and Ti on grain refinement in 70-mm-diameter billets from an AA7475 alloy: (a) at 0.05% Zr and (b) at 0.16%–0.18% Zr; (1) without sonication and (2) with sonication. The dashed lines show the dendrite arm spacing in billets cast without sonication.

The growth restriction by Ti remaining in the liquid aluminum and by other alloying elements plays an additional role in hindering the growth of the grains, which is vividly demonstrated by very strong grain refinement observed in concentrated commercial alloys. The plot in Figure 5.17c shows that in 7XXX-series alloys, the ultrasonic treatment causes the multiplication of active nucleating particles (lower intercept) and increases their potency (changed slope) [53].

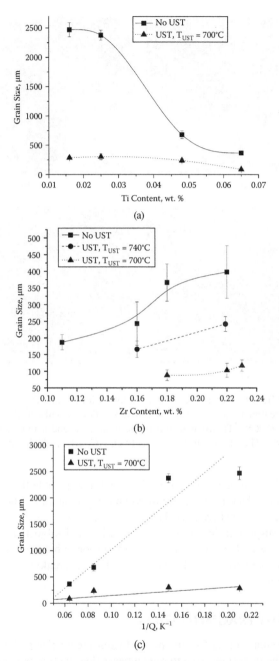

FIGURE 5.15 Effect of ultrasonic processing (17.5 kHz, 40-μm amplitude) and the concentration of Zr and Ti on grain refinement of commercial aluminum: (a) at 0.18% Zr and ultrasonic processing at 700°C; (b) at 0.065% Ti and at different temperatures of ultrasonic processing; (c) replot of (a) showing the relationship between the grain size and inverse growth-restriction factor in Al–Zr–Ti alloys. Growth-restriction factor Q is calculated from the Ti concentration in the alloys.

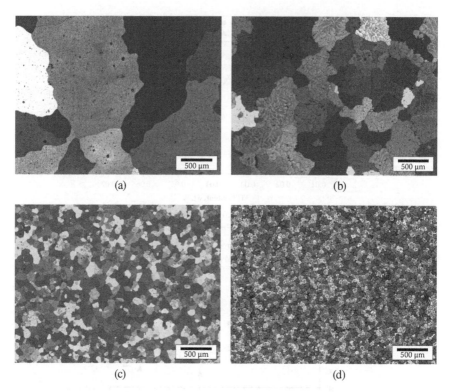

FIGURE 5.16 Microstructures of aluminum alloys with Zr and Ti: (a) Al–0.22% Zr cast with ultrasonic treatment; (b) Al–0.22% Zr–0.065% Ti cast without ultrasonic treatment; (c) Al–0.22% Zr–0.065% Ti cast with ultrasonic treatment; and (d) an AA7075 alloy with 0.18% Zr and 0.08% Ti cast with ultrasonic treatment. (Courtesy of T.V. Atamanenko [46].)

The mechanisms of primary-phase refinement may include the enhanced nucleation on activated substrates (e.g., alumina) and fragmentation. The former mechanism was described elsewhere [54]. The latter mechanism was observed on faceted calcite crystals suspended in a transparent solution and subjected to 17-W 42-kHz sonication [55]. It was shown that cavitation (implosion of bubbles and related microjet impingement onto the crystal surface) resulted in breaking and indenting the suspended crystals.

5.5 DENDRITE FRAGMENTATION

Dendrite fragmentation is one of the most obvious mechanisms of grain refinement by cavitation. The important issue to remember is that the fragmentation of the solid phase can only happen when this solid phase is present in the cavitation field. Other mechanisms that we have discussed before are not usually in play in the semisolid state of the solidifying alloy. Fracture by collapsing bubbles can happen to the primary intermetallics in the range of their formation, as we have discussed in the previous section. In this case, the alloy would be considered liquid from a technological

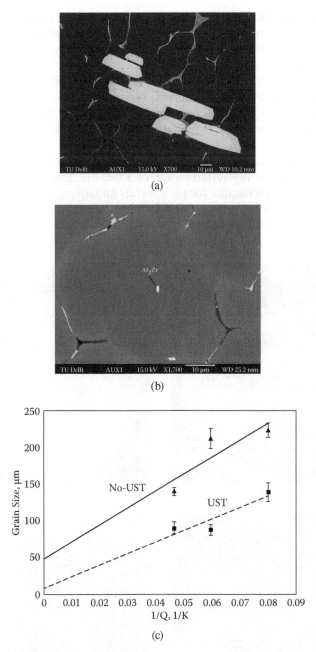

FIGURE 5.17 Effect of ultrasonic processing on primary Al₃Zr intermetallics in an AA7075 alloy with 0.6% Zr and 0.06% Ti: (a) without sonication and (b) with sonication at 710°C; and (c) grain size versus inverse growth-restriction factor in 7XXX-series alloys with 0.2% Zr and 0.06% Ti. Q is calculated based on the concentration of Zn, Mg, and Cu.

point of view, as the formation of these particles as well as the ultrasonic processing occur well above the liquidus temperature of the matrix solid solution. The fragmentation of dendrites, however, is typical of ultrasonic processing in the solidification range of the matrix.

Figures 5.2b and 5.4 demonstrate the fragmentation by oscillating and collapsing cavitation bubbles on transparent dendrites. Two mechanisms are acting: explosion-type fragmentation by a collapsing bubble and fatigue-type fragmentation by an oscillating bubble. The former is typical of fragmentation during solidification of metallic alloys.

A lot of research has been done under conditions when the ultrasonic oscillations are applied during solidification of a metallic alloy. The results were always good and reproducible for pure metals and aluminum and magnesium alloys, but were limited in the volume treated. Figure 5.18 illustrates that the grain can be dramatically refined in a small crucible (400 g of an Al–4% Cu

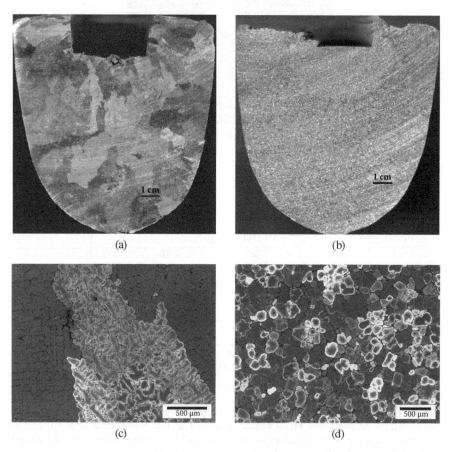

(a)   (b)

(c)   (d)

FIGURE 5.18 Effect of ultrasonic treatment during solidification on macro- (a, b) and grain (c, d) structure of an Al–4% Cu alloy (17.5 kHz, 30-μm amplitude, 400 g): (a, c) no ultrasonic processing, inserted sonotrode; (b, d) ultrasonic processing until the end of solidification.

melt treated in this case). The effect is always greater immediately under the sonotrode and decays with the distance and the deviation from the vertical direction [56].

The feasibility of breaking out dendrite branches by collapsing cavitation bubbles can be estimated using a Pilling–Hellawell model for the effect of a concentrated melt flow acting normal to the dendrite arm [20]. In the case of an imploding cavitation bubble, the jet velocity gives the momentum acting on the dendrite root. The estimates performed by Shu et al. [26] give the stresses at the dendrite root as 12.8 MPa at a jet velocity of 10 m/s. This stress should be sufficient to break aluminum, whose yield strength at the melting point can be taken as 6.5 MPa [20]. The fracture is assisted by high strain rates, the shear mode of the stress, and liquid-metal embrittlement.

Despite the very good grain-refining effect of fragmentation, the practical application of this mechanism is limited to small volumes. There might be a potential to use this mechanism in direct-chill casting or other continuous processes (e.g., arc remelting), where the position of the solidification front is fixed in space and the cavitation can be applied throughout the process in the locations below the liquidus isotherm. The limitation in this case would be the lateral spread of the effect; hence multiple cavitation sources would be required for processing of larger cross sections.

5.6 NONDENDRITIC SOLIDIFICATION

The dependence of dendrite arm spacing (d or DAS) on the cooling rate or solidification rate is now a matter of textbook knowledge. This dependence can be written as a unique function of the local solidification time (t_f):

$$d = Bt_f^n.$$ (5.5)

As $t_f = \Delta T_0/Gv_s$ (here ΔT_0 is the solidification range, G is the temperature gradient, and v_s is the velocity of the solidification front) and Gv_s is the cooling rate (v_c) in the solidification range, we can write the same expression in a well-known form:

$$d = Cv_c^{-n},$$ (5.6)

where n, a so-called coarsening exponent, varies between 0.4 and 0.2 for various light alloys.

It is important to note that the cooling rate v_c and the solidification rate v_s are, in general, proportional to each other, but not always linearly. In the case of DC casting, the solidification rate is equal to the velocity of the solidification front and depends on the casting speed, as shown by the following relationship [57]:

$$v_s = v_c \cos \alpha,$$ (5.7)

where α is the angle between the billet axis and the normal to the solidification front.

The dependence was apparently first reported by Dobatkin in 1948 [58] for the dependence of the dendrite arm spacing on the solidification rate (casting

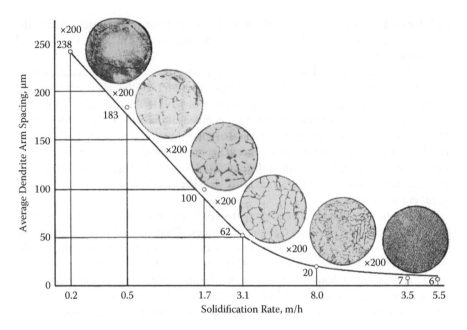

FIGURE 5.19 Dependence of dendrite arm spacing on the solidification rate upon DC casting. (Adapted from Dobatkin [58].)

speed) during DC casting (see Figure 5.19). This plot was reproduced in 1964 by Chalmers [3]. The dependence of dendrite arm spacing on the cooling rate was studied by Alexander and Rhines [59] but was not formalized. In the 1960s, the relationships (5.5) and (5.6) were firmly established for metallic alloys [16, 60–62], also at very high cooling rates [63].

The dependence of grain size on the cooling rate is more complicated and depends on both the nucleation rate and growth conditions (cooling rate). Under certain conditions, however, the grain size may be a function of the cooling rate. In 1967, Kattamis, Holmberg, and Flemings [62] showed that, for magnesium alloys with Zr additions, the structure may become nondendritic, and then the size of these nondendritic grains will depend on the local solidification time in the same way as dendrite arm spacing (see Equation 5.5). "For a given local solidification time, grain diameter in the grain-refined alloy is approximately the same as dendrite-arm spacing in a similar nongrain-refined alloy" [62]. Comparable behavior was observed for Al–Cu alloys slowly solidified within a vibrating crucible with a water-cooled titanium rod submerged in the melt [64]. Fragmentation of dendrite branches formed on the titanium rod, and their spread over the solidifying volume was reported to be responsible for the formation of nondendritic grain structure.

This preliminary evidence for the formation of nondendritic structure in the 1960s–1970s was confirmed and thoroughly studied in the laboratory and at the pilot and industrial scale during systematic research performed in the USSR under the supervision of Dobatkin and G.I. Eskin [65–67]. Various aluminum alloys containing small additions of Zr (and apparently Ti, see Section 5.4) were direct-chill cast in billets ranging from 134 to 1000 mm in diameter. This work showed that the application of ultrasonic cavitation treatment to the liquid bath during DC casting promoted multiplication of solidification nuclei and resulted in the formation of nondendritic grain structure.* Such a structure, where the sole control parameter of its fineness is the cooling (or solidification) rate, represents the finest structure that can be obtained at the given cooling rate. The formation of nondendritic grain structure was demonstrated in Al, Mg, Ni, and Fe alloys. It was suggested that the formation of this structure requires an excess of active solidification sites ahead of the solidification front. The nature of the multiplication of solidification sites does not play any decisive role. Under typical casting conditions, it can be heterogeneous nucleation assisted by additions of powerful grain refiners (Zr in Mg, Sc in Al), activation and multiplication of substrates by cavitation, fragmentation of dendrites, etc. It can also be an enhanced heterogeneous or even homogeneous nucleation at very high undercooling achievable upon rapid solidification.

Cavitation-induced mechanisms of grain refinement are powerful enough to facilitate the formation of nondendritic grains in a wide range of alloying systems and solidification conditions. The excess of active solidification sites (substrates) in the vicinity of the solidification front is the cornerstone of the nondendritic solidification. Each of the undercooled (thermally or constitutionally) volumes at any given cooling rate will contain active nuclei. The presence of excessive nuclei means that their activation can only be achieved by increasing the undercooling, which in most cases means an increased cooling rate. The formed grains do not have an opportunity to initiate branching, as almost immediately after formation they found themselves in the solute–temperature field of the neighboring grains. Thus, with excessive nuclei, the cooling rate becomes a factor determining the grain size in an ingot.

The wealth of fundamental and applied research in the formation of nondendritic structure allowed the formulation of a physical law: "The formation of excess solidification nuclei in the melt, for example by combination of cavitation melt treatment with grain refining additions, results in the formation of nondendritic structure with the grain size solely controlled by the cooling rate" [68]. Figure 5.20 illustrates the dependence of dendrite arm spacing in dendritic structures and grain size in dendritic and nondendritic structures on cooling rate for various groups of alloys.

Let us take a closer look at the nondendritic structure (see Figure 5.16d as an example) and its features. Microsegregation profiles underlying the consecutive stages of grain growth may shed some light on the growth mode. Concentration isolines given in Figure 5.21 for solute copper in nondendritic and dendritic

---

* Note that the term *nondendritic structure* is also used in semisolid processing of metals, where the nondendritic structure is usually formed either by isothermic coarsening of dendritic grains or by their shear deformation. In both cases, the resultant grain structure is indeed nondendritic, but the mechanisms of its formation are completely different from those discussed here.

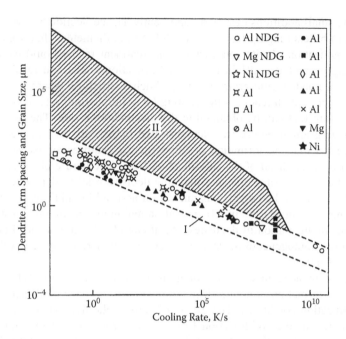

FIGURE 5.20 The effect of cooling rate on nondendritic grain size (NDG, I), dendritic grain size (II), and dendritic arm spacing (I) for aluminum, magnesium, and nickel alloys.

structures obtained in an AA2024-type alloy ingot show some peculiar features. It is easy to see that the nondendritic grain has developed concentrically in all directions without segmenting into dendritic parts. As the nondendritic grain grows, its shape can deviate somewhat from the spherical one because of the influence from neighboring grains (Figure 5.21a). Similar isolines constructed for a large dendritic branch (Figure 5.21b) indicate that the development of this branch also occurs by its growth in all directions. The degree of copper segregation in dendritic and nondendritic structures is virtually the same and is equal to 5 and 6, respectively. However, it should be noted that the nondendritic grain has a larger fraction of its volume occupied by strongly depleted solid solution, which reflects the different pattern of growth. In the case of nondendritic structure or spherical grain growth, the solute accumulation in the liquid occurs at a lower rate than in the case of the cylindrical grain, which may be considered as a model for a dendritic branch. The diffusion distances between concentration maxima become less in nondendritic structure, as illustrated in Figure 5.21c for an AA7055-type alloy, which facilitates the diffusion upon solution treatment and allows for a shorter homogenization time.

Eutectic colonies or particles of eutectic phases appear at the nondendritic grain boundaries in the same manner as between dendrite branches. At the same time, the thickness of these eutectic colonies and the size of eutectic constituents decrease by an order of magnitude in the nondendritic structure (Figure 5.22).

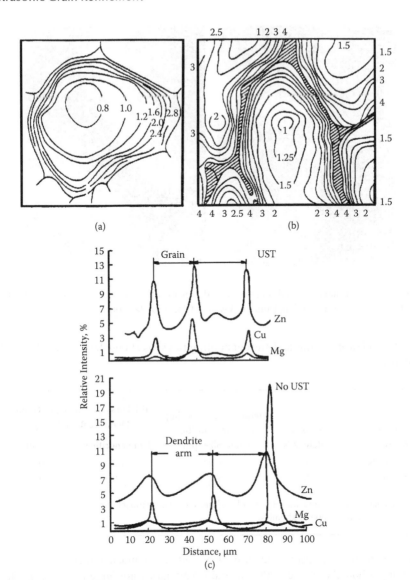

FIGURE 5.21 Microsegregation isoconcentrates for copper in (a) nondendritic and (b) dendritic structures of an AA2024-type alloy billet (eutectic areas in (b) are marked by crosshatching; lines are marked by wt% Cu) and (c) microsegregation profiles in nondendritic (UST) and dendritic grains (no UST) in an AA7055-type billet.

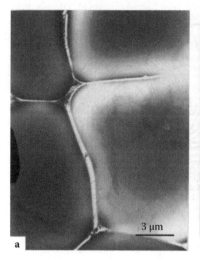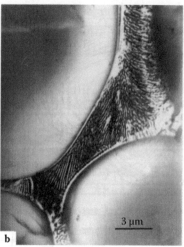

FIGURE 5.22 Eutectic colonies at the boundaries of (a) nondendritic grains and (b) dendrite branches in 70-mm billets from an AA7055-type alloy (×3000).

This dependence is valid at different cooling rates, as shown here for an AA7055-type alloy:

Type of Casting	Cooling Rate, K/s	Dendrite Arm Spacing or Nondendritic Grain Size, μm	Thickness of Eutectic at Grain or Dendrite Boundary, μm	
			Nondendritic [a]	Dendritic
70-mm billet	60	20–30	1.0	30
50-μm granule	10^4	5.0	0.01	0.1

[a] Nondendritic structure was obtained by ultrasonic melt treatment during DC casting of billets or atomization of granules.

The nondendritic grain structure also contains a much larger proportion of large-angle boundaries, i.e., 90% of grain boundaries are large-angle in nondendritic structure as opposed to 90% of small-angle boundaries typical of the dendritic structure. This is illustrated in Figures 5.23a,b for both types of structures obtained in ingots from an AA7055-type alloy. The nondendritic structure is also more uniform with regard to the size distribution of structural features, as demonstrated in Figure 5.23c for size distribution of dendritic and nondendritic grains as well as dendrite arm spacing.

The formation of nondendritic structure is linked to the change in the morphology and extent of the transition region in DC casting (and generally in any type of casting). The transition region can be conventionally subdivided into a slurry zone (between the liquidus isotherm and the coherency isotherm) and a mushy zone (between the coherency isotherm and the solidus) [69]. The grain refinement was

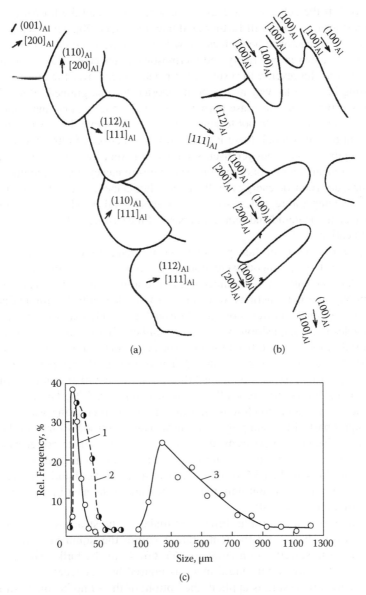

FIGURE 5.23 Distribution of large- and small-angle boundaries in (a) nondendritic and (b) dendritic grain structures (crystallographic orientations are shown) and (c) size distribution of nondendritic (1) and dendritic (3) grains and dendrite arm spacing in the latter (2). Data are given for 70-mm billets from an AA7055-type alloy. Nondendritic structure is obtained by ultrasonic melt treatment during DC casting.

shown to shift the coherency to higher fractions solids, and the formation of very fine, nondendritic grains will further facilitate this effect [70]. This means that the nondendritic grains, once being formed, will have some freedom to move through the slurry zone driven by gravity and thermosolutal convective flows. The fineness of these grains, however, will result in their wide spread over the cross section of the casting in a similar way as has been reported for fine-grained alloys [71, 72]. Such a structure would increase permeability of the mushy zone and improve the resistance of the casting to hot tearing as well as equilibrate the macrosegregation profile. On the other hand, there is an important influence of ultrasonic cavitation processing that we have not yet discussed but that is very important for understanding the ultrasonic-aided solidification. The high-frequency, high-amplitude vibrations introduce acoustic energy into the treated liquid medium (melt) and effectively increase its temperature, unless special measures are taken. This changes the thermal balance in the liquid and semiliquid parts of the casting and affects the kinetics of solidification.

To quantify this important phenomenon, the temperature fields in the sump of DC-cast billets and ingots were studied by "freezing" thermocouples into the casting. A frame with six K-thermocouples in ceramic insulation was placed in the liquid pool. Their hot junctions were bare spark-welded butts 2.0 mm in diameter, and their cold junctions were connected through relatively long (>1.0 m) compensation cables to an acquisition system. In a number of cases, direct measurements by a Ti rod of the distance from the melt surface to the solidification front (about the coherency isotherm) were conducted. The results for the temperature fields in the liquid pool of 270-mm-diameter billets from an AA1070-grade commercial aluminum and an AA7055-type alloy are shown in Figure 5.24. These billets were cast under steady-state conditions, and the ultrasonic processing was performed in the liquid part of the sump by a Nb ultrasonic sonotrode 40 mm in diameter. According to acoustic measurements, about 1.0 kW of acoustic power was transferred to the melt. Our estimates for commercially pure aluminum give the contribution of ultrasound as 3% of the total heat input in the liquid pool. However, when one takes into account only the superheating of the melt in the liquid pool, this contribution rises to 40%.

The sump of the aluminum billet cast under in the steady-stage regime has a conventional shape, and the temperature within the liquid pool varies from 660°C near the solidification front to 665–667°C in the bulk, which is characteristic of the case when liquid metal is poured by a concentrated jet. When the ultrasonic processing is applied, the bottom of the sump becomes straighter and the melt temperature in the liquid part of the sump rises by more than 10°C (Figure 5.24a). As a result of the heat input by ultrasonic cavitation as well as due to acoustically induced flows, the liquidus isotherm is shifted toward the solid part of the billet. Consequently, the liquid pool becomes deeper (by 23% in this case).

In alloys, when a transition region is formed between liquidus and solidus, the vertical size of the transition region decreases under the action of ultrasound and induced additional melt superheat (Figure 5.24b). The liquidus shifts downward due to the acoustic energy, and the slurry zone, where fine grains nucleate, grow, and

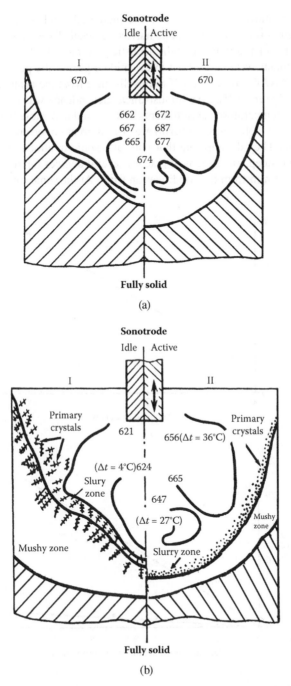

FIGURE 5.24 Variation in measured melt temperature in the liquid pool of 270-mm billets from (a) AA1070 and (b) AA7055 under conventional casting conditions with idle sonotrode (I, left-hand parts of the diagrams) and subjected to ultrasonic melt treatment (II, right-hand sides of the diagrams).

have freedom to move, narrows. Due to the increased solid fraction at coherency with the refined grains, the mushy zone narrows as well. To estimate the superheating in the ingot pool as a function of the acoustic power supplied to the melt, temperature measurements were carried out in the middle part of the pool of the 270-mm billet from an AA7055 alloy at a half-radius distance from the center and at a depth of 65–70 mm (Figure 5.25a). An almost linear dependence was observed in these measurements. Temperature measurements in the liquid pool close to the continuous solidification front (coherency isotherm) are also of great importance and are shown in Figure 5.25b. They demonstrate that with ultrasonic treatment, the temperature gradient is quite steep, and superheat reaches several degrees (up to 8°C) already at a distance of 2–3 mm from the solidification front. At the same time, the superheat

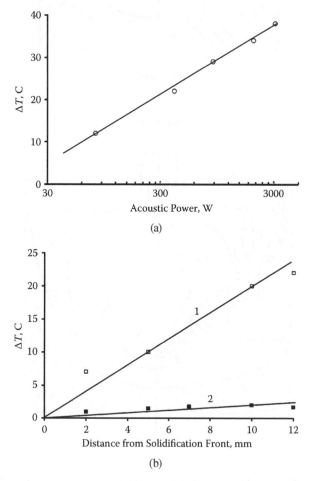

(a)

(b)

FIGURE 5.25 (a) Effect of acoustic power on the melt superheat in a liquid pool of an AA7055 alloy billet and (b) the temperature gradient in the liquid pool of the same ingot measured from the solidification front (coherency isotherm): (1) with sonication and (2) without sonication.

does not exceed 2°C at a distance of 15 mm from the solidification front under conventional casting conditions.

Similar measurements were performed for large billets 650 to 1200 mm in diameter. The results showed that ultrasonic treatment always led to a noticeable superheat of the melt and a 15%–20% increase in the liquid pool volume. The melt superheat and the steep temperature gradient ahead of the solidification front are important features of ultrasonic melt processing that prevent excessive growth of nondendritic grains and intermetallics in the slurry zone.

We can summarize that the ultrasonic melt processing creates conditions for the formation of an excess of active solidification nuclei (through activation of substrates and their multiplication and dispersion) that by itself prevents dendritic growth through interaction of solutal and thermal fields of neighboring grains. In addition, the ultrasonic melt processing hinders the growth of the formed grains by narrowing the slurry zone and creating a steep temperature gradient at the solidification front. The changed geometry of the transition region and the acoustic flows also have consequences for the formation of such defects as macrosegregation and cracks, as we will show in Chapter 7.

Cavitation is an important parameter required for the formation of nondendritic structure, so a certain acoustic power is required to achieve the developed cavitation conditions, as shown in Figure 5.26. When the required conditions are achieved, the transition from dendritic to nondendritic structure occurs almost

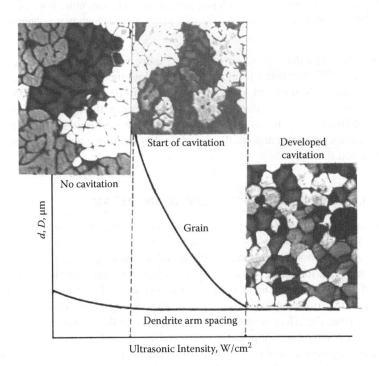

FIGURE 5.26 Effect of ultrasonic intensity on the formation of nondendritic structure. Micrographs show the grain structure of an AA7055 alloy.

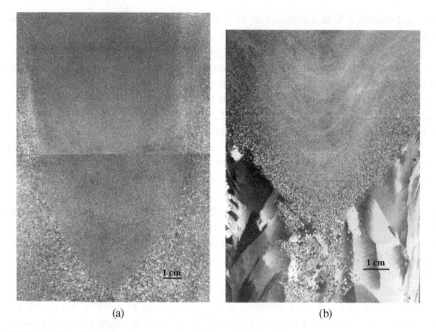

FIGURE 5.27 Transition between dendritic and nondendritic structure as a result of ultrasonic melt processing performed in a liquid bath of (a) a 112-mm billet from an AA7055 alloy and (b) 75-mm billet from an AA7050 alloy prepared using high-purity aluminum.

immediately over a distance of several mm. Figure 5.27 illustrates this transition for a DC-cast 112-mm billet from an AA7055 alloy (the grain size changes from 500–1000 μm to 20 μm) and for a DC-cast 75-mm billet from an AA7050 alloy prepared using high-purity aluminum. In the latter case, the transition from very coarse columnar structure to very fine nondendritic grains is striking. In both cases, the sonication was done in the liquid pool of the billet, and both alloys contained Zr additions.

5.7 ULTRASONIC MELT PROCESSING IN THE MOLD

Melt processing in a mold or in stationary volume is typical for small-scale experiments and small-scale shape casting, as well as for DC casting. In the latter case, the increasing mold cross section requires additional sonotrodes.

It is important to distinguish experiments when the processing is performed in the liquid phase (in a crucible or mold) with subsequent solidification without sonication and the processing that involves sonication during solidification as well. In the former case, the effect of ultrasound is primarily on the activation of substrates and nucleation and growth of primary intermetallics (as well as in degassing). In the latter case, fragmentation of the growing solid phase is most likely to be responsible for the observed grain refinement, while other effects can be related to the altered solidification kinetics.

Experiments in a limited volume are useful for establishing basic relationships and parameters of ultrasonic treatment. Figures 5.28a–c demonstrate the effect of treatment time, treated volume, and the holding time after the end of treatment on the grain size of an Al–0.18% Zr–0.07% Ti alloy [36]. Evidently, the longer the processing time and the smaller the volume, the greater is the grain refinement. The effect can survive for several minutes after the end of ultrasonic processing,

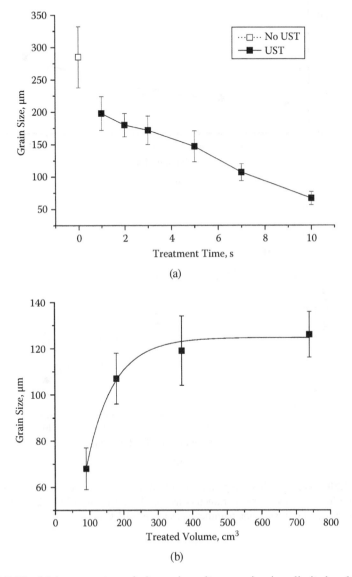

FIGURE 5.28 Main parameters of ultrasonic melt processing in a limited melt volume: (a) treatment time at 700°C, treated volume 90 cm³, amplitude 40 μm, frequency 17.5 kHz, Al–0.18% Zr–0.07% Ti; (b) treated volume, same parameters as (a), treatment time 10 s. (*continued*)

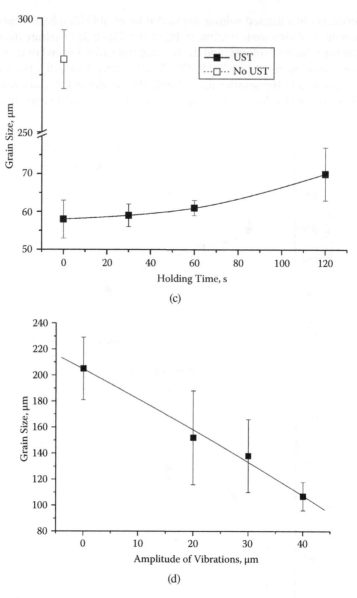

FIGURE 5.28 (*continued*) Main parameters of ultrasonic melt processing in a limited melt volume: (c) holding time, same parameters as (b), Al–2.5% Cu–0.22% Zr–0.06% Ti; and (d) amplitude of vibrations, same parameters as (a), Al–4% Cu.

before fading starts. These several minutes are quite sufficient for most casting processes. The decisive role of the intensity of ultrasonic processing has been demonstrated throughout this chapter (see Figures 5.8a and 5.26). Figure 5.28d further illustrates this for an Al–4% Cu alloy treated in a small volume [36]. Similar results, albeit on a different scale, have been demonstrated upon ultrasonic treatment in a DC casting mold. This chapter gives sufficient evidence of that.

Working with a limited volume allows for isothermal processing. Such experiments give an interesting insight into the ultrasonic processing. We have already discussed the mechanism of dendritic fragmentation that leads to a considerable grain-refining effect when the ultrasonic vibrations are applied to the solidifying alloy. In this case, the treatment started in the liquid state and proceeded through the onset of solidification until a significant solid fraction was formed (see, for example, Figure 5.18). Different results are obtained when the ultrasonic treatment is applied isothermally to the semisolid material [73, 74]. Ultrasonic cavitation is limited to a small volume close to the sonotrode tip and generates quite a lot of heat, while acoustic streams cannot develop in the dense semisolid structure. As a result, instead of fragmentation of the solid phase with dispersion of fragments in the liquid phase, the solidification effectively slows down, and the grain coarsening follows. This is demonstrated in Figure 5.29. Note that grain coarsening is accompanied by an increase in dendrite arm spacing. For alloys with a considerable amount of eutectics, ultrasonic treatment in the semisolid state causes coarsening of eutectics, as shown in Figure 5.30 [75, 76], while when performed above the liquidus it may result in eutectic refinement [77, 78].

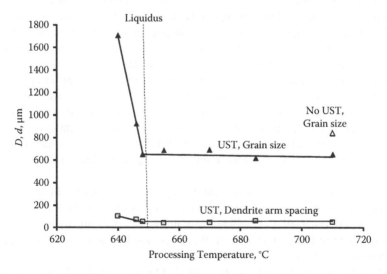

FIGURE 5.29 Effect of isothermal ultrasonic processing (UST) on the grain size in an Al–4% Cu alloy: amplitude 40 µm, frequency 17.5 kHz, processing time 15 s, volume 90 cm^3.

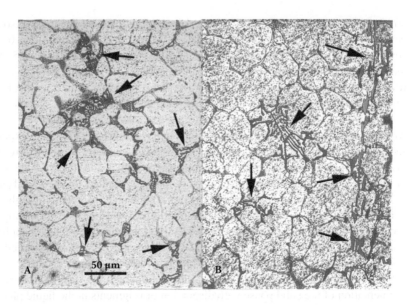

FIGURE 5.30 Effect of ultrasonic treatment during solidification on the eutectic phases in an Al–Cu–Mg alloy with a high level of impurities: (a) without sonication and (b) with sonication until the end of solidification.

5.8   ULTRASONIC MELT PROCESSING IN THE MELT FLOW

Sonication of melt flow—in the melt transport or melt feeding system—is a promising way of ultrasonic melt processing, as we discussed previously in Section 3.4 (see also Figure 3.17). This type of processing is characterized by large throughput of melt, which means that the time available for processing is necessarily short but also that the instantaneously treated volume is rather small.

This scheme of sonication has not been well studied. The earlier applications for degassing (see Section 3.4) and filtration (see Section 4.5) were successful but required the use of several sonotrodes to process the amount of melt needed for DC casting. It is obvious that some form of flow management is necessary to achieve a requisite ratio between treatment time and treated volume, as is done now for in-line filtration. Computer simulations and physical modeling may be very useful.

The processing of the melt in the launder (apart from degassing) invokes the activation of substrates and refinement of primary intermetallics. The introduction of grain-refining master alloys combined with ultrasonic melt processing in a launder is one of the promising technological directions based on activation and dispersion of substrates, as illustrated in Figures 5.12 and 5.13.

Relatively low casting temperatures typical of DC casting also allow for the mechanism that involves primary intermetallics into solidification. As we have already

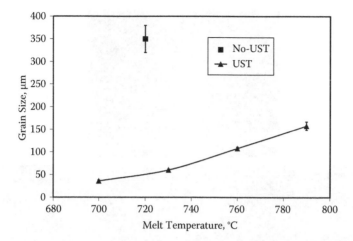

FIGURE 5.31 Effect of melt temperature on the grain size in AA7075-type alloy with addition of 0.2 wt% Zr and 0.07 wt% Ti cast through a launder pouring system. (Courtesy of L. Zhang [79].)

shown in Section 5.4, a combination of Zr and Ti triggers efficient grain refinement under cavitation conditions. This effect can be used during melt processing in a launder, when the melt temperatures vary between 720°C and 690°C, which is within the formation range of primary Al$_3$Zr.

Laboratory-scale experiments with a small launder (80 cm long with 6 × 5-cm cross section) were performed with a melt volume of 1 kg and an AA7075-type alloy with Zr and Ti additions. The melt temperature was varied between 700°C and 790°C. In such a small system, the introduction of a sonotrode resulted in a temperature drop of up to 75°C over the length of the launder. The results in Figures 5.31 and 5.32 clearly demonstrate the possibility of grain refinement with sonication in the launder, but also confirm that the melt temperature plays a crucial role for the involved mechanism of primary intermetallic refinement. The grain size was reduced from 300–350 μm (dendritic) to 40 μm (nondendritic) at a pouring melt temperature of 700°C. The cooling rate in the mold also plays a role in grain refinement. In these series of experiments, the decrease in the cooling rate during solidification from 5 to 2 K/s resulted in an increase in the grain size obtained without sonication from 300 to over 800 μm, while the grain size after sonication remained virtually the same, 40–45 μm.

Although these laboratory-scale results show the potential of ultrasonic treatment in the melt flow for grain refinement, large-scale trials are yet to confirm this, and more research in the flow management and optimization of time–volume ratio for cavitation melt processing is needed.

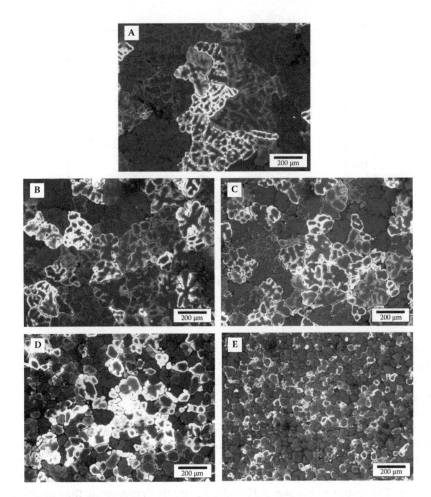

FIGURE 5.32 Grain structure of castings produced with ultrasonic melt treatment in the launder (see Figure. 5.31): (a) no UST, poured at 720°C; (b) UST, poured at 790°C; (c) UST, poured at 760°C; (d) UST, poured at 730°C; and (e) UST, poured at 700°C. (Courtesy of L. Zhang [79].)

REFERENCES

1. Kapustin, A.P. 1962. *Effect of ultrasound on the kinetics of crystallization.* Moscow: Nauka.
2. Walker, J.L. 1958. In *Proc. ASM Seminar Liquid Metals and Solidification*, 319–36. Metals Park, OH: ASM Publication.
3. Chalmers, B. 1964. *Principles of solidification.* New York: John Wiley & Sons.
4. Hunt, J.D., and K.A. Jackson. 1966. *J. Appl. Phys.* 37:254–57.
5. Frawley, J.J., and W.J. Childs. 1968. *Trans. Met. Soc AIME.* 242:256–63.
6. Eskin, G.I. 1961. *Ultrasound in metallurgy.* Moscow: Metallurgizdat.
7. Eskin, G.I. 1965. *Ultrasonic treatment of molten aluminum.* Moscow: Metallurgiya.
8. Danilov, V.I. 1956. *Constitution and crystallization of liquids.* Kiev: Izd. Akad. Nauk SSSR.
9. Danilov, V.I., and V.E. Neimark. 1949. *Zh. Eksper. Teor. Fiz.* 19:235–41.

10. Danilov, V.I., and D.E. Ovsienko. 1951. *Zh. Eksper. Teor. Fiz.* 21:879–87.
11. Kazachkovsky, O.D. 1948. In *Collective papers of the Laboratory of Metals Physics*, ed. V.I. Danilov, 103–112. Kiev: Izd. Akad. Nauk UkrSSR.
12. Swallowe, G.M., J.E. Field, C.S. Rees, and A. Duckworth. 1989. *Acta Mater.* 37:961–67.
13. Chow, R., R. Blindt, R. Chivers, and M. Povey. 2005. *Ultrasonics* 43:227–30.
14. Chvorinov, N.I. 1954. *Crystallization and non-homogeneity of steel*. Prague: Publishing House of Czechoslovak Academy of Sciences (NČSAV).
15. Balandin, G.F. 1973. *Formation of crystalline structure of castings*. Moscow: Mashinostroenie.
16. Flemings, M.C. 1974. *Solidification processing*. New York: McGraw-Hill.
17. Campbell, J. 1981. *Int. Met. Rev.*, no. 2:71–108.
18. Abramov, O.V. 1972. *Crystallization of metals in an ultrasonic field*. Moscow: Metallurgiya.
19. Vogel, A., R.D. Doherty, and B. Cantor. 1979. In *Solidification and casting of metals*, 518–25. London: TMS.
20. Pilling, J., and A. Hellawell. 1996. *Metall. Mater. Trans. A* 27A:229–32.
21. Jackson, K.A., J.D. Hunt, D.R Uhlmann, and T.P. Seward. 1966. *Trans. Metall. Soc. AIME* 236:149–60.
22. Chernov, A.A. 1956. *Kristallografiya* 1:589–93.
23. Kattamis, T.Z., J.C. Coughlin, and M.C. Flemings. 1967. *Trans. Metall. Soc. AIME* 239:1504–11.
24. Ruvalcaba, D., R.H. Mathiesen, D.G. Eskin, L. Arnberg, and L. Katgerman. 2007. *Acta Mater.* 55:4287–92.
25. Abramov, O.V., and I.I. Teumin. 1970. Crystallization of metals. In *Physical principles of ultrasonic technology*, ed. L.D. Rozenberg, 427–514. Moscow: Nauka. (Translated by New York: Plenum, 1973.)
26. Shu, D., B. Sun, J. Mi, and P.S. Grant. 2012. *Metall. Mater. Trans. A* 43A:3755–66.
27. Grandfield, I.F., D.G Eskin, and I.F. Bainbridge. 2013. *Direct-chill casting of light alloys: Science and technology*. Hoboken, NJ: Wiley.
28. Dobatkin, V.I., R.M. Gabidullin, B.A. Kolachev, and G.S. Makarov. 1976. *Gases and oxides in aluminum wrought alloys*. Moscow: Metallurgiya.
29. Šebo, P., J. Ivan, L. Táborský, and A. Havalda. 1973. *Kovové Materiály* 11 (2): 173–80.
30. Eskin, D.G. 1996. *Z. Metallkde.* 87:295–99.
31. Fan, Z., and Y. Wang. 2011. Brunel University, BCAST. Private communication.
32. Fan, Z., Y. Wang, M. Xia, and S. Arumuganathar. 2009. *Acta Mater.* 57:4891–4901.
33. Chen, Y-J., W.-N. Hsu, and J.-R. Shih. 2009. *Mater. Trans.* 50:401–8.
34. StJohn, D.H., M. Qian, A. Easton, P. Cao, and Z. Hildebrand. 2005. *Metall. Mater. Trans. A* 36A:1669–79.
35. Bhingole, P.P., and G.P. Chaudhari. 2012. *Mater. Sci. Eng. A* 556:954–61.
36. Atamanenko, T.V., D.G. Eskin, L. Zhang, and L. Katgerman. 2010. *Metall. Mater. Trans. A* 41A:2056–66.
37. Eskin, G.I. 1988. *Ultrasonic treatment of molten aluminum*. Moscow: Metallurgiya.
38. Dantzig, J.A., and M. Rappaz. 2009. *Solidification*. Lausanne, Switzerland: EPFL, 51–56; Boca Raton, FL: CRC Press, 268–75.
39. StJohn, D.H., M. Qian, M.A. Easton, and P. Cao. 2011. *Acta Mater.* 59:4907–21.
40. StJohn, D.H., M.A. Easton, M. Qian, and J.A. Taylor. 2013. *Metall. Mater. Trans. A* 44A:2935-49.
41. Rumpf, H. 1962. In *Agglomeration*, ed. W.A. Knepper, 382–403. New York: John Wiley.
42. Agranat, B.A., M.N. Dubrovin, N.N. Khavsky, and G.I. Eskin. 1987. *Fundamentals of physics and technology of ultrasound*, 297–300. Moscow: Vysshaya Shkola.
43. Bochvar, S.G., and G.I. Eskin. 2012. *Tekhnol. Legk. Spl.*, no. 1:9–17.
44. Eskin, G.I., A.A. Rukhman, S.G. Bochvar, V.I. Yalfimov, and D.V. Konovalov. 2008. *Tzvetn. Met.*, no. 3:105–110.

45. Eskin, G.I., and D.G. Eskin. 2010. In *Proc. Intern. Conf. Aluminum Alloys ICAA12*, 638–45. Tokyo: Jpn. Inst. Light Metals.
46. Atamanenko, T.V. 2010. Cavitation-aided grain refinement in aluminium alloys. PhD thesis, Delft, Delft University of Technology.
47. Zhang, L., D.G. Eskin, and L. Katgerman. 2011. *J. Mater. Sci.* 46:5252–59.
48. Han, Y., K. Li, J. Wang, D. Shu, and B. Sun. 2005. *Mater. Sci. Eng. A* 405:306–12.
49. Han, Y., D. Shu, J. Wang, and B. Sun. 2006. *Mater. Sci. Eng. A* 430:326–31.
50. Ramirez, A., M. Qian, B. Davis, T. Wilks, and D.H. StJohn. 2008. *Scr. Mater.* 59:19–22.
51. Atamanenko, T.V., D.G. Eskin, L. Zhang, and L. Katgerman. 2010. *Metall. Mater. Trans. A* 41A:2056–66.
52. Atamanenko, T.V., D.G. Eskin, M. Sluiter, and L. Katgerman. 2011. *J. Alloys Compd.* 509 (1): 57–60.
53. Zhang, L., D.G. Eskin, A. Miroux, and L. Katgerman. 2012. In *Proc. 13th Intern. Conf. Aluminum Alloys ICAA13*, ed. H. Weiland, A.D. Rollet, and W.A. Cassada, 1389–94. Warrendale, PA: TMS.
54. Zhang, L., D.G. Eskin, A. Miroux, and L. Katgerman. 2011. *IOP Conf. Series: Mater. Sci. Eng.* 27: paper 012002.
55. Wagterveld, R.M., L. Boels, M.J. Mayer, and G.J. Witkamp. 2011. *Ultrason. Sonochem.* 18:216–25.
56. Ramirez, A., and M. Qian. 2007. In *Magnesium technology*, ed. R.S. Beals, A.A. Luo, N.R. Neelameggham, and M.O. Pekguleryuz, 127–32. Warrendale, PA: TMS.
57. Livanov, V.A. 1945. In *Proc. First Technological Conference of Metallurgical Plants of Peoples' Commissariat of Aviation Industry*, 5–58. Moscow: Oborongiz.
58. Dobatkin, V.I. 1948. *Continuous casting and casting properties of alloys*. Moscow: Oborongiz.
59. Alexander, B.H., and F.N. Rhines. 1950. *Trans. AIME* 188:1267–73.
60. Spear, R.E., and G.R. Gradner. 1963. *Trans. AFS* 71:209–15.
61. Bower, T.F., H.D. Brody, and M.C. Flemings. 1966. *Trans. AIME* 236:624–34.
62. Kattamis, T.Z., U.T. Holmberg, and M.C. Flemings. 1967. *J. Inst. Met.* 95:343–47.
63. Matyia, H., B.C. Giessen, and N.J. Grant. 1968. *J. Inst. Met.* 96:30–32.
64. Kattamis, T.Z., and R.B. Williamson. 1968. *J. Inst. Met.* 96:251–52.
65. Dobatkin, V.I., G.I. Eskin, and S.I. Borovikova. 1971. *Tekhnol. Legk. Spl.*, no. 6:9–17.
66. Dobatkin, V.I., G.I. Eskin, S.I. Borovikova, and Yu.G. Golder. 1976. In *Processing of light and high-temperature alloys*, 151–62. Moscow: Nauka.
67. Dobatkin, V.I., and G.I. Eskin. 1991. *Tsvetn. Met.*, no. 12:64–67.
68. Dobatkin, V.I., A.F. Belov, G.I. Eskin, S.I. Borovikova, and Yu.G. Golder. 1983. *Bull. Discoveries, Inventions*, no. 37:1.
69. Eskin, D.G. 2008. *Physical metallurgy of direct chill casting of aluminum alloys*, 79–124. Boca Raton, FL: CRC Press.
70. Dobatkin, V.I., and G.I. Eskin. 1995. *Izv. Ross. Akad. Nauk., Met.*, no. 4:36–41.
71. Nadella, R., D.G. Eskin, and L. Katgerman. 2008. *Metall. Mater. Trans. A* 39A:450–61.
72. Eskin, D.G., R. Nadella, and L. Katgerman. 2008. *Acta Mater.* 56:1358–65.
73. Atamanenko, T.V., D.G. Eskin, and L. Katgerman. 2008. In *Proc. 11th Intern. Conf. Aluminium Alloys: Their Physical and Mechanical Properties, ICAA11*, ed. J. Hirsch, B. Skrotzki, and G. Gottstein, 316–320. Weinheim, Germany: Wiley-VCH.
74. Li, J.-W., T. Momono, Y. Fu, Z. Jia, and Y. Tayu. 2007. *Trans. Nonferrous Met. Sci. China.* 17:691–97.
75. Eskin, G.I., and D.G. Eskin. 2004. *Z. Metallkde.* 95:682–90.
76. Jain, X., H. Xu, T.T. Meek, and Q. Han. 2005. *Mater. Lett.* 59:190–93.
77. Jian, X., T.T. Meek, and Q. Han. 2006. *Scr. Mater.* 54:893–96.
78. Zhang, S., Y. Zhao, Z. Cheng, G. Chen, and Q. Dai. 2009. *J. Alloys Comp.* 470:168–72.
79. Zhang, L., 2013. Ultrasonic Processing of Aluminum Alloys, PhD thesis, Delft, Delft University of Technology.

6 Refinement of Primary Particles

6.1 FORMATION OF PRIMARY PARTICLES DURING SOLIDIFICATION

Primary crystals (other than grains of Al- or Mg-based solid solution) in aluminum and magnesium alloys are formed in peritectic or eutectic systems (in the latter case, the amount of alloying elements should be hypereutectic) at melt temperatures higher than the formation temperature of the aluminum or magnesium solid solution. In hypereutectic alloys, where the concentration of the second main component is above the eutectic point, e.g., Al–Si alloys containing more than 12 wt% Si, primary crystals are the main structural constituent. Due to the high-temperature nature of these phases, they nucleate and grow in the liquid sump of a billet or an ingot as individual particles, whereas the main solid-solution phase or eutectics mostly grow progressively with a continuous solidification front. After some time when reaching a certain size, suspended particles either sediment onto the solidification front or become captured by the moving solidification front. Frequently these crystals grow to a considerable size, have elongated shape, and may form agglomerates. In addition, these crystals are usually hard and brittle and can act as stress concentrators, decreasing the ductility, toughness, and deformability of as-cast metal. On the other hand, these particles can act as grain refiners, improve wear and thermal resistance, decrease thermal expansion coefficient, and increase elastic modulus of the alloy. The significant refinement of these particles can essentially improve the structure, properties, and workability of cast material.

Although in most cases primary particles are harmful, there are two classes of alloys where they are welcome. The first type has been considered in Chapter 5, where the grain-refining potential of some primary intermetallics is discussed. The second class is represented by so-called natural composites, with Al–Si hypereutectic alloys as the best-known representatives. Here primary crystals of Si are the main structural constituents responsible for useful properties (reduced thermal expansion, increased thermal stability, improved wear resistance). In these alloys, primary crystals (e.g., Si) nucleate and grow in the melt until the beginning of the eutectic reaction. The volume fraction of Si determines such properties as elastic modulus, linear thermal expansion coefficient, and wear resistance—the more, the better. At the same time, the size of primary silicon particles affects ductility, fracture toughness, tensile strength, and general work-ability—in this case, the smaller, the better. If the size of primary silicon is less than 40–50 μm, then the plasticity of cast material is sufficient for crack-free casting and further deformation [1].

According to widely accepted views on the solidification, the size of primary intermetallic or silicon particles that nucleate and grow in the liquid volume depends on the number of active solidification nuclei and the time period during which these crystals grow until they are captured by the continuous solidification front of solid solution or eutectic. The solidification or cooling rate affects the morphology and sometimes the crystal structure of growing crystals and can influence the nucleation rate by increasing bulk undercooling.

Several ways are possible for controlling the process of spatial solidification of such free-growing crystals. One group of mechanisms relates to heterogeneous nucleation on specially added or natively existing substrates. The mechanisms of ultrasonic-aided nucleation are generally the same as those described in Chapter 5 for grain refinement, i.e., activation of nucleation sites and multiplication of crystals. The second group of mechanisms concerns the growth of particles. Ultrasonic processing introduces additional energy into the treated volume, effectively increasing the temperature in the liquid part of the casting (see Figure 5.24). As a result, the slurry zone where the primary crystals can grow narrows, limiting the time available for particle growth.

The effects of ultrasonic melt treatment on the formation of primary intermetallics have been studied since the 1960s [2], and the main control mechanisms were formulated in the 1980s–1990s [3, 4]. We have already discussed in Chapter 5 that cavitation treatment of the melt can be efficiently used to control the grain size in aluminum and magnesium alloys. Here we will look at the application of ultrasonic processing to the refinement of primary-phase particles.

6.2 EFFECT OF ULTRASOUND ON SOLIDIFICATION OF INTERMETALLIC COMPOUNDS

The refining effect of ultrasound on primary crystals has been known since the 1960s [5–7]. It was demonstrated that the maximum effect could be observed when ultrasound was applied during solidification of an alloy.

We have studied several binary systems of aluminum with Fe, Mn, Zr, Ti, or Cr in which intermetallic compounds are formed either in eutectic or peritectic systems. Most of the experiments were performed in small volumes (up to 1 kg of melt), with ultrasound introduced from the top of the melt by an immersed Nb sonotrode. The cooling rates during solidification varied between 0.1 and 2 K/s. The sonication was usually performed at temperatures above 660°C. In some cases, the alloys were produced by DC casting with sonication in the sump of a round billet 60 to 270 mm in diameter.

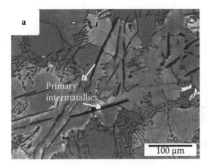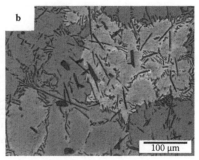

FIGURE 6.1 Effect of ultrasound on the size of primary intermetallics in Al–2.1% Fe alloys: (a) without ultrasonic processing and (b) with ultrasonic processing in the temperature range 700–660°C.

In eutectic systems at nearly eutectic concentration of Fe (2.1 wt%) or Mn (1.9 wt%), the ultrasonic treatment results in refinement of primary particles, coarsening of eutectics, and a general shift of the eutectic point to lower concentrations of the alloying elements [8]. Figures 6.1 a,b show the structures for an Al–2.1% Fe alloy treated in a crucible for 10 s in the temperature range 700°C to 660°C, with subsequent solidification in a copper mold at 2 K/s. The refinement of primary intermetallics along with eutectic coarsening are clearly seen. It is suggested that fragmentation of primary (AlFe) particles results in their refinement, while the extra heat introduced in the system by ultrasound slows down the cooling rate during eutectic solidification, causing eutectic coarsening.

Ultrasonic processing performed below the liquidus of a complex hypereutectic Al–20% Si alloy containing 2% Fe results in acicular β (AlFeSi) particles being substituted with blocky δ (AlFeSi) particles with the refinement of the latter [9]. Homogeneous solute and temperature distribution resulting from slurry sonication were suggested as possible reasons for the observed effects.

Experiments on hypereutectic alloys with 3–4% Mn showed the importance of cooling rate and processing temperature. Figure 6.2 demonstrates that ultrasonic processing reduces the average size of intermetallic compounds 45–50 times at 0.1 K/s and 5–9 times at 0.5 K/s.

In order to demonstrate the efficiency of ultrasonic processing at different stages of solidification, two series of experiments were performed. In the first series, an Al–4% Mn alloy was treated with ultrasound during primary solidification of the Al₆Mn compound, during eutectic reaction or during the entire solidification. The experiments were performed at a cooling rate of 0.1 K/s. The results are shown in Table 6.1. In the second series, a binary Al–4% Mn alloy was first slowly solidified to allow for large intermetallics to form. Then the alloy was remelted and treated for 3 min isothermally at temperatures of 710°C, 690°C, and 680°C, i.e., at temperatures that are, respectively, 50°C, 30°C, and 20°C above the eutectic point but still below

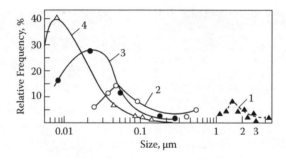

FIGURE 6.2 Particle size distributions of intermetallic compound Al_6Mn in an Al–3% Mn alloy as functions of the cooling conditions of the solidifying melt and ultrasound: (1) graphite mold (0.1 K/s), without sonication; (2) same as (1), but with sonication; (3) metallic mold (0.5 K/s) without sonication; and (4) same as (3), but with sonication.

the liquidus (717°C for this alloy). Cooling conditions were the same as in the first series. The results are given in Table 6.2.

These results show that the processing during solidification, when nucleation, fragmentation, and growth are affected (Table 6.1, primary solidification), is more efficient than fragmentation only (Table 6.2). In the latter case, the lower temperature of sonication may assist in fragmentation, as there are more crystals available.

Similar results were obtained for magnesium alloys. Figure 6.3 shows the distribution curves and microstructures for Mn-containing intermetallics in commercial flat ingots of an AZ31B-type alloy.

It is noteworthy that cavitation melt processing does not only refine primary particles, but also changes the phase composition of Mn-containing compounds in Mg alloys. In particular, the number of grain-refining, stable, hexagonal-phase Al_4Mn (η) particles increases, substituting metastable, tetragonal Al_6Mn (τ). As a result, the Mg grain size becomes smaller upon ultrasonic treatment. The formation of the stable phase is due to the enhanced nucleation under cavitation conditions that prevents the high undercooling required for the formation of the metastable phase. At high

TABLE 6.1

Effect of Ultrasonic Processing at Different Stages of Solidification on the Dimensions of Primary Al_6Mn Particles in an Al–4% Mn Alloy

Stage of Solidification	Duration of Sonication, min	Length, mm	Width, mm	Area, mm²
No sonication	0	1.83	0.075	0.137
Primary	0.5	0.016	0.016	0.00026
Eutectic	3	0.121	0.07	0.084
Throughout	5	0.04	0.016	0.00064

TABLE 6.2

Effect of Isothermal Ultrasonic Processing During Remelting to the Semisolid State on the Size of Primary Al_6Mn Particles in an Al–4% Mn Alloy

Temperature of Solidification	Duration of Sonication, min	Length, mm	Width, mm	Area, mm²
No sonication	0	1.3	0.075	0.137
710	3	0.76	0.075	0.057
690	3	0.194	0.06	0.0116
680	3	0.131	0.038	0.00498

Mn concentrations, ultrasonic processing facilitates the formation of fine particles of the high-temperature βMn phase.

In other, more complex magnesium alloys (Mg–Y–Ce–Mn, Mg–Y–Cd–Mn–Ce–Al, Mg–Zn–Nd–Cd–Zr, etc.), the ultrasonic cavitation treatment results in the formation of fine grain structure and refinement of intermetallics. Figure 6.4 demonstrates this effect for a ML19 Russian Grade (Mg–Nd system).

Practically important intermetallics are formed in peritectic systems, i.e., Al–Ti and Al–Zr. The refinement of primary particles in these systems during solidification with ultrasonic processing was reported in 1965 [6]. Recently, these data were confirmed by laboratory experiments with controlled processing temperature, as demonstrated in Figure 6.5. Obviously, the ultrasonic processing performed in the temperature range of primary solidification of intermetallic phases results in their refinement. We have already discussed this in Chapter 5 for Al–Zr–Ti alloys (see Section 5.4, Figure 5.17).

Interesting data were obtained in studies of the formation of intermetallic compounds in direct-chill casting when the cavitation region is localized near the radiating face in the upper part of the liquid pool.

Let us consider the solidification of intermetallic Al–Zr–Ti compounds as a function of the main process factors that define the formation of high-quality ingots of light alloys. These results are of great significance in the production of ingots of aluminum alloys with nondendritic structure and a large cross section (up to 1200-mm diameter). In this case, the use of high-purity-grade aluminum is mandatory. For example, an addition of <0.20% (Zr + Ti) to high-strength aluminum alloys such as AA7475 and AA2324 prepared using high-purity 99.99% Al ensures the formation of nondendritic structure in billets 65–960 mm in diameter that are direct-chill cast with ultrasonic processing in the sump of the billet. In this case, no coarse primary intermetallics are formed despite the considerable amount of added Zr and Ti. Table 6.3 illustrates the efficiency of adding Zr and Ti to high-purity, high-strength aluminum alloys, while Figure 6.6 gives the size distribution of primary intermetallics in DC-cast billets of these alloys. The formation of nondendritic structure as a result of ultrasonic processing during DC casting can be achieved in a wide range of casting temperatures, making the billet structure

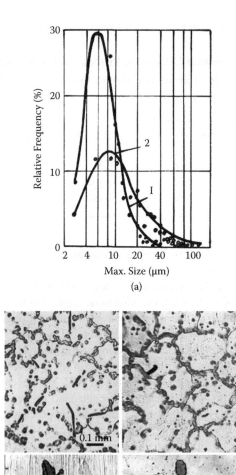

(a)

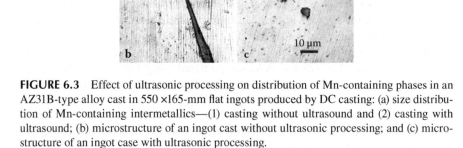

FIGURE 6.3 Effect of ultrasonic processing on distribution of Mn-containing phases in an AZ31B-type alloy cast in 550 ×165-mm flat ingots produced by DC casting: (a) size distribution of Mn-containing intermetallics—(1) casting without ultrasound and (2) casting with ultrasound; (b) microstructure of an ingot cast without ultrasonic processing; and (c) microstructure of an ingot case with ultrasonic processing.

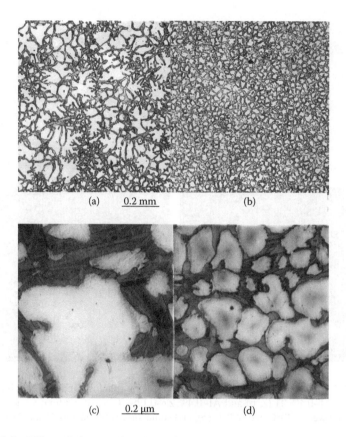

(a)   0.2 mm   (b)

(c)   0.2 µm   (d)

FIGURE 6.4 Effect of ultrasound on the grain structure (a, b) and intermetallic inclusions (c, d) in 174-mm-diameter billets of a Mg–Nd–Y alloy cast without (a, c) and with (b, d) ultrasonic processing.

insensitive to melt superheating. For example, 65 to 285-mm billets from Russian Grade 1960 and AA7055 alloys of the Al–Zn–Mg–Cu system with 0.15%–0.18% Zr were produced by DC casting with ultrasonic treatment at different casting temperatures ranging from 720°C to 800°C and demonstrated nondendritic grain structure [10].

There are two principal ways of enhancing nucleation and inducing refinement of primary phases. The first is to clean the melt of active coarse substrates and facilitate deep undercooling of the melt before nucleation. This can be done by melt filtration or melt superheating [4, 11]. The second way is to create an excessive number of active solidification sites so that there will always be available substrate in the thermally or constitutionally undercooled volume. These methods were applied to DC casting of billets 143–270 mm in diameter from an AA7075-type alloy. The filtration (F) was performed using multilayered mesh filters with small cell size, allowing one to filter out particles larger than 1 µm (see Chapter 4 for details). Alternatively, the melt was superheated at 1000°C for more than

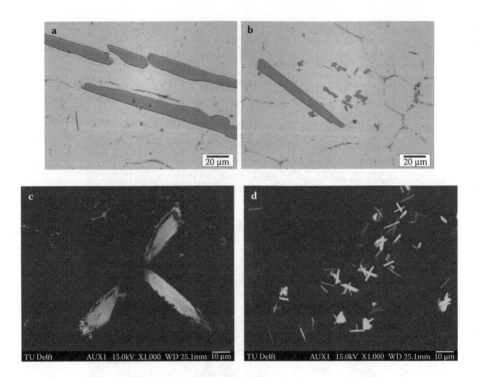

FIGURE 6.5 Effect of ultrasonic processing on the morphology and size of primary inter-metallics in Al–0.4% Ti (a, b) and Al–0.4% Zr–0.12% Ti (c, d): without sonication (a, c), sonication between 720°C and 680°C (b), and sonication between 790°C and 750°C (d).

3 h (SH). The multiplication of active solidification sites was done by increasing Zr concentration to 0.3% and using different processing schemes, i.e., ultrasonic treatment in a launder (UST1), ultrasonic treatment in the billet sump (UST2), melt transport through a water-cooled launder without UST (L), and a combination of L with ultrasonic processing. Combinations with filtration and melt superheating were also investigated. Ultrasonic treatment was done using a standard 5-kW magnetostrictive transducer working at 18 kHz. Three parameters were used for the comparison: average size of intermetallics (d_i), volume fraction of Al$_3$Zr particles (V), and fraction of nondendritic grains (A). The results are given in Figure 6.7 for two billet diameters.

It is obvious that the change from melt superheating (SH) to melt undercooling (L) and to ultrasonic processing leads to the progressive refinement of primary particles. At the same time, the nondendritic structure becomes dominant only with ultrasonic processing (Figure 6.7a). Cleaning of the melt of nonmetallic inclusions larger than 1 μm using a seven-layered filter prevents the formation of nondendritic structure, even upon ultrasonic processing in the sump. Larger undercooling is required to trigger the formation of nucleating substrates. The use of a five-layered filter in combination with melt undercooling and ultrasonic treatment allows one to obtain predominantly

TABLE 6.3

Effect of Ultrasonic Processing and Concentration of Zr and Ti Additions on the Grain Size in 70-mm Billets Produced by DC Casting

Ti, %	Zr, %	Grain Size, µm	
		UST	No UST
	AA2324		
0.03	...	300	550
0.03	0.12	120	280
0.03	0.16	82	190[b]
0.03	0.22	28[a]	115[b]
0.08	0.22	27[a]	187[b]
0.05	...	93	270
0.08	...	87	190
0.14	...	50	150
0.13	0.06	50	130[b]
0.12	0.11	50	100[b]
0.11	0.16	30[a]	100[b]
	AA7475		
0.02	0.05	190	300
0.02	0.10	150	250
0.02	0.15	50	180
0.02	0.17	29[a]	150
0.07	0.17	36[a]	150[b]
0.11	0.17	30[a]	144[b]

[a] Nondendritic structure.
[b] Primary intermetallic compounds are visible.

nondendritic structure in DC-cast billets (Figure 6.7b). The size of the intermetallic particles decreases to less than 3 µm at a volume fraction of 0.13%. Figure 6.8 shows the corresponding structures.

This series of experiments allows us to conclude the following. Fine filtration is more efficient in removing substrates than their deactivation by melt superheating. Melt undercooling is a productive way of controlling the grain structure (apparently by allowing small substrates to become available for heterogeneous nucleation of aluminum). Ultrasonic processing is very efficient in refining primary intermetallics and making them available as solidification substrates. A combination of ultrasonic processing that refines substrates with sufficient melt undercooling is important for getting optimum grain refinement [12]. It was recently

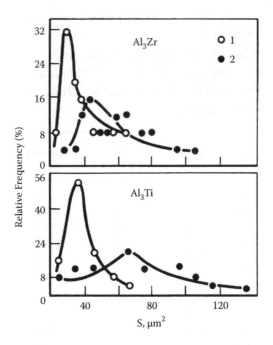

FIGURE 6.6 Size distributions of Al_3Zr and Al_3Ti primary particles in 70-mm billets of an AA2324 alloy with 0.18% Zr and 0.15% Ti: (1) DC casting with ultrasonic processing and (2) DC casting without ultrasonic processing.

suggested that alumina particles activated by cavitation may become substrates for Al_3Zr particles [13].

It is important that the use of Zr for grain refinement does not mean that its amount available for recrystallization control (in the supersaturated aluminum solution) becomes less. In fact, the optimum melt processing results not only in grain refinement, but also increases the concentration of Zr in the aluminum solid solution after the end of solidification, as shown here:

Fraction of Nondendritic Structure, %	Average Size of Al_3Zr, μm	Volume Fraction of Al_3Zr, %	Zr in Solid Solution, %
30	6.8	0.855	0.067
80	3.65	0.290	0.073
100	3.25	0.130	0.110

Our industrial experience shows that the same trends are observed upon casting large billets and ingots (up to 800–1200 mm in cross section) and at even smaller Zr concentrations of 0.15–0.16%.

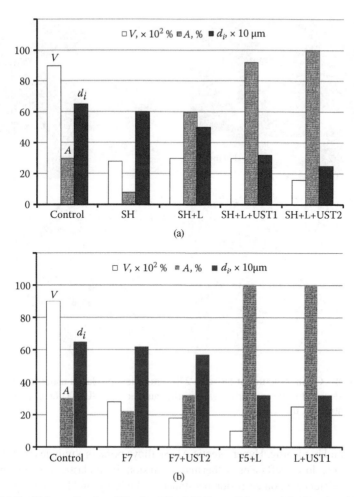

FIGURE 6.7 Volume fraction (V), size (d_i) of Al$_3$Zr particles, and the fraction (A) of non-dendritic grain structure in (a) 145-mm and (b) 270-mm DC-cast billets of an AA7075-type alloy with 0.3% Zr: C, control billet; SH, melt superheat; L, melt undercooling in a water-cooled launder; F7, filtration through a 7-layered filter; F5, filtration through a 5-layered filter; UST1, ultrasonic processing in the launder; and UST2, ultrasonic processing in the billet sump.

6.3 EFFECT OF ULTRASOUND ON THE SOLIDIFICATION OF HYPEREUTECTIC AL–SI ALLOYS

Hypereutectic Al–Si alloys represent alloys where the main primary phase is not aluminum. These alloys are widely used in industry for making thermally stable and wear-resistant castings, mainly as parts of engines. Commercial Al–Si alloys typically contain 12–18% Si, but this amount may be as high as 23% in special cases. In addition, Cu, Mg, Ni, Fe, and some other elements are added to provide additional heat

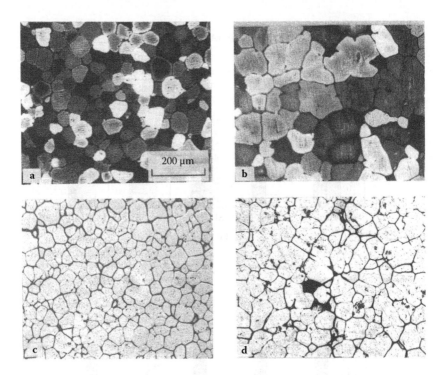

FIGURE 6.8 Grain structure of DC-cast 270-mm billets produced using filtration, water-cooled launder, and ultrasonic processing in the launder (F+L+UST1) (a, c) and a standard process (control) (b, d).

resistance and hardening. Hypereutectic Al–Si alloys possess a unique combination of properties, i.e., low coefficient of thermal expansion, high elastic modulus, hardness and wear resistance, good corrosion resistance, weldability, and low density.

Hypereutectic Al–Si alloys can be considered as natural composites having a ductile eutectic matrix and hard particles of primary silicon. Homogeneous distribution of fine Si particles is therefore a desired feature of these alloys. Without taking care about the refinement of primary crystals, Si particles can grow to more than 100 μm in size, which is detrimental to the alloys' mechanical and casting performance. The refinement of primary particles, their uniform distribution in the volume, and the refined or modified structure of the eutectic matrix are the issues that are dealt with by alloying and melt processing.

The solidification of hypereutectic Al–Si alloys is illustrated in Figure 6.9 based on experimental measurements and observations [3]. An intense heat transfer from the liquid metal to the solid shell of a casting, observed in real casting conditions, brings the liquid pool's temperature close to that of the binary eutectic, that is, far below the liquidus temperature for hypereutectic alloys (line 2). This means that the bulk solidification of primarily Si in the sump of a billet is unavoidable. Therefore, the crystal size is defined by the number of active solidification nuclei formed in the unit volume of liquid metal and by the time taken by the primary crystals to reach

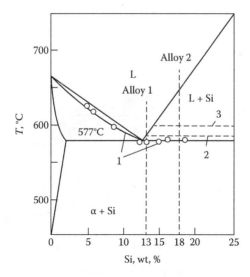

FIGURE 6.9 Temperature conditions on the solidification front and in the liquid pool of 280-mm-diameter billets of Al–Si alloys: (1) temperature at the solidification front (depends on the cooling conditions and does not depend on sonication); (2) temperature in the liquid pool 50 mm from the solidification front, casting without ultrasound; (3) temperature in the liquid pool 50 mm from the solidification front, casting with ultrasound.

the solidification front. As we have already discussed, the introduction of acoustic energy into the liquid bath of a casting considerably increases the temperature. This temperature increment can be as high as 15–20°C, as shown in Figure 6.9 by line 3. At the same time, ultrasonic processing can potentially increase the amount of active solidification sites.

Let us analyze the conditions of nucleation and growth of Si crystals in two hypereutectic alloys solidified with and without sonication. In the first case (alloy with 13% Si), melt superheat (by 20°C) induced by ultrasound prevents the formation of primary Si crystals in the liquid pool at a conditionally chosen distance of 50 mm from the solidification front. Without sonication, at the same distance, we would find primary Si in a form of large faceted crystals. The solidification of the second alloy (18% Si) is accompanied by the formation of primary Si, irrespective of ultrasonic treatment. However, acoustic cavitation considerably increases the number of active solidification nuclei in the liquid pool. Therefore, more Si crystals will be nucleated per unit volume in the melt. In addition, the increased melt temperature in the liquid pool will effectively decrease the solidification range of primary crystals, allowing them less time for growth.

The number of active nuclei (substrates) and the degree of melt superheat are directly proportional to the ultrasonic intensity in the developed cavitation mode. Thus, the acoustic cavitation during the primary solidification of Si (and in fact any primary compounds) affects both the processes of crystal nucleation and growth. Figure 6.10 demonstrates the importance of acoustic power (proportional to the squared amplitude) for the refining of primary Si [3].

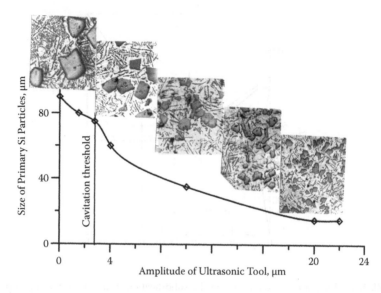

FIGURE 6.10 Effect of ultrasonic intensity on the refinement of primary silicon in an unmodified binary Al–17% Si alloy.

The temperature range where the ultrasonic processing is applied is very important. Zhang et al. [14] showed that the best result for refinement of primary Si in a binary Al–17% Si alloy can be achieved when the processing is performed in the temperature range 690–720°C (10 s for 400 g), which is above the liquidus (647°C). The cavitation conditions were achieved at 17.5 kHz with an amplitude of 40 μm. Ultrasonic processing close to the liquidus did not show significant refinement of primary particles; and the continuation of ultrasonic treatment until the mushy state of the alloy resulted in significant coarsening of the eutectics. Feng et al. [15] used a less efficient scheme of a vibrating crucible attached to a waveguiding system. Primary Si in Al–23% Si alloy was considerably refined after processing at 680–700°C (around the liquidus temperature). However, the amplitude was only 4 μm at 20 kHz, and the treated volume was 200 g.

There are three main technological approaches to producing castings from hypereutectic Al–Si alloys, i.e., gravity casting (sand and permanent mold), pressure-assisted casting (high-pressure die and squeeze casting), and rapid solidification. Special additions of Si refiners are required in the first two groups of technologies, but considerable scatter in cooling rates across the casting frequently prevents a uniform refining effect. The third technology requires special equipment and is expensive.

Plastic deformation by forging and extrusion of DC-cast billets is an alternative to traditional casting routes. However, this approach can be feasible only in the case of efficient refining of primary Si at the stage of DC casting.

Extended research performed in the 1980s–2000s showed that ultrasonic melt processing before and during solidification can be very efficient in refining primary Si, especially when combined with optimum additions of Si refiners [1, 3, 4, 16–18]. The AlP phase, formed as a result of the interaction of phosphorus with

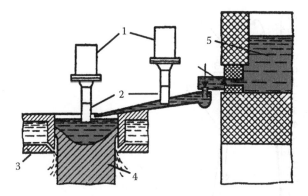

FIGURE 6.11 Schematic diagram of DC casting with ultrasonic processing: (1) transducer, (2) sonotrode, (3) mold, (4) billet, and (5) holding furnace.

aluminum melt, acts as a solidification site for Si [19]. Phosphorus is added in a form of P-containing master alloys like Cu–P, Al–Cu–P, and Al–Fe–P, the latter being the most efficient [19, 20]. Ultrasonic cavitation processing of melt becomes an efficient controlling means of primary Si formation, as it activates—through mechanisms explained in Chapter 5—nonmetallic (mostly alumina and aluminum phosphide) inclusions that act as substrates for Si.

Figure 6.11 gives a scheme of ultrasonic melt processing during DC casting that was used. Two ways of processing, in the sump of a billet and in a launder, were applied. The use of 4.5 kW of generator power at 18 kHz enabled the working amplitude of a sonotrode larger than 15 μm.

It was found that the quality and type of Si-refining master alloys are as important and sometimes more important than their amount. Ternary Al–Fe–P and Al–Cu–P master alloys demonstrated enhanced performance when being introduced in a launder with simultaneous ultrasonic melt treatment. The activation of AlP particles and prevention of their agglomeration were suggested as the main reasons for this performance [17]. In addition, the cavitation treatment promotes better dissolution of compounds from master alloys [21] and higher recovery of phosphorus. Figure 6.12 shows the size distribution of primary Si particles in a 114-mm round DC-cast billet, demonstrating the advantage of ultrasonic processing in the melt flow with simultaneous introduction of a P-containing master alloy [4].

It was found that hydrogen dissolved in the melt deteriorates the nucleating ability of AlP particles. Therefore, degassing of the melt becomes an essential part of success. Ultrasonic processing is accompanied by degassing, as was discussed in Chapter 3, so the treatment becomes combined, including degassing and grain-refining functions.

We also studied the possibility of using Al–40% Si master alloys produced by direct carbon reduction of cheap nepheline, alunite, kaolin, and other minerals and slags. Such silico-aluminum was commercially produced at Zaporizhia Aluminum Works (Ukraine).

Billets 98 to 178 mm in diameter were cast from hypereutectic Al–18% Si alloys (18% Si, 1% Fe, 0.2% Cu, 0.3% Ti, 0.15% Zr, impurities of Mg and Mn) made using

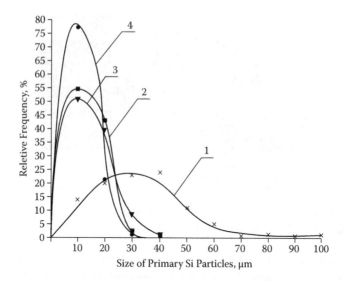

FIGURE 6.12 Size distribution of primary Si particles in a 114-mm DC-cast billet of a commercial Al–18% Si alloy: (1) no P addition, no UST; (2) no P addition, UST in the melt flow in a launder; (3) P addition in the furnace, UST in the launder; and (4) P addition in the furnace and addition of an Al–Fe–P master alloy rod in the launder with UST.

pure elements and master alloys (charge 1) and by dilution of an Al–40% master alloy obtained by carbon reduction (charge 2). The chemical analysis of the latter showed that it contained about 40 impurities on a ppt level, including Ni, Bi, Hf, Cs, Na, C, Be, Cr, Pb, V, Mo, Ce, B, Y, Sb, La, etc. [17, 18]. Many of these impurities may be useful due to their modifying action on the Al–Si eutectics. Addition of phosphorus was made using an Al–Fe–P master alloy (0.02% P added) in a launder. Ultrasonic melt processing was also performed in the launder.

Typical structures of the Al–40% master alloys and billets produced with and without melt processing are shown in Figure 6.13. The initial master alloy contained coarse silicon particles up to 3–5 mm in size and modified eutectics. After dilution to the nominal 18% Si, the size of primary Si particles was reduced to 150–250 μm, addition of phosphorus further refined Si particles to 40–50 μm, and a combination of P addition with ultrasonic treatment resulted in particles 20–30 μm in size. The size distribution of primary Si particles in DC-cast billets produced with and without ultrasonic processing is given in Figure 6.14. Such a structure allows extrusion and forging of billets to produce various shapes, as illustrated in Figure 6.15. The physicomechanical properties of extruded plates (7×100 mm) from hypereutectic Al–18% Si alloys are listed in Table 6.4 in comparison to the properties of a typical extrusion alloy AA6063. It is evident that DC-cast and extruded hypereutectic Al–Si alloys produced using combined phosphorus and ultrasonic treatment demonstrate mechanical properties similar to AA6063 while having advantage in elastic modulus and thermal expansion coefficient.

Hypereutectic Al–Si alloys have excellent weldability with all welding technologies. For example, the hot tearing susceptibility during MIG welding at a rate of

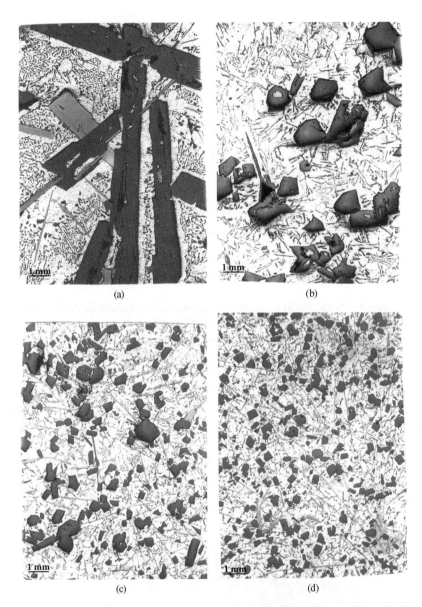

FIGURE 6.13 Microstructures of (a) carbon-reduced Al–40% Si master alloy; (b) diluted Al–18% Si alloy; (c) Al–18% Si alloy with an addition of 0.02% P; and (d) Al–18% Si alloy with an addition of 0.02% P and ultrasonic processing of the melt.

12 m/h was lower than for an Al–6% Mg (AMg6) alloy. The weld strength was 140–160 MPa at a bending angle of 40°. These alloys also exhibit very high corrosion resistance under different climatic conditions. Deformed hypereutectic Al–Si alloys also show sufficiently high ductility at low temperatures, which makes them suitable for Arctic applications as pipes and profiles.

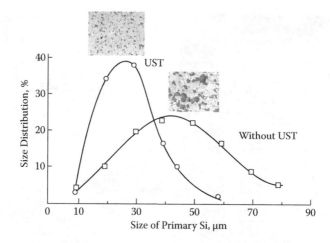

FIGURE 6.14 Size distribution of primary Si particles in a DC-cast 112-mm billet of a 01390 alloy with phosphorus addition: (1) without UST; (2) with UST.

As a result of these studies a series of commercial hypereutectic Al–Si alloys has been developed as heat-treatable (Grade 01392) and nonhardening (Grades 01390 and 01391) alloys. The 01391 alloy is made using cheap carbon-reduced master alloys. The chemical composition of these alloys is given here:

Grade	Si, %	Cu, %	Mg, %	Mn, %	Ti, %	Zr, %	Fe, %	P, %	Other, %
01390	17–19	<0.5	<0.5	0.3-0.8	<0.15	–	<1.5	<0.1	<0.1
01391	17–19	<0.5	<0.2	<0.3	<0.5	<0.3	<1.5	<0.1	<0.1
01392	17–19	3.0-3.5	0.3-0.5	0.1-0.3	<-0.15	–	<0.7	<0.1	<0.1

Physicomechanical properties of new alloys are illustrated in Figure 6.16 for extruded plates.

FIGURE 6.15 Extruded and forged shapes made from billets of Al–18% Si alloys produced by DC casting with ultrasonic melt treatment.

TABLE 6.4

Properties of Hot-Extruded Plates (7 × 100 mm) from Al–18% Si and AA6063 Alloys

Alloy	UTS, MPa		YS, MPa		El, %		E, GPa	HB, MPa	KCU, J/cm²		σ_{-1}, MPa	LTEC, 10^{-6} K^{-1}
	20°C	–50°C	20°C	–50°C	20°C	–50°C			20°C	–50°C		
Al–18% Si, charge 1, O	192	184	156	155	5.9	6.2	80.6	630	5.3	5.0	…	…
Al–18% Si, charge 2, F	210	210	120	145	4.0	3.9	79	800	4.6	4.6	90	16
Al–18% Si, charge 2, O	190	182	100	130	4.4	4.4	79	730	5.1	5.9	90	15
AA6063, T5	240	…	200	…	10	…	71	800	5.0	…	90	26.7

Note: UTS is ultimate tensile strength; YS is 0.2% proof yield strength; El is tensile elongation; E is Young's modulus; HB is Brinell hardness; KCU is impact toughness; σ_{-1} is the fatigue limit; and LTEC is the linear thermal expansion coefficient.

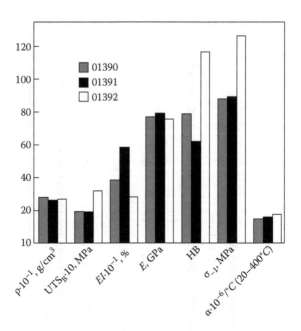

FIGURE 6.16 Physicomechanical properties of new hypereutectic Al–Si alloys.

Further improvement of structure and performance of hypereutectic Al-Si alloys can be achieved by helical (cross) rolling of billets. This deformation technology allows further refinement of silicon particles (both primary and eutectic) with improved homogeneity of their distribution, as illustrated in Figure 6.17 and in Table 6.5. The complex technology of ultrasonic-assisted DC casting with subsequent helical (cross) rolling and extrusion or forging was developed and thoroughly

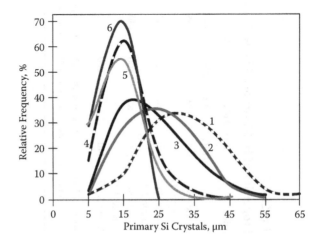

FIGURE 6.17 Evolution of size distribution of primary Si particles during helical rolling of DC-cast cross-rolled rods: (1) 112, (2) 92, (3) 80, (4) 50, (5) 28, and (6) 14 mm in diameter.

TABLE 6.5

Effect of Helical Rolling in the Size of Silicon Crystals in Billets of a 01392 Alloy

Product	Si Crystals	Si Shape	Si size, μm Range	Average	Refinement Effect
Billet	Primary	Plates	30–80	50	…
Rolled rod	Primary	Faceted	10–20	15	3–4 times
Billet	Eutectic	Plates	<10	8	…
Rolled rod	Eutectic	Globules	<1	<1	10 times

TABLE 6.6

Mechanical Properties of DC Cast Billets, Cross-Rolled Rods, Forgings, and Extruded Plates from a 01392 Alloy

Product	Condition	UTS, MPa	El,%
Billet	Homogenized	177	2.6
Rod	Cross rolled	185	2.6
Rod	T6	325	2.0
Forged	From billet, T6	330	1.7
Forged	From rod, T6	335	2.0
Plate	Extruded, T6	320	3.0

studied in 2000–2011 [22–25]. As a result, significant improvement of mechanical properties and technological performance has been achieved, as summarized in Table 6.6.

REFERENCES

1. Eskin, G.I., and D.G. Eskin. 2003. *Ultrason. Sonochem.* 10:297–301.
2. Eskin, G.I., and I.N. Fridlyander. 1961. *Izv. Akad. Nauk SSSR, Otdel. Tekhn. Nauk, Metall. Topl.* no. 5:110–112.
3. Dobatkin, V.I., and G.I. Eskin. 1985. In *Metals science of aluminum alloys*, 163–71. Moscow: Nauka.
4. Eskin, G.I., and D.G. Eskin. 2004. *Z. Metallkde.* 95:682–90.
5. Pogodin-Alekseev, G.I., and V.M. Gavrilov. 1966. *Izv. Akad. Nauk SSSR, Met.*, no. 1:27–38.
6. Eskin, G.I. 1965. *Ultrasonic treatment of molten aluminum.* Moscow: Metallurgizdat.
7. von Seemann, H.-J., and K.-G. Pretor. 1966. *Z. Metallkde.* 57:347–49.
8. Zhang, L. 2013. Ultrasonic processing of aluminum alloys. PhD thesis, Delft, Delft University of Technology.
9. Zhong, G., S. Wu, H. Jiang, and P. An. 2010. *J. Alloy Comp.* 492:482–87.

10. Dobatkin, V.I., and G.I. Eskin. 1991. *Tsvetn. Met.*, no. 12:64–67.
11. Eskin, G.I., G.S. Makarov, and Yu.P. Pimenov. 1995. *Adv. Perform. Mater.* 2:43–50.
12. Zhang, L., D.G. Eskin, and L. Katgerman. 2011. *J. Mater. Sci.* 46:5252–59.
13. Zhang, L., D.G. Eskin, A.G. Miroux, and L. Katgerman. 2011. *IOP Conf. Series: Mater. Sci. Eng.* 27:paper 012002.
14. Zhang, L., D.G. Eskin, A. Miroux, and l. Katgerman. 2012. *Light metals 2012*, ed. C.E. Suarez, 999–1004. Warrendale, PA: TMS.
15. Feng, H.K., S.R. Yu, Y.L. Li, and L.Y. Gong. 2008. *J. Mater. Process. Technol.* 208:330–35.
16. Altman, M.B., G.I. Eskin, and I.S. Gotsev. 1981. In *Alloying and processing of light alloys*, 30–35. Moscow: Nauka.
17. Pimenov, Yu.P., V.I. Tararyshkin, and G.I. Eskin. 1997. *Tekhnol. Legk. Spl.*, no. 3:17–23.
18. Eskin, D.G., M.L. Kharakterova, G.I. Eskin, and Yu.P. Pimenov. 1998. *Izv. Ross. Akad. Nauk, Met.*, no. 2:91–98.
19. Kyffin, W.J., W.M. Rainforth, and H. Jones. *Mater. Sci. Technol.* 17:901–5.
20. Eskin, G.I., G.S. Makarov, and Yu.P. Pimenov. 1997. *Mater. Sci. Forum* 242:65–70.
21. Brodova, I.G., P.S. Popel, and G.I. Eskin. 2002. *Liquid metal processing: Applications to aluminium alloy production*. London and New York: Taylor and Francis.
22. Eskin, G.I., and E.I. Panov. 2002. *Tsvetn. Met.*, no. 9:85–89.
23. Panov, E.I., G.I. Eskin, A.A. Voskan'yants, and O.Yu. Il'in. 2002. *Metallurgist* 46 (7–8): 237–40.
24. Panov, E.I., and G.I. Eskin. 2004. *Metalloved. Termich. Obrab. Mater.*, no. 9:7–13.
25. Eskin, G.I., E.I. Panov, L.B. Ber, B.N. Stepanov, S.G. Bochvar, and V.I. Yalfimov. 2008. *Metallurgist* 52 (7–8): 388–94.

7 Ultrasonic Processing during Direct-Chill Casting of Light Alloys

Wrought-aluminum and -magnesium alloys are mostly produced by a semicontinuous casting method called direct-chill (DC) casting. Invented in the late 1930s, it quickly became the dominant technology of producing round (billets) and flat (ingots) large-scale castings intended for further deformation (extrusion and rolling). Sometimes large billets are made for forgings and flat ingots, intended for further remelting (as master alloys).

The principle of the casting is illustrated in Figure 7.1. The melt is poured into an open-ended, water-cooled mold that is initially closed from the bottom by a starting block (dummy). As the solid shell starts to form at the walls of the mold and onto the starting block, the latter is moved downward at a certain withdrawal (casting) speed, extracting the solid part of the billet from the mold. As soon as the shell exits the mold from the bottom, a direct spray of cooling water hits the surface (hence the term *direct chill*) and facilitates further solidification of the metal in the interior of the billet. The melt is continuously poured into the mold and the billet continues to move downward until the prescribed length is cast. There are modifications of this process such as hot-top casting that helps with maintaining the permanent melt level in the mold, gas- and oil-assisted mold cooling that controls the cooling conditions and lubrication in the mold, and horizontal DC casting when the casting is performed in the horizontal direction. The reader is referred to a recent treatise on the technology of DC casting for further details [1].

Direct-chill casting has a unique feature that makes it very distinct from shape casting and permanent-mold ingot casting. The solidification occurs in a narrow layer of the casting inside and below the relatively short mold. During the steady-state stage of casting, the shape and the dimensions of this region remain constant and reproducible from one heat to another. By controlling the melt distribution and cooling conditions inside the mold, direct cooling below the mold, and controlling the casting speed, the shape and dimensions of the solidification region can be maintained within optimum limits that ensure the repeatable quality of the casting. Specifics of structure formation during DC casting are well described elsewhere [2].

DC casting, on the one hand, simplifies the application of ultrasonic melt treatment, as the melt containers (launder, mold, melt distribution systems) are simple and constant in shape with well-defined temperature profiles. On the other hand, DC casting requires processing of larger melt volumes in a continuous manner, which

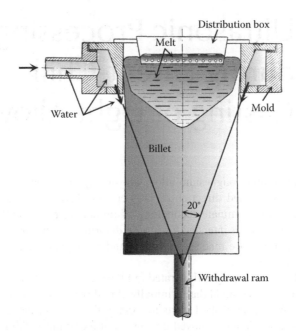

FIGURE 7.1   Principal scheme of direct-chill casting.

creates some challenges for the technology of ultrasonic processing. In this chapter we will consider the technological approaches to ultrasonic melt processing during DC casting and its benefits for the quality and properties of the cast metal.

7.1   DIRECT-CHILL CASTING OF ALUMINUM ALLOYS

7.1.1   Ultrasonic Processing in the Sump of a Billet

The first industrial DC-casting installation with ultrasonic melt treatment (UST) was built in the 1970s at a Soviet metallurgical plant. A standard DC caster with a 10-t holding furnace was equipped with an ultrasonic processing station that could be controlled remotely. Thyristor generators, each with 5.4 kW in power, were located some 30 m away from the casting pit in a separate room and linked to water-cooled 18-kHz, 4.5-kW magnetostrictive transducers by cables. Figure 7.2a shows the principal scheme of the installation, and Figures 7.2b,c show photos of the casting process. The ultrasonic processing was performed by dipping several sonotrodes (each fed by an individual transducer) into the sump of a billet or an ingot. The horns were made of a Nb alloy that ensured the stable and continuous operation in the melt during the entire casting process. The choice of Nb alloys as the most suitable material for ultrasonic horns for liquid aluminum processing was first made and proved by G. Eskin in the 1960s [3] (see also Section 13.3). Extended casting trials upon DC casting of billets 650 to 1200 mm in diameter and up to 6 m in length demonstrated that Nb-alloy horns sustained minor cavitation erosion that did not exceed 1.5 mm

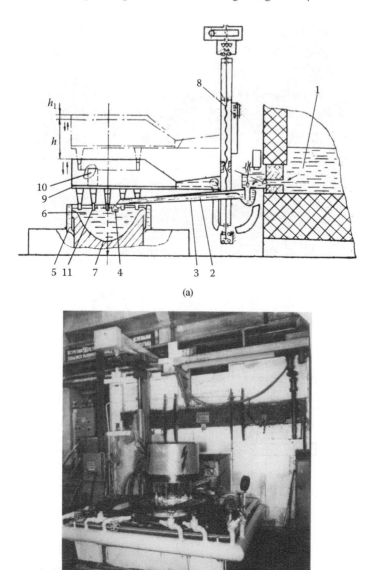

(a)

(b)

FIGURE 7.2 Industrial DC casting installation for ultrasonic processing of large billets (ingots): (a) principal scheme [(1) holding furnace with melt, (2) melt flow, (3) launder, (4) distribution box, (5) mold, (6) billet sump, (7) billet, (8) ultrasonic equipment support with water and power lines, (9) platform with transducers, (10) ultrasonic transducer, (11) sonotrode)]; (b) general view of the DC caster with ultrasonic transducers in the mold. (*continued*)

(c)

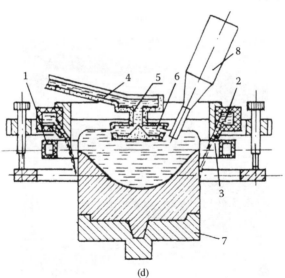

(d)

FIGURE 7.2 (*continued*) Industrial DC casting installation for ultrasonic processing of large billets (ingots): (c) actual casting with sonotrodes in the melt; and (d) a scheme of ultrasonic melt processing during electromagnetic casting [(1) induction coil, (2) water cooling, (3) electromagnetic screen, (4) launder, (5) spout, (6) melt distributor, (7) starting block, (8) ultrasonic transducer with sonotrode)].

TABLE 7.1

Main Processing Parameters for DC Casting of Round Billets with Ultrasonic Melt Processing in the Sump

Billet Diameter, mm	Casting Speed, mm/min	Acoustic Power, kW [a]
70–100	180–240	0.6–0.8
100–200	90–180	0.8–1.0
200–300	36–90	1.0
300–400	24–36	1.0–3.0
400–500	18–24	3.0–7.0
600–1200	12–18	7.0–10.0

[a] The maximum acoustic power produced by a single 18-kHz source used is 0.6–1.0 kW.

after 100-h continuous work in the melt. Taking into account the resonance length of 90–100 mm of the horn, this did not affect much the resonance characteristics of the sonotrode.

The main casting parameters of DC casting with ultrasonic melt processing are given in Table 7.1. Depending on the size of the billets, one to ten transducers with sonotrodes were used in a single casting in order to achieve uniform nondendritic structure throughout the entire billet.

With ultrasonic processing in the melt, the presence of substrates that can be activated or refined is important, as we have discussed in Chapter 5. With regard to aluminum alloys, additions of Zr and Ti are essential for getting the ultimate grain-refining effect as a result of ultrasonic melt treatment—formation of nondendritic grain structures.[*] This type of structure and the mechanisms involved in its formation were considered in detail in Chapter 5. Figure 7.3 gives examples of this structure for DC-cast billets. Similar results were obtained for AA2124-type and AA5182-type alloys, with the nondendritic grain size ranging from 25 to 55 μm for billets 74 to 370 mm in diameter, respectively. Billets and ingots with such a structure exhibit higher mechanical properties, improved casting properties, and better response to heat treatment and deformation.

Table 7.2 demonstrates that the formation of the nondendritic structure, in addition to refining the grain proper, considerably refines the eutectics (see also Figure 5.22), which has consequences for both mechanical properties (Table 7.3) and homogenization efficiency (see Section 7.5).

In addition to structure refinement, the casting contraction (shrinkage) was found to be lesser due to a higher ductility of the solid shell. As a result, the air gap becomes smaller and the surface quality is greatly improved, as demonstrated in Figure 7.4. The casting shrinkage decreased from 2.2%–2.5% to 1.8%–2.0% for a 285-mm billet. The depth of cold shuts also decreased substantially, as shown in Figure 7.5 and

---

[*] These experiments were performed in the 1970–1980s when common grain-refining practice was to add Ti in the furnace. The modern grain refining with AlTiB master-alloy rods was adopted later.

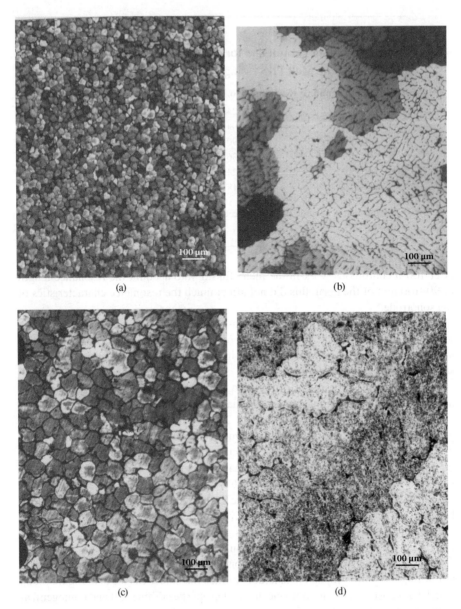

FIGURE 7.3 Microstructure of 70-mm (a, b) and 285-mm (c, d) billets from an AA7055 alloy cast with (a, c) and without (b, d) ultrasonic processing in the sump.

Table 7.4. The improvement of surface quality was such that 65–74-mm billets could be extruded without surface scalping or turning.

An important consequence of ultrasonic melt processing in a DC casting mold and the formation of nondendritic structure is the increased homogeneity of chemical composition on the scale of a billet, i.e., decreased macrosegregation. There are several reasons for this effect. According to the modern views on

TABLE 7.2

Effect of Ultrasonic Processing on the Structure Features of AA7055 Billets

Billet Diameter, mm	Structure Type	Structure parameters, µm		
		Grain	Dendrite Arm Spacing	Divorced Eutectic Thickness
70	Nondendritic	20–30	…	0.1
	Dendritic	400–500	20–40	1–3
285	Nondendritic	50–60	…	0.3
	Dendritic	≤1000	50–70	3–5

Note: AA7055 billets contained 0.14% Zr.

TABLE 7.3

Mechanical Properties of Homogenized AA7055 Billets (Transverse Direction)

Billet Diameter, mm	Structure Type	Testing Temperature, °C	Ultimate Tensile Strength (UTS), MPa	Elongation (El), %
70	Nondendritic	20	325	3.2
		400	35	118
	Dendritic	20	260	2.0
		400	30	65
285	Nondendritic	20	220	2.0
		400	35	131
	Dendritic	20	180	1.8
		400	36	98

Note: AA7055 billets contained 0.14% Zr.

FIGURE 7.4 Surface appearance of 285-mm DC-cast billets from AA7055 alloys: (left) without ultrasonic melt processing and (right) with ultrasonic melt processing.

FIGURE 7.5 Macrostructure of the cross section of an 830-mm billet from an AA7474 alloy showing cold shuts: (a) casting with ultrasonic melt processing and (b) conventional casting.

TABLE 7.4

Effect of Ultrasonic Melt Processing during DC Casting on the Depth of Cold Shuts on the Surface of Large Billets from an AA7475-Type Alloy

Diameter, mm	UST	Cold-Shut Depth, mm	Average Inclination Angle, Deg
845	+	10–30	65
845	−	48–54	45
960	+	10–30	75
960	−	52–56	50

macrosegregation [4], the main acting mechanisms are convection of the melt in the liquid pool with penetration of the flows into the upper part (slurry) of the transition region; shrinkage-induced flow acting in the lower part (mush) of the transition region; and transport of floating crystals by the melt flow. It was shown on the pilot scale and confirmed by computer simulations [5] that the downward flow along the centerline of the billet would decrease the typical negative centerline segregation due to the suppression of the natural upward flow of melt enriched in solute elements. Sonotrodes immersed into the liquid bath of a billet will create strong downward flows (see Figures 2.14–2.16) and, thereby, affect the macrosegregation pattern [6, 7]. Depending on the arrangement of the sonotrodes, the macrosegregation can be effectively controlled.

Grain refinement has a dual effect on permeability of the transition region. On the one hand, it decreases the coherency temperature and effectively widens the slurry region, thereby increasing its permeability. As a result, convective flows have more opportunity to penetrate into the slurry zone and bring solute-rich liquid to

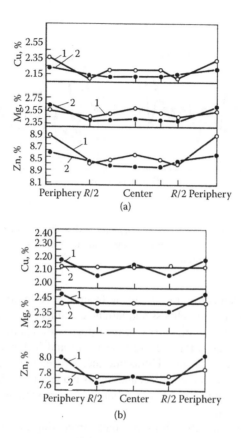

FIGURE 7.6 Effect of ultrasonic melt treatment on macrosegregation of alloying elements in (a) 270-mm-diameter billets and (b) 112-mm-diameter billets from an AA7055 alloy: (1) casting without ultrasound and (2) casting with ultrasound.

the center. On the other hand, the smaller grains decrease the permeability of the coherent mushy zone. This results in less potent shrinkage-induced flow, hence in less negative centerline segregation [8]. By making the slurry zone wider, the grain refinement also allows for more floating grains. However, these grains are smaller and less prone to settling down and distributing evenly in the billet cross section [8, 9]. Figure 7.6 illustrates the beneficial effect of ultrasonic melt treatment on the macrosegregation in AA7055 billets of two diameters.

The uniformity in structure and chemical composition translates to the uniformity and high level of mechanical properties, as shown in Figure 7.7 for large 830-mm billets from an AA2324 alloy. As a consequence, the susceptibility of the cast metal to hot and cold cracks decreases. It is well known that higher ductility of semisolid and solid metals plays a decisive role in the occurrence of hot and cold cracks, respectively [1, 10–12]. In the solid state, the ductility at temperatures below 300°C should be larger than 1.0% in order to prevent cold cracks (as the cast metal is subjected to the residual tensile strains on the order of 0.55%–0.6% [13]).

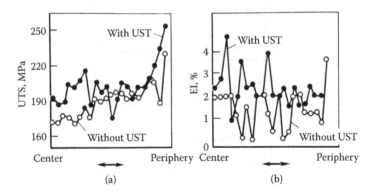

FIGURE 7.7 Effect of nondendritic structure on the distribution of mechanical properties in the cross section of annealed 830-mm billets from an AA2324 alloy: (a) ultimate tensile strength and (b) tensile elongation.

Large deformed items for aircrafts made from high-strength alloys require high characteristics of fracture toughness and fatigue endurance. These requirements are typically met by increasing the purity of the alloys with regard to Fe and Si. This, however, results in coarsening the grain structure and in higher susceptibility to cracking upon and after casting. Ultrasonic melt processing and the formation of nondendritic structure in larger billets and ingots made it possible to meet the challenge and solve the problem of producing large castings without cracking and with uniform fine structure [6, 14–18].

Figure 7.7 shows how the ultrasonic melt processing and nondendritic structure improve the tensile properties and, especially, ductility in a cross section of an 830-mm billet from an AA2324 alloy. Under conventional DC casting conditions, the ductility drops to 0.5%, with the resultant high probability of cold cracks. Nondendritic structure assures that the ductility stays above 2%, which guarantees crack-free billets. It is also important that the mechanical properties are improved throughout the cross section of a casting, as additionally illustrated in Figures 7.8a,b. Actually, the ductility is only an approximate parameter of fracture resistance. Impact toughness is a more appropriate property to characterize the toughness of the material. Figure 7.8c shows that the fracture toughness is considerably increased in a large billet of an AA7050-type alloy.

In addition to the effects of ultrasonic melt processing on structure refinement, chemical homogeneity, and cast-metal properties, a relatively low casting speed during DC casting of large billets (less than 15 mm/min) and intense cavitation processing of liquid metal in the sump of a billet result in efficient cleaning of the melt from nonmetallic impurities. The data show that the hydrogen concentration decreases by 25%–30%, and the amount of defects (porosity, inclusions) detected at technological fracture samples is lower by an order of magnitude.

As a result of experience gained in industrial DC casting with ultrasonic melt processing of various alloys and different-scale billets, it became possible in the 1980s to commercially produce large-scale crack-free billets with diameters of

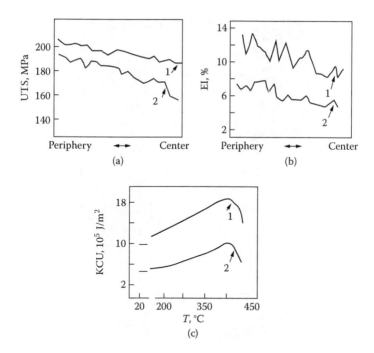

FIGURE 7.8 Effect of nondendritic structure on the mechanical properties of homogenized 960-mm billets from an AA7050 alloy: (1) nondendritic and (2) dendritic structure: (a) ultimate tensile strength, (b) tensile elongation, and (c) impact toughness.

960 mm (AA7055) and 1200 mm (AA2324) (Figure 7.9). These billets were used for special forgings and extrusions for transport airplanes [19, 20]. This experience was extended to flat ingots of an AA2324 alloy where the nondendritic structure was successfully obtained in ingots 450 × 1200 mm in cross section.

The role of Zr and Ti in the formation of nondendritic structure in commercial alloys has already been mentioned on several occasions. In high-strength alloys of 2XXX and 7XXX series, these elements are frequently present as nominal alloying elements, so only slight adjustment of their concentration may be required. In other alloys, where Zr is not an alloying element, it can be added in small amounts. For example, addition of 0.15% Zr to an AA2628-type alloy (Al–Cu–Mg–Ni–Fe) and its ultrasonic processing during DC casting allowed one to produce a nondendritic structure with a grain size less than 200 μm across the entire cross section of a 1200-mm billet, while the grain size in a conventionally cast billet was nonuniform and exceeded 1000–1500 μm.

It is worth noting that billets of such dimensions from high-strength alloys were not commercially produced in the West until 2003, when Vista Metals Corp. reported a 42-inch (1068 mm) billet successfully cast from an AA7174 alloy [21]. In this case, a sophisticated mold with a hot top, graphite inserts, ceramic rings, and gas-assisted lubrication was used, and the cold cracking was prevented by stress relief using specially designed wipers. The melt quality and

FIGURE 7.9 One of the largest ingots (1200-mm diameter, 10-ton weight) from an AA2324 alloy with nondendritic structure obtained by casting with ultrasonic melt processing in the billet sump.

structure were controlled by an in-line refining system. A grain size of 390 μm was achieved.

The ultrasonic melt processing in the billet sump can be efficient in casting special alloys such as Russian Grade 1420 (Al–Mg–Li–Zr). In these alloys the addition of Zr can result in the formation of nondendritic structure even without ultrasonic processing [22]. However, the concentration of Zr should be maintained at 0.12%–0.14%, as the structure coarsens and become dendritic at lower concentrations. The application of ultrasound allows for a wider range of Zr concentrations at which the nondendritic structure can be obtained, as shown in Figure 7.10.

Similar effects are observed in other systems where the nondendritic structure can be formed under conventional DC casting conditions, e.g., Al–Mg–Li–Sc–Zr (Russian Grades 1570, 1421) and Al–Cu–Li–Sc–Zr (Russian Grades 1461, 1424) [23]. These alloys, however, contain very expensive Sc that acts as a grain refiner, anti-recrystallizing agent, and hardener. Ultrasonic melt processing allows one to achieve nondendritic structure in these alloys at a lower concentration of Sc compensated by an increased Zr content.

One of the interesting applications of ultrasonic melt processing is combining it with electromagnetic casting. Billets cast in electromagnetic molds are characterized by a smooth surface and macroscopically fine grains [12]. The latter is the result of melt stirring by induced flows and, as a consequence, colder melt in the billet sump. The internal structure of dendritic grains is, however, coarser than in conventional DC casting, also

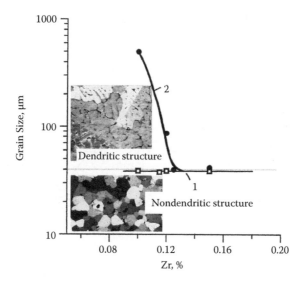

FIGURE 7.10 Grain refinement in 178-mm billets from a 1420 (Al–Li–Mg–Zr) alloy (1) with and (2) without ultrasonic melt processing in dependence on Zr concentration.

due to the induced melt flows and longer time available for grain coarsening. A combination of electromagnetic casting with ultrasonic melt processing compensates for melt cooling from stirring by melt heating due to acoustic energy (see Figure 5.24). In addition, multiplication of solidification sites leads to additional grain refinement. This process was tested upon casting 370-mm round billets from 7XXX- and 5XXX-series alloys containing up to 0.15% Zr. A principal scheme is shown in Figure 7.2d. Three ultrasonic sources were arranged at an angle around the melt distributor and submerged into the liquid bath by 5–10 mm. The melt temperature was 720–730°C, and casting speed varied from 46 to 36 mm/min. After casting 1 m of a billet, ultrasonic processing was performed for the casting length 4–5 m, and then the last 1 m was cast without ultrasound. The billets sections cast without ultrasound had columnar structure at the periphery and fine dendritic grains in the center. The billet section cast with ultrasonic treatment demonstrated uniform nondendritic grain structure [24].

7.1.2 Ultrasonic Melt Processing in the Launder

Ultrasonic melt processing can be applied to the melt flow during the transport of the melt from the holding durance to the mold. This technological approach is more versatile, as the processed melt can be directed to several molds, but it also poses a number of challenges related to the treatment time–melt volume–acoustic power ratios. We have already discussed these challenges and a possible solution with regard to ultrasonic degassing (see Section 3.4) and grain refinement (see Section 5.8). A general scheme of melt processing in the melt flow is shown in Figure 7.11a. A combination of melt processing in an intermediate vessel with ultrasonication in the launder proper could be a plausible solution (see, e.g., Figure 4.8b).

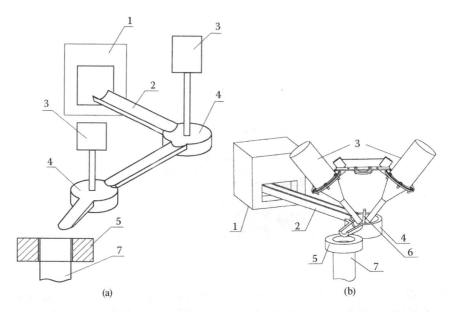

FIGURE 7.11 Principal schemes of ultrasonic melt processing in melt flow: (a) ultrasonication in an intermediate vessel and in the launder and (b) with introduction of a grain-refining rod in the cavitation zone during DC casting: (1) holding furnace, (2) launder, (3) ultrasonic transducer with sonotrode, (4) intermediate vessel, (5) DC casting mold, (6) grain refining rod with a feeder, and (7) billet.

Table 7.5 compares the efficiency of ultrasonic processing in the mold and in the launder for a 270-mm billet of an AA7055 alloy with 0.16% Zr. The melt temperature in the launder was 80–100 K above the liquidus, while in the sump the superheat was limited to 5–10 K. The grain refinement can be achieved by ultrasonic melt treatment in the flow, but its efficiency for grain refinement is less than that for ultrasonic processing in the mold. In this case, the mechanisms of substrate activation and refinement of primary particles are acting. The efficiency may be improved by managing the melt flow (see Figure 3.17).

TABLE 7.5

Comparison of the Efficiency of Ultrasonic Melt Treatment in the Launder and in the Mold During DC Casting of a 270-mm Billet from an AA7055 Alloy

Ultrasonic Processing	Melt Temperature, °C	Number of US Sources	Grain Size, μm	H_2, cm³/100 g
No	720	...	<1000 (dendritic)	0.24
In the launder	720	1	190 (dendritic)	0.07
In the mold	650	1	60 (nondendritic)	0.12
In the mold	650	3	50 (nondendritic)	0.07

We have already discussed in Section 5.3 that the application of ultrasonic cavitation to the introduction of grain-refining master alloys is very promising (see Figure 5.12). The introduction of an Al–Ti–B grain-refining rod in the melt flow during DC casting was suggested in the 1970s [25], but it was not widely adopted until later in the 1980s [26]. One of the deficiencies of grain-refining rod introduction is the agglomeration of TiB_2 particles, wide size distribution of these particles and, as a consequence, low efficiency, i.e., only several percent of particles are acting as nucleation substrates [26, 27]. Ultrasonic melt processing improves the performance of standard AlTiB master alloys by dispersing agglomerates and activating substrates, as has been confirmed by a number of studies [28–30].

Figure 7.11b gives a principal scheme of ultrasonic processing of melt flow with introduction of a grain-refining rod. The best results are obtained when the rod is introduced into the cavitation zone. In this case, the deagglomeration effect is most pronounced.

This technology was tested using pilot-scale DC casting of flat (90 × 230 mm) and round (178 mm) billets from an AA7050-type alloys. Experimental Al–5% Zr and Al–5% Ti and commercial Al–5% Ti–1% B master-alloy rods were used. Table 7.6 shows that the optimum concentration of alloying additions and ultrasonic melt processing of the melt flow promote the formation of nondendritic structure in DC-cast billets (see also Figure 5.12).

7.2 DIRECT-CHILL CASTING OF MAGNESIUM ALLOYS

Direct-chill casting of magnesium alloys is relatively smaller, both in the dimensions of ingots/billets and in annual production, as compared to aluminum. Nevertheless, this is the main technology of producing wrought Mg alloys that face similar challenges. Ultrasonic melt processing has been applied to DC casting of Mg since the 1980s based on extensive research [31–34].

An industrial-scale DC casting installation with ultrasonic equipment was built in the 1980s at a Soviet casting plant. This DC casting rig was able to produce flat 550 × 165-mm ingots up to 6 m long. Two magnetostrictive ultrasonic transducers with sonotrodes were fixed on a platen above the casting mold. The parameters of

TABLE 7.6
Effect of Combined Grain Refining and Ultrasonic Melt Processing of the Melt Flow on the Grain Size of 178-mm Billets from an AA7050 Alloy

Ultrasonic Processing	Grain Refining Rod	Zr, wt%	Ti, wt%	Grain Size, μm
No	...	0.10	0.02	>600 dendritic
No	Al–5% Zr	0.10 + 0.04 [a]	0.02	120–130 dendritic
In the launder	Al–5% Zr	0.10 + 0.04 [a]	0.02	40–50 nondendritic
In the launder	Al–5% Ti	0.10	0.02 + 0.02 [a]	40–50 nondendritic
In the launder	Al–5% Ti–1% B	0.10	0.02 + 0.01 [a]	40–50 nondendritic

[a] Additional concentration introduced from the master alloy rod.

generators and transducers were the same as for Al DC casting (see Section 7.1). The sonotrodes made from a Nb alloy were submerged into the melt in the liquid part of the ingot. The DC casting installation allowed one to cast simultaneously two ingots. Therefore, two ingots were typically produced, with and without ultrasonic processing. The melt from a resistance holding furnace was supplied to the distribution box by an electromagnetic pump and then fed into the mold. A gas mixture was used to protect the melt from oxidation.

Figure 7.12 shows a principal scheme of the installation and a casting procedure for an AZ31B-type alloy. Similar to DC casting of aluminum, the ultrasonic

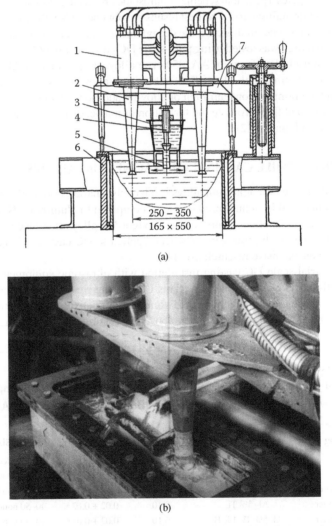

(a)

(b)

FIGURE 7.12 DC casting of magnesium alloys with ultrasonic melt processing: (a) a diagram of the process [(1) ultrasonic transducer, (2) sonotrode, (3) melt holder, (4) stopper, (5) melt distribution box, (6) mold, (7) frame]; and (b) casting process.

melt processing in the ingot sump resulted in the increased melt temperature in the liquid pool of the ingot (by 20°C for a 550 × 165-mm ingot) and flattening of the sump profile [24]. These effects should potentially decrease macrosegregation and hot tearing as well as prevent formation of coarse primary intermetallics.

Note that an AZ31B alloy is a Zr-free Mg–Al–Mn alloy. Grain refinement of these alloys is a challenge. The ingots produced with ultrasonic processing, however, had significantly refined grain structure, as illustrated in Figure 7.13 by fracture surfaces; but the nondendritic structure was not achieved. It is quite possible that the acoustic power was not sufficient for the maximum grain-refining effect. Nevertheless, the columnar structure was eliminated, and the grain size was refined from over 1000 μm to 70–100 μm, which significantly decreased the propensity to hot tears.

As we have already discussed, the grain refinement in AZ31B-type alloys is accompanied with refinement of primary intermetallics (see Figure 6.3) and degassing (see Figure 3.16). As a result of this complex improvement of ingot quality, the technological plasticity of the ingots upon hot deformation is significantly increased, as demonstrated in Figure 7.14.

It appeared possible to produce the nondendritic grain structure in AZ31B-type alloys when smaller billets were cast, i.e., 118 to 178 mm. A pilot-scale DC caster was constructed in the All-Union Institute of Light Alloys (VILS). The special features were a steel-lined induction furnace, a melt dosing system when the melt was transferred from a closed pressurized vessel to the mold through a tundish, and 1:5 SO_2:Ar gas mixture melt protection in the tundish and the mold. Billets from AZ31B-type alloys were produced with nondendritic grains 30–40 μm in size as compared to 120-μm dendritic grains in conventionally cased billets. A grain structure that was close to nondendritic, with the grain size 30–40 μm (as compared to 60 μm), was obtained in billets from a Russian Grade VMD9 (Mg–Mn–Nd–Y–Zn–Al) (see Figures 7.15c,d). Casting of magnesium alloys of other Zr-free systems, such as

(a)

(b)

FIGURE 7.13 Fracture surfaces and microstructures of a 550 ´ 165-mm ingot from an AZ31B-type alloy: (a) DC cast with ultrasonic melt processing and (b) conventional DC casting.

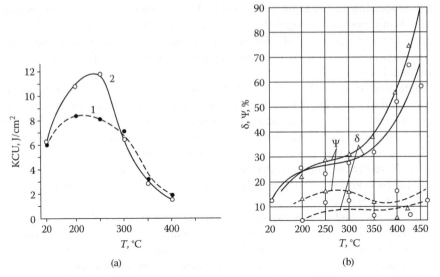

FIGURE 7.14 Mechanical properties of as-cast ingots from an AZ31B-type alloy: (a) impact toughness KCU and (b) elongation δ and reduction in area ψ [(1) without ultrasonic processing and (2) with ultrasonic processing].

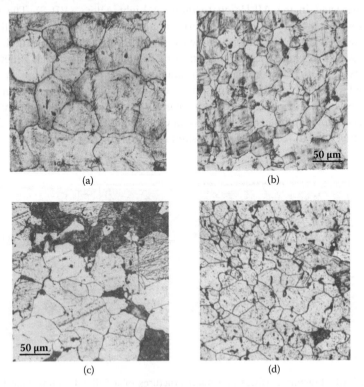

FIGURE 7.15 Grain structure of homogenized 174-mm billets from MA14 (a, b) and VMD9 (c, d) alloys: (a, c) without ultrasonic processing and (b, d) with ultrasonic processing.

Mg–Zn–Mn and Mg–Mn–Ce, showed that ultrasonic treatment produced the non-dendritic structure only at high ultrasound intensities in small ingots (98–118-mm diameter). Evidently, high ultrasonic intensities are required to activate the substrates that are present in these alloys.

Recently, a combination of ultrasonic processing and electromagnetic stirring during DC casting of an AZ80 (Mg–Al–Zn–Mn) alloy was attempted [35]. The alloy was cast in a 160-mm mold at a casting speed of 150 mm/min. In addition to ultrasonic processing using a single source (20 kHz, 2-kW consumed power), low-frequency electromagnetic stirring was applied to the melt in the mold through electromagnetic coils. The ultrasonic processing resulted in grain refinement from 400 to 280 μm, and a combination of stirring and sonication refined the grains somewhat further, to 260 μm. The limited effect of grain refinement may be due to the specific temperature conditions in the sump subjected to low-frequency electromagnetic stirring. The temperature throughout the sump becomes very uniform and even somewhat below the liquidus (which is opposite to melt superheating by ultrasonic cavitation). As a result, the ultrasonic processing occurs actually in the semiliquid state, and fragmentation (refinement) is accompanied by coarsening of floating crystals. The important outcome of the combined processing was the uniformity of grain structure in the cross section. Electromagnetic stirring usually causes grain coarsening in the center of the billet, while ultrasonic processing mitigates this effect. The optimization of the structure led to improvement of mechanical properties. In particular, the elongation was increased from 1.7% to 4% after ultrasonic processing and to 4.2% after the combined treatment. The tensile strength improved from 157 MPa to over 185 MPa.

Another group of Mg alloys contains Zr and rare-earth metals. We studied Russian Grades MA14 (Mg–Zn–Zr), VMD10 (Mg–Y–Zn–Zr–Cd), and VMD7 (Mg–Zn–Y–Cd–Zr). The modifying effect of Zr is defined by the peritectic reaction and the formation of primary particles of Zr or ZnZr (in the Mg–Zn–Zr system) that act as substrates for Mg nucleation. Even without sonication, a grain structure close to nondendritic can be obtained in these alloys, with high Zr concentrations (up to 0.9%) cast to billets up to 270 mm in diameter and flat ingots 100 × 300 mm in cross section. Reduction of Zr concentration to 0.4% results in the formation of dendritic grains.

Experiments with larger billets, up to 370 mm in diameter, showed a lower probability of nondendritic grain structures, even for the maximum Zr content. The main cause of the transition to a dendritic structure seems to be the scaling factor of larger billets with longer holding times and slower casting, which increases the probability of precipitation of zirconium in the holding furnace and during melt transport to the mold.

Casting with sonication in the mold under conditions of developed cavitation ensures the formation of a nondendritic structure in billets of commercial alloys with zirconium concentration within the entire nominal range, as shown in Figures 7.15a,b for an MA14 alloy.

The analysis of cooling condition in the sump showed that sonication in the mold somewhat increases the cooling rate due to a positive temperature gradient in the liquid pool. The cooling rate was estimated using the measured temperature gradient and casting speed [24]. The results are given here:

	Billet Diameter, mm			
	270	204	174	118
Cooling rate with sonication, K/s	41	54	63	93
Cooling rate without sonication, K/s	39	52	61	89

Taking these cooling rates into account, the average nondendritic grain size in magnesium alloys with Zr fits well with the plot in Figure 5.20. In addition to grain refinement, ultrasonic processing results in refinement of intermetallic particles, as described in Chapter 6 (see Figures 6.3 and 6.4).

The structure refinement and modification improves the mechanical properties of complex magnesium alloys with rare earths, as illustrated in Table 7.7 for 174-mm billets from Russian Grade VMD10.

Finally, the typical casting parameters used in DC casting of magnesium alloys are listed in Table 7.8.

TABLE 7.7
Mechanical Properties of Homogenized 174-mm Billets from a VMD10 Alloy

Ultrasonic Processing	UTS, MPa	YS, MPa	El, %	RA, %	KCU, J/cm^2
In the mold	240	155	15.2	17.5	6
No ultrasound	230	150	13.4	15.1	5.1

Note: UTS = ultimate tensile strength; YS = yield strength; El = elongation; RA = reduction in area; KCU = impact toughness.

TABLE 7.8
Typical Casting Parameters for DC Casting and Ultrasonic Processing of Magnesium Alloys

Casting Dimensions, mm	Casting Speed, mm/min	Acoustic Power,[a] kW
Diam. 98–118	30	0.6–0.8
Diam. 118–178	12	0.8–1.0
Cross section 550 × 165	6	1.6–2.0

[a] The acoustic power of a single 18-kHz source used is 0.6–1.0 kW.

7.3 EFFECT OF ULTRASONIC DEGASSING AND FILTRATION ON STRUCTURE AND PROPERTIES OF WROUGHT PRODUCTS

7.3.1 ULTRASONIC DEGASSING

Ultrasonic degassing performed during DC casting (see Chapter 3) improves the mechanical properties of billets/ingots and wrought products produced from the processed metal. Tables 7.9 and 7.10 show that the better metal quality achieved through ultrasonic degassing increases ductility at room and elevated temperatures and improves impact toughness.

Elimination of porosity, cleaning of the metal from solid inclusions, and reducing the amount of hydrogen remaining in the solid solution benefit the technological behavior and properties of deformed metal [36–38]. A Russian Grade AMg6 (Al–6% Mg–0.6% Mn) is a good example of an alloy whose properties can be significantly improved by ultrasonic degassing. Billets 370 mm in diameter from this alloy were cast with ultrasonic degassing of the melt and then worked into hot-extruded 110-mm bars, 20-mm-thick plates, and 6- and 2.5-mm-thick sheets.

First of all, the decreased amount of hydrogen in the solid metal and reduced porosity result in much less sensitivity of the metal to delamination during hot deformation. This sensitivity was tested using the so-called heat test, where the sample is quickly heated to a temperature slightly above the solidus (583°C for AMg6) and then held at this temperature for 15 min. The laminations are then ranked from I (no lamination) to III (severe lamination). Table 7.11 and Figure 7.16 illustrate this effect.

TABLE 7.9
Mechanical Properties of 460-mm Billets from an AA2117-Type Alloy

Degassing Technology	Hydrogen Content, $cm^3/100$ g	UTS, MPa	El, %
Regular technology	0.40	160–185	4.8–6.4
Ultrasonic degassing	0.21	180–210	6.5–9.5

TABLE 7.10
Ductility of 1700 × 300-mm Ingots from an AMg6 Alloy at 400°C

Degassing Efficiency, %	24		33		48	
Property	Conventional Technology	Ultrasonic Degassing	Conventional Technology	Ultrasonic Degassing	Conventional Technology	Ultrasonic Degassing
El, %	54	69	58	68	55	67
KCU, J/cm^2	69	77	62	76	73	82

TABLE 7.11

Effect of Ultrasonic Degassing on Delamination during Deformation of Hot-Extruded and Hot-Rolled Items from an AMg6 Alloy

	110-mm Bars		6-mm Sheet	
Degassing Technology	H_2, cm³/100 g	Heat Test Number	H_2, cm³/100 g	Heat Test Number
Regular technology	0.36	II–III	0.39	II–III
	0.30	II	...	...
Ultrasonic degassing	0.16	I	0.23	I
	0.13	I	0.20	I
	0.10	I	0.18	I

The higher quality of the metal translates to better quality of welded joints. Special tests after MIG (metal inert gas) welding showed the absence of porosity in the weld (Figure 7.16) and in the heat-affected zone as well as increased He-tightness, as shown in Table 7.12.

Another "rupture" probe was developed for testing wrought alloys for delamination [36, 37]. The technique allows one to test flat samples by rupturing them in the height direction using standard tensile testing equipment. The samples were tested as received and after the heat test. The difference in rupture stresses or their ratio gives the susceptibility of the material to cracking during welding. Table 7.13 gives the properties of 10-mm plates rolled from large 1700 × 300-mm ingots from an AMg6 alloy DC cast using conventional technology and with ultrasonic degassing

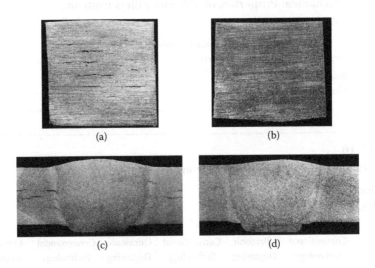

(a)   (b)

(c)   (d)

FIGURE 7.16 Effect of ultrasonic degassing on delamination behavior of plates (a, b) and welds (c, d) from an AMg6 alloy subjected to the heat test: (a, c) conventional DC casting technology; (b, d) DC casting with ultrasonic degassing.

TABLE 7.12

Effect of Ultrasonic Degassing on Gas Tightness of Welded Joints from an AMg6 Alloy

	Fraction of Tight Samples, %	
Sample Thickness, mm	Conventional Technology	Ultrasonic Degassing
4.0	100	100
1.8	27	93
1.0	0	86

in the melt flow. These results demonstrate that the decrease in hydrogen concentration in the deformed metal to 0.2 cm³/100 g leads to twofold improvement in the fatigue endurance.

The comparison of different degassing techniques testifies to the advantages of ultrasonic degassing, as illustrated here for the quality of welded joints of AMg6 sheets:

Weld Property	Ar Lancing	Vacuum Degassing	Ultrasonic Degassing
UTS, MPa	300	320	325
Bending angle, deg.	57	65	87

Note: UTS = ultimate tensile strength.

7.3.2 ULTRASONIC FILTRATION

The effect of the Usfirals process (see Chapter 4) on the properties of wrought metal was studied using billets and ingots produced in pilot and industrial casting facilities

TABLE 7.13

Effect of Ultrasonic Degassing on the Quality and Properties of 10-mm-Thick Hot-Rolled Plates from an AMg6 Alloy

Property	Conventional DC Casting	DC Casting with Ultrasonic Degassing
H_2, cm³/100 g in melt	0.6	0.3–0.33
H_2, cm³/100 g in ingot	0.33–0.37	0.20–0.25
H_2, cm³/100 g in plate	0.3–0.34	0.18–0.22
Heat-test number	II–III	I
Cycles to failure, kc	70–261	196–645
Rupture test of as-received plate, MPa	193–235	239–247
Rupture test after heat test, MPa	69–80	170–206

TABLE 7.14

Effect of Ultrasonic Filtration on the Properties of Extruded Plates from an AA2024-Type Alloy

Filtering	Inclusions, mm²/cm²	UTS, MPa	YS, MPa	El, %	RA, %	K_{IC}, MPa·m$^{1/2}$	LCF, kc at 180 MPa
1 layer	0.32	420	330	6.4	6.7	26.5–35 (av. 30)	70–151 (av. 97)
3 layers	0.004	415	315	7.3	8.6	32–35 (av. 33.5)	67–209 (av. 110)
5 layers	0.0013	435	330	7.6	8.4	39–44 (av. 40.7)	104–151 (av. 116)

(see Figure 4.8). The use of fiberglass multilayered screen filters allowed the efficient cleaning of fine nonmetallic inclusions from the melt. Ultrasonic filtration in the setting depicted in Figure 4.8 is accompanied by degassing. Tables 4.4 and 4.5 give the capacity of ultrasonic filtering for different filter designs. The impact of ultrasonic filtration on the properties of deformed metal can be illustrated with the data in Table 7.14 for extruded 200 × 65-mm plates from 370-mm DC-cast billets of an AA2024-type alloy [39]. The plates were quenched from 500°C in water and then naturally aged. It is noteworthy that the structural strength (toughness, fatigue endurance) is significantly improved, whereas the tensile properties are less responsive to the metal purity.

7.4 EFFECT OF NONDENDRITIC STRUCTURE ON STRUCTURE AND PROPERTIES OF WROUGHT PRODUCTS

The effect of billet/ingot as-cast structure on the deformation behavior and final properties of wrought semifinished products is generally known. Various optimal deformation schemes have been developed to deal with coarse and inhomogeneous grain structure of the casting. It is generally accepted that grain refinement and resultant small grains and their uniformity across the casting benefit the properties of the billet/ingot (less cracking and macrosegregation) and the wrought products (uniform deformation, better structure). On the other hand, the deformation does transform the original as-cast grain structure and, therefore, it is assumed that with deformation ratio of 80%–85% (extrusion ratio 6–8), one can obtain adequate properties in the deformed metal regardless of the initial ingot structure. There are not many detailed studies on the actual effect of the as-cast structure on the structure and properties of rolled, forged, and extruded products.

In this section, we will demonstrate that fine, nondendritic grain structure of the DC-cast billet or ingot has significant impact on the technological and mechanical properties of wrought products.

7.4.1 ALUMINUM WROUGHT ALLOYS

High-strength aluminum alloys suffer from insufficient ductility in the as-cast condition, which is one of the main reasons for their high susceptibility to hot and cold cracking. The reduced ductility also makes their deformation difficult and troublesome. These alloys are usually used in high-fidelity and sensitive applications such as aircraft, automotive, and aerospace structures, where endurance to fatigue and delayed fracture is important. The improvement of their ductility by controlling the grain structure already at the casting stage has great practical value. In the following discussion, we will illustrate the benefits of nondendritic grain structure for the properties of deformed products.

Let us first look at small forgings from an AA7055-type alloy. Billets 67–70 mm in diameter had a dendritic structure with 350–500-μm grains after conventional DC casting and a nondendritic structure with 20–30-μm grains after casting with ultrasonic melt processing in the mold. Forgings from nondendritic billets exhibited a homogeneous structure, whereas those from dendritic billets had uneven, fiberlike structure, as illustrated in Figure 7.17. A study of microstructure clearly indicated that the forged items from nondendritic billets had much finer deformed grains with uniform distribution of excess phases (Figure 7.18) [40]. Table 7.15 demonstrates how this difference in structure translates to the improvement in mechanical properties [40]. Evidently the forgings from nondendritic billets have an advantage in mechanical properties both at room and warm temperatures. In addition, impact tests revealed that the forgings from nondendritic billets had better isotropy of properties, with the impact toughness in the critical, chord direction increased by 15% from 58–63 J/cm^2 to 65–75 J/cm^2.

An examination of the fracture surfaces of impact samples showed that crack propagated to a great extent intergranular (up to 40%) in coarse-grained samples, whereas 80% transgranular fracture was observed in fine-grained samples. The fracture surface showed more signs of plastic deformation in the latter case. Particles of

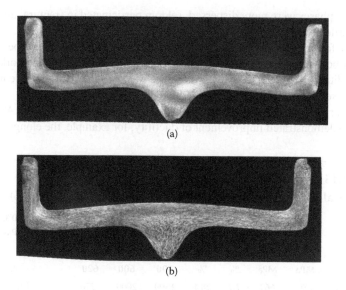

(a)

(b)

FIGURE 7.17 Macrostructure of AA7055-type alloy forgings obtained from DC-cast billets with (a) nondendritic grain structure and (b) dendritic grain structure.

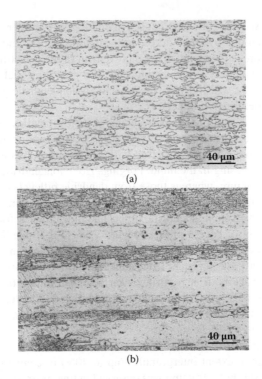

FIGURE 7.18 Microstructure of AA7055-type alloy forgings obtained from DC-cast billets with (a) nondendritic grain structure and (b) dendritic grain structure.

secondary phases were more dispersed, seldom participated in fracture, and seemed to have little effect on it.

Larger forgings produced from 315-mm billets from an AA7055-type alloy had a thicker bottom part with less worked structure and a thinner, well-worked shell. The hereditary effect of nondendritic structure was particularly noticeable in the less deformed sections, as shown in Table 7.16.

Even larger forgings made from 830-mm DC-cast billets of an AAA7175-type alloy also demonstrated improvement of ductility; for example, the elongation in the

TABLE 7.15

Mechanical Properties of Forgings from an AA7055-Type Alloy (T6)

As-cast Structure	Tensile Properties at 20°C				Time to Rupture at 50°C at Load (MPa), h			Creep at 50°C at 4000 h at Load (MPa), µm/mm h
	UTS, MPa	YS, MPa	El, %	RA, %	590	600	620	450
Nondendritic	680	667	7.7	28.1	1300	2000	770	1.0×10^{-4}
Dendritic	675	640	6.2	26.9	1200	815	90	2.1×10^{-4}

TABLE 7.16

Mechanical Properties of Forgings from AA7055-Type Alloy Billets with Dendritic (Numerator) and Nondendritic (Denominator) Structure

Part of a Forging	Sampling Direction	UTS, MPa	YS, MPa	El, %
Bottom	Chord	590/640	570/630	6.8/11.8
	Radial	570/600	550/590	6.8/7.2
	Height	540/600	530/580	3.2/5.1
Shell	Longitudinal	640/660	610/640	8.0/10.5
	Transversal	650/660	630/640	9.0/11.2

height direction increased from 5.6% to 8.8%, reduction in area from 10% to 30%, and fracture toughness from 41.4 to 54.9 MPa·m$^{1/2}$ upon transition from dendritic to nondendritic billets [24].

The reduced anisotropy of properties can be demonstrated with extruded plates from an AA7055 alloy produced from DC-cast 270-mm billets with either dendritic or nondendritic grains. Plates of two cross sections 10 × 120 mm and 6 × 90 mm were extruded (extrusion ratio 19.6). The plates from nondendritic billets demonstrated a noticeably better structure. Tests of heat-treated plates showed reduced anisotropy and a higher level of mechanical properties in the fine-grained samples, as seen in Figure 7.19.

The same tendency is observed in long hot-extruded panels from an AA7175-type alloy. The original billets, 830 mm in diameter, were first swaged and then extruded. The effect of nondendritic structure was pronounced in all testing directions, but in the height direction in particular (Table 7.17). Also, the number of defects detected by nondestructive ultrasonic testing decreased from 150 to 20 per panel.

Hollow 285-mm billets with dendritic and nondendritic grain structures (Figure 7.20) were extruded into pipes with 20-mm wall thickness. It appeared that the maximum extrusion speed could be increased 1.5–2 times for nondendritic billets. Figure 7.21 shows that the as-cast nondendritic grain structure has a clear effect on the deformed structure of thin-walled 120-mm pipes from an AA7055-type alloy. The mechanical properties of the pipes were improved, especially the time to rupture at warm temperatures, as shown here:

Billet Structure (Grain Size)	Time to Rupture at 60°C at Load (MPa), h		
	600	630	640
Nondendritic (60–70 μm)	>1000	450	380
Dendritic (>800 μm)	700	200	<20

The nondendritic ingot structure also has a positive effect on rolled sheets. This was demonstrated using flat DC-cast ingots, 550 × 165 mm, from AA2324, AA7155, and AA2319-type alloys cast with and without ultrasonic melt processing.

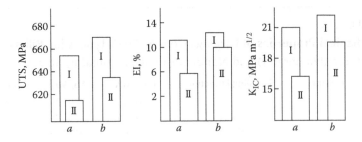

FIGURE 7.19 Anisotropy of the mechanical properties of heat-treated (water quenched from 470°C; aged 10 h at 100°C and 3 h at 160°C), hot-extruded AA7055-type plates: (a) plates obtained from dendritic billets, (b) plates obtained from nondendritic billets. I, longitudinal direction; II, transverse direction.

TABLE 7.17
Mechanical Properties of Hot-Extruded Panels from an AA7175-Type Alloy

Billet Structure	Sampling Direction	UTS, MPa	YS, MPa	El, %
Nondendritic	Longitudinal	560	480	11.0
	Transversal	550	470	10.6
	Height	475	450	7.0
Dendritic	Longitudinal	525	460	8.6
	Transversal	530	460	8.3
	Height	470	425	3.5

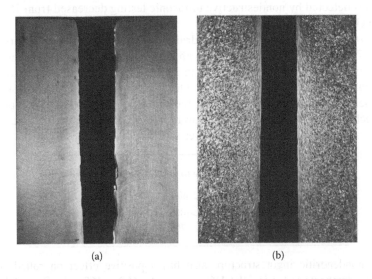

(a)   (b)

FIGURE 7.20 Macrostructure of hollow 285-mm billets from an AA7055-type alloy: (a) nondendritic structure and (b) dendritic structure.

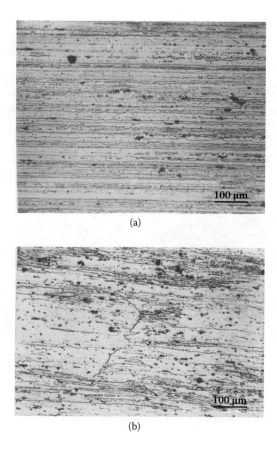

FIGURE 7.21   Microstructure of pipes from an AA7055-type alloy extruded from DC-cast billets with (a) nondendritic structure and (b) dendritic structure.

Figure 7.22 shows the AA2319 ingot structures as well the structure of sheets with different thicknesses.

The improvement in the properties of hot-rolled plates from AA2324-type and AA7155-type alloys is demonstrated in Table 7.18 for the most sensitive height direction. The hot-rolled plates were characterized by more-isotropic mechanical properties, which was confirmed also on 40–80-mm thick plates rolled from larger ingots measuring 1200 × 450 mm in cross section.

The effect of ultrasonic melt processing on the anisotropy of properties in cold-rolled sheets from AA7055-type alloys of different purity is illustrated in Table 7.19. These data suggest that the ultrasonic melt treatment and the formed as-cast nondendritic structure reduce anisotropy of properties in cold-rolled sheets, especially with the increased purity of the starting materials.

The improved ductility in wrought products made of nondendritic ingots was also observed at cryogenic temperatures, as shown in Table 7.20 for plates and sheets of different thicknesses from an AA2319-type alloy.

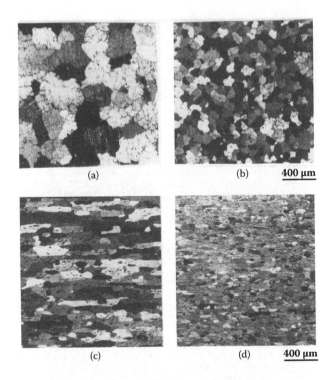

FIGURE 7.22 Grain structure of flat 550×165-mm ingots from an AA2319-type alloy produced (a) without and (b) with ultrasonic melt processing and structure of (c) 6-mm and (d) 2-mm sheets rolled from an ingot with the nondendritic structure.

The ability of deformed metal to inhibit cracks is of special interest. Extruded 6×90-mm plates produced from 134-mm billets of a high-strength 7055-type alloy were used as an object of the study. Flat test pieces ($85 \times 300 \times 6$ mm in size; a central hole of 2.0-mm diameter with a through-notch 1-mm long and 0.3-mm wide) were used to test the fracture toughness K_C and crack propagation rate (CPR). Three

TABLE 7.18

Mechanical Properties in the Height Direction of Hot-Rolled 60-mm Plates from AA2324- and AA7175-Type Alloy Ingots with Dendritic and Nondendritic Structure

Alloy	Ingot Structure	UTS, MPa	El, %	Fatigue Limit at 120 Hz, 10^7 Cycles, MPa
AA2324	Dendritic	350–380	0.8–4.4	130
	Nondendritic	390–395	4.4–4.8	150
AA7175	Dendritic	440–445	0.5–1.0	...
	Nondendritic	530–535	1.8–2.0	...

TABLE 7.19

Mechanical Properties of 2-mm Cold-Rolled Sheets from an AA7055-Type Alloy in the Longitudinal and Transversal Directions After T6 Heat Treatment

DC-Casting Technology	Impurities, wt%		Longitudinal Direction			Transversal Direction		
	Fe	Si	UTS, MPa	YS, MPa	El, %	UTS, MPa	YS, MPa	El, %
With UST	0.04	0.01	595	550	10.1	590	550	10.8
Without UST	0.04	0.01	610	585	8.3	580	535	9.8
With UST	0.14	0.01	600	555	10.3	610	565	10.0
Without UST	0.14	0.01	605	565	9.6	590	550	6.8
With UST	0.34	0.01	600	555	10.3	580	540	9.8
Without UST	0.34	0.01	595	550	9.7	600	560	9.4

TABLE 7.20

Effect of the Nondendritic Structure on the Mechanical Properties of Ingots, Plates, and Sheets from an AA2319-Type Alloy

Ingot Structure	Mechanical Properties at			
	20°C		−196°C	
	UTS, MPa	El, %	UTS, MPa	El, %
550 × 165-mm Ingot				
Nondendritic	185	10.2	...	...
Dendritic	175	6.4	...	...
60-mm-Thick Plate (Height Direction)				
Nondendritic	420	4.2	540	5.2
Dendritic	420	2.0	545	3.1
6-mm Sheet				
Nondendritic	420	13.8	505	14.0
Dendritic	420	7.1	510	7.0
2-mm Sheet				
Nondendritic	480	19.1	565	26.3
Dendritic	450	12.7	580	15.9

TABLE 7.21

Effect of Billet Structure on Fracture Toughness of Extruded Plates from an AA7055-Type Alloy

As-cast Structure	Grain Size, µm	K_C, MPa·m$^{1/2}$
Dendritic	300	33.6
Nondendritic	40	49.8
Nondendritic	20	52.5

Note: Quenched in water from 470°C and aged for 16 h at 140°C.

grain sizes were obtained in the DC-cast billets with and without ultrasonic melt processing, as shown in Table 7.21.

The effect of the nondendritic structure on the crack propagation rate was especially pronounced for small nondendritic grains when the CPR decreased by a factor of 10 at the same stress intensity. The fatigue fracture surface always exhibits two well-defined zones. One is smooth with traces of a fatigue crack, and the other shows a morphology typical of brittle or ductile fracture under static loading. The transition region between these zones allows one to determine the critical crack opening H_c that can be used to characterize crack bluntness. The H_c values were found to be different (80 ± 1.5 and 200 ± 8 µm) for the materials with low K_C (33.6 MPa·m$^{1/2}$) and high K_C (52.5 MPa·m$^{1/2}$), respectively (see Table 7.21). These values reflect the level of plastic deformation sustained by the material in the tip of a static crack, characterizing the fracture toughness.

We may conclude that the nondendritic structure of billets and ingots favorably affects the structure of deformed products from high-strength aluminum alloys and improves their ductility and service properties (fracture and impact toughness, time to fracture).

There are also positive effects on corrosion behavior [41]. The increased purity of starting materials in Al–Zn–Mg–Cu alloys with additions of up to 0.18% Zr as well as the structural factors (grain size and dendrite arm spacing) play an essential role in the corrosion behavior of deformed semifinished products. Metallographic and X-ray studies of the sheets indicated that the ultrasonic melt processing performed during DC casting considerably improves the uniformity of distribution for $Al_2Mg_3Zn_3$ particles as well as for the primary intermetallic compounds of chromium and zirconium. Billets produced by conventional DC casting frequently have coarse particles of these phases, which facilitates recrystallization of hot-rolled sheets and stress-corrosion cracking. Conversely, ultrasonic melt processing refines primary crystals, slows down recrystallization, and impedes stress-corrosion cracking. The corrosion behavior depends on the purity of the metal. The positive effect of ultrasonic melt processing is most pronounced in high-purity alloys with iron concentration below 0.1%. In the alloys with iron concentration larger than 0.2%, the difference in grain size in the as-cast and deformed material as a result of ultrasonic melt processing becomes less, and therefore the application of ultrasonic melt processing is more applicable to high-purity alloys. The stress-corrosion cracking tests of sheets from an AA7055-type alloy prepared from high-purity starting materials indicated that ultrasonic melt processing increased the endurance of the sheets in a corrosive environment almost twofold, i.e., 55 days for sheets from ultrasonically treated billets versus 30 days for sheets from conventionally cast billets.

7.4.2 MAGNESIUM WROUGHT ALLOYS

Ultrasonic melt treatment performed during DC casting of magnesium alloys has positive consequences for properties and structure of wrought items in the same manner as has been demonstrated for aluminum alloys.

In Al-containing alloys like AZ31B, ultrasonic melt treatment produces a grain structure that is almost ten times finer than after conventional DC casting. This structure significantly improves the plasticity in hot and cold rolling. In addition to increased plasticity, the refined and uniform grain structure in the ingot results in reduced anisotropy of ductility in plates and sheets.

Technological properties such as deformability are improved, as demonstrated in Figure 7.23 for hot-deformed ingots from an AZ31B-type alloy (Mg–Al–Zn). In this case, hot forging at 425°C of wedged samples showed that the ingot cast with ultrasonic processing had significantly fewer cracks than the conventionally cast ingot, even though the nondendritic structure was not achieved in these ingots. The fracture surfaces of ingots and hot-rolled plates from the same alloy show much more ductility in the ultrasonically processed products (Figure 7.24). Rolled plates and sheets of different thicknesses produced from ultrasonically treated ingots demonstrate better

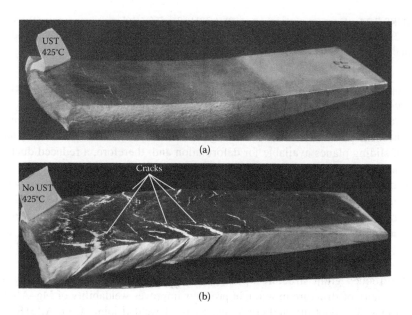

(a)

(b)

FIGURE 7.23 Wedged ductility test pieces from an AZ31B-type alloys after forging at 425°C, obtained from ingots cast (a) with and (b) without ultrasound.

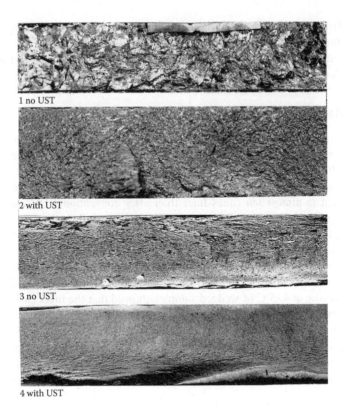

FIGURE 7.24   Fracture surfaces of (1, 2) AZ31B flat 550 ´ 165-mm ingots cast (1) without and (2) with ultrasonic melt processing and (3, 4) hot-rolled 30-mm-thick plates produced from these ingots, respectively.

isotropy of mechanical properties that can be traced back to more uniform ductility of the as-cast ingot, as illustrated in Figure 7.25 and Table 7.22 [31, 34, 42].

Grain refinement in as-cast ingots and billets from magnesium alloys is important because plastic deformation of magnesium is associated with a limited number of sliding planes available for deformation and, therefore, a reduced ductility (technological plasticity). The grain refinement of ingots and billets suppresses deformation by twinning and increases the share of deformation by sliding, which can improve the plasticity and reduce the anisotropy of wrought metal. This is illustrated by Table 7.23, which gives the minimum and maximum values for the mechanical properties of 8-mm-thick plates obtained from ingots with large grains solidified without ultrasound and ingots with refined grains solidified with ultrasound.

The refined grain structure of DC-cast ingots and the inherited ductility and homogeneity of structure in wrought products improves weldability of Mg–Al–Zn–Mn alloys. The ductility and crack sensitivity of welded joints from AZ31B-type alloys also depend on the iron concentration both in the base metal sheet and the filling wire. Purer alloys are less prone to cracking. It turned out that the combination of

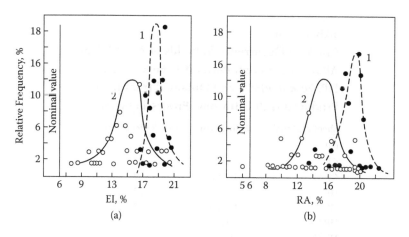

FIGURE 7.25 Distribution of the elongation (El) and reduction in area (RA) in 20-mm rolled plates from an AZ31B-type alloy produced from ultrasonically treated (1) and conventionally cast (2) ingots.

TABLE 7.22

Influence of Ultrasonic Melt Processing during DC Casting on Mechanical Properties of Hot-Rolled Plates and Cold-Rolled and Annealed Sheets from an AZ31B-Type Alloy

			Longitudinal/Transversal		
Product	Thickness, mm	UST	UTS, MPa	YS, MPa	El, %
Plate	30	−	277/282	161/172	14/14.7
	30	+	280/284	162/175	18/17.1
	8	−	266/292	188/171	8.9/21.1
	8	+	294/289	188/169	19.8/21.4
Sheets	2.5	−	286/324	228/267	3.1/14.3
	2.5	+	307/322	243/257	10.9/17.2
	1.5	−	308/314	216/247	13.7/16.6
	1.5	+	308/313	231/248	16.2/18.9
	0.8	−	302/315	206/237	17.0/18.0
	0.8	+	297/305	208/224	17.9/19.4

TABLE 7.23

Mechanical Properties of Plates from an AZ31B-Type Alloy in the Height Direction

Casting Technology	Grain Size, mm	UTS, MPa	YS, MPa	El, %	RA, %
With UST	0.2–0.3	302–327	300–322	15.5–18.6	17.2–25.2
Without UST	2–5	282–322	271–311	9.8–18.7	11.1–21.2

TABLE 7.24

Cracking Occurrence in Welds of AZ31B-Type Alloys of Different Purity (CP: Fe < 0.04%; HP: Fe < 0.005%) and Obtained from Ingots Produced with Ultrasonic Processing (US)

Sheet Material	Filling Wire	Crack Occurrence, %
CP	CP	85
CP	HP	45
CP	HP US	37
HP	HP	22
HP	HP US	7
HP US	HP	17
HP US	HP US	1–3

improved purity with ultrasonic treatment substantially reduces fracturing in welding, as shown in Table 7.24 [43].

In practical terms, these results show that the grain refinement of an ingot cast with ultrasonic melt processing is retained in the deformed metal (sheet, wire) and improves its weldability. To achieve improved weldability, a filling wire made from such a metal would suffice.

Mechanical tests of the metal treated by ultrasound during casting revealed a shift of the zero-ductility temperature to higher temperatures as well as an increase of ductility in this temperature range (Figure 7.26). It appeared that the alloy cast with ultrasonic melt processing also exhibited a smaller linear thermal-expansion coefficient in the temperature range 300°C–400°C, i.e., 26.4×10^{-6} K^{-1} versus 30.7×10^{-6} K^{-1} [24]. The tested alloy (AZ31B) can form a nonequilibrium eutectic at

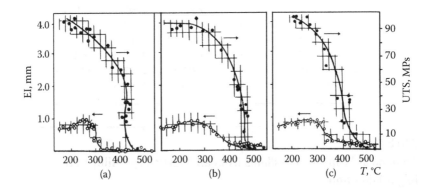

FIGURE 7.26 Effect of iron concentration in an AZ31B-type alloy and ultrasonic melt processing during casting on mechanical properties (absolute elongation and ultimate tensile strength) in the brittle temperature range relevant to welding: (a) commercial purity (Fe ≤ 0.04%), (b) high purity (Fe ≤ 0.005%), and (c) high purity and ultrasonic melt processing.

338.5°C, and the increase of the (apparent) linear thermal-expansion coefficient in the conventionally cast metal in the temperature range 300–400°C can be associated with the formation of a liquid phase, which spreads solid grains apart because of its larger specific volume. The grain refinement induced by ultrasonic treatment decreases the volume of this liquid phase, thus reducing the linear expansion coefficient.

Nondendritic crystallization in Al-free magnesium alloys and fine nondendritic grains may yield an even greater hereditary effect. This can be illustrated by the properties of forged pieces from a MA14 alloy (Mg–Zn–Zr) DC cast into 204-mm billets with either dendritic (conventional casting) or nondendritic (ultrasonic melt processing) structure. The billets were upset with a deformation degree of 85%, and the mechanical properties are listed in Table 7.25. The mechanical properties in forgings produced from nondendritic billets improved considerably.

Ultrasonic melt processing of high-strength and heat-resistant magnesium alloys with rare-earth metals like VMD7 (Mg–Nd–Zn–Zr–Cd) not only refines the grain structure, but also greatly refines intermetallic particles, which increases the ductility of the as-cast metal and improves the properties of the deformed metal. This is illustrated in Table 7.26. The impact toughness increased from 3.5 to 4.5 J/cm^2 at 20°C, from 6.6 to 8.8 J/cm^2 at 200°C, and from 10.8 to 11.8 J/cm^2 at 250°C [34].

TABLE 7.25

Mechanical Properties of MA14 T5 Forgings Produced from Dendritic and Nondendritic Billets

Casting Technology	Grain Size, µm	Sample Direction	UTS, MPa	YS, MPa	El, %	RA, %
With UST	30	Height	287	130	12.6	29.0
		Radial	305	255	21.3	38.1
Without UST	>100	Height	260	100	5.0	...
		Radial	300	250	19.0	...

TABLE 7.26

Mechanical Properties of 65-mm Rods from 174-mm VMD7 Billets

	Mechanical Properties at Testing Temperature								
	20°C			200°C			300°C		
Casting	UTS, MPa	YS, MPa	El, %	UTS, MPa	YS, MPa	El, %	UTS, MPa	YS, MPa	El, %
With UST	320	240	11.6	245	140	20.9	205	133	32.6
Without UST	265	170	11.0	240	145	20.0	195	140	28.4

7.5 EFFECT OF NONDENDRITIC GRAIN STRUCTURE ON HOMOGENIZATION

Homogenization or solution treatment of large-scale billets and ingots is a lengthy and energy-expensive procedure. The purpose of homogenization is to: dissolve nonequilibrium phases that have been formed as a result of solidification; transfer solute elements into the solid solution for future precipitation hardening; precipitate dispersoids of transition metals for recrystallization control; change the morphology of Fe-, Si-, and Mn-containing intermetallics to a more compact shape; and improve the ductility during deformation. All the processes occurring during homogenization are diffusion controlled, and therefore their duration depends directly on the diffusion length, which is a function of structure fineness.

The ultrasonic melt processing that results in the formation of nondendritic grain structure also allows for a considerably shorter homogenization time without a loss of ductility. This is the direct result of the increased specific surface area of grains, providing more paths for faster diffusion and finer eutectic colonies and phases.

Table 7.27 shows the advantage of nondendritic structure with an example of 960-mm billets from an AA7175-type alloy [19, 20]. The homogenization time can be reduced by up to five times as compared to the standard regime.

The production of high-quality extrusions from 6XXX-series alloys requires control of (AlFeSi) intermetallics. There are two main forms of these intermetallics occurring in aluminum alloys: $\beta(Al_5FeSi)$ that forms needles and plates and $\alpha(Al_8Fe_2Si)$ that forms "Chinese script"-like particles [44]. High-temperature annealing can transform β into α, with a corresponding improvement in mechanical properties, deformation behavior and, eventually, surface quality [45]. The ultrasonic melt processing refines the intermetallic particles, especially those that form at high temperatures and grow in the liquid environment (see Chapter 6).

We performed experiments with an A6063 alloys containing 0.5% Si, 0.3% and 0.7% Mg, and 0.02% Ti. Billets 60 mm in diameter were cast in a steel mold. Ultrasonic melt treatment was performed while also introducing 0.01%–0.07% Ti from an Al5Ti1B grain-refining master alloy. After such processing, the grain

TABLE 7.27
Effect of Nondendritic Structure on Homogenization Duration of 960-mm Billets from an AA7175-Type Alloy

	Elongation in the Temperature Range of Hot Deformation (350°C–400°C), %	
Duration of Homogenization, h	Nondendritic	Dendritic
6	75–80	60–65
16	75–80	60–65
30 (standard regime)	>80	70–75

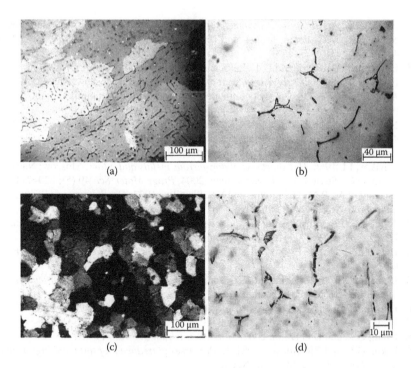

(a)

(b)

(c)

(d)

FIGURE 7.27 Grain structure and morphology of the α(AlFeSi) phase in an AA6063 alloy after high-temperature annealing at 570°C for 2 h: (a, b) casting without ultrasonic melt processing and (c, d) casting with ultrasonic melt processing.

structure was nondendritic with a grain size of 60 μm. The control billet cast without ultrasonic processing and Al5Ti1B master alloy had a dendritic grain structure with a grain size of 500–1000 μm.

The formation of nondendritic structure with the significantly increased specific intergrain surface changes the conditions for the formation of high-temperature eutectics containing (AlFeSi) phases. The particles become thinner and discontinuous, i.e., α(AlFeSi) particles represented by long veins and Chinese script larger than 10 μm in size are substituted with globules 1–7 μm in size. The effect resembles that produced by an increased cooling rate. It is estimated that the use of ultrasonic melt processing to cast 100–300-mm billets equals a two- to threefold increase in the cooling rate with regard to the refinement of eutectic phases. In this case, high-temperature annealing (570°C–580°C, 2 h) does not change the morphology and size of the particles, apart from some fragmentation (Figure 7.27).

REFERENCES

1. Grandfield, J.H., D.G. Eskin, and I.F. Bainbridge. 2013. *Direct-chill casting of light alloys: Science and technology*. Hoboken, NJ: John Wiley and Sons.
2. Eskin, D.G. 2008. *Physical metallurgy of direct-chill casting of aluminum alloys*. Boca Raton, FL: CRC Press.

3. Eskin, G.I. 1965. *Ultrasonic treatment of molten aluminum.* Moscow: Metallurgiya.
4. Nadella, R., D.G. Eskin, Q. Du, and L. Katgerman. 2008. *Progr. Mater. Sci.* 53 (3): 421–80.
5. Eskin, D.G., A. Jafari, and L. Katgerman. 2011. *Mater. Sci. Technol.* 27:890–96.
6. Eskin, G.I. 1988. *Ultrasonic treatment of molten aluminum.* Moscow: Metallurgiya.
7. Atamanenko, T.V. 2010. Cavitation-aided grain refinement in aluminium alloys. PhD thesis, Delft, Delft University of Technology.
8. Nadella, R., D.G. Eskin, and L. Katgerman. 2008. *Metall. Mater. Trans.* A 39 (2): 450–61.
9. Eskin, D.G., R. Nadella, and L. Katgerman. 2008. *Acta Mater.* 56 (6): 1358–65.
10. Novikov, I.I. 1966. *Hot shortness of non-ferrous metals and alloys.* Moscow: Nauka.
11. Eskin, D.G., Suyitno, and L. Katgerman. 2004. *Progr. Mater. Sci.* 49 (5): 629–711.
12. Livanov, V.A., R.M. Gabidullin, and V.S. Shepilov. 1977. *Continuous casting of aluminum alloys.* Moscow: Metallurgiya.
13. Teitel, I.L. 1961. In *Wrought aluminum alloys*, ed. I.N. Fridlyander, V.I. Dobatkin, and E.D. Zakharov, 200–9. Moscow: Oborongiz.
14. Dobatkin, V.I., G.I. Eskin, and S.I. Borovikova. 1973. *Fiz. Khim. Obrab. Mater.*, no. 6:37–41.
15. Dobatkin, V.I., G.I. Eskin, and S.I. Borovikova. 1976. In *Light and high-temperature alloys*, 151–61. Moscow: Nauka.
16. Dobatkin, V.I., and G.I. Eskin. 1991. *Tsvetn. Met.*, no. 12:64–67.
17. Eskin, G.I. 2002. *Z. Metallkde.* 93:503–7.
18. Eskin, G.I., and D.G Eskin. 2002. *Mater. Sci. Forum* 396–402:77–82.
19. Eskin, G.I., and P.R. Silayev. 1981. In *Processes of treatment of light and high-temperature alloys*, 118–22. Moscow: Nauka.
20. Yunyshev, V.K., G.I. Eskin, and P.R. Silayev. 1981. *Tekhnol. Legk. Spl.*, no. 2:26–30.
21. Anonymous. 2003. *Light Metal Age*, February:47–49.
22. Turkina, N.I., and E.V. Semenova. 1992. *Technol. Legk. Spl.*, no. 1:57–59.
23. Yelagin, V.I., V.V. Zakharov, and T.D. Rostova. 1995. In *Metals science, casting and processing of alloys*, 6–16. Moscow: VILS.
24. Eskin, G.I. 1998. *Ultrasonic melt treatment of light alloy melts.* Amsterdam: Gordon and Breach Science Publ.
25. Hoff, J.C. 1970. US Patent 3,634,075.
26. Granger, D.A. 1998. In *Light metals*, ed. B. Welch, 941–52. Warrendale, PA: TMS.
27. Greer, A.L, A.M. Bunn, A. Tronche, P.V. Evans, and D.J. Bristow. 2000. *Acta Mater.* 48:2823–35.
28. Eskin, G.I., A.A. Rukhman, S.G. Bochvar, V.I. Yalfimov, and D.V. Konovalov. 2008. *Tsvetn. Met.*, no. 3:105–10.
29. Eskin, G.I., S.G. Bochvar, and V.I. Yalfimov. 2010. *Tekhnol. Legk. Spl.*, no. 1:38–43.
30. Bochvar, S.G., and G.I. Eskin. 2012. *Tekhnol. Legk. Spl.*, no. 1:9–17.
31. Guryev, I.I., G.I. Eskin, and A.E. Ansyutina. 1978. In *Magnesium alloys*, 136–40. Moscow: Nauka.
32. Guryev, I.I., G.I. Eskin, and A.E. Ansyutina. 1986. In *Metals science: Casting and processing of light alloys*, 114–26. Moscow: VILS.
33. Eskin, G.I. 1994. In *Technology of processing light and special alloys*, 136–48. Moscow: VILS.
34. Eskin, G.I. 2003. *Metallurgist* 47:265–72.
35. Shao, Z., Q. Le, Z. Zhang, and J. Cui. 2011. *Mater. Design* 32:4216–24.
36. Shvetsov, P.N., V.I. Lyakhova, and T.A. Vlasova. 1973. *Tekhnol. Legk. Spl.*, no. 9:13–17.
37. Eskin, G.I., and P.N. Shvetsov. 1977. In *Metals science and casting of light alloys*, 8–17. Moscow: Metallurgiya.

38. Eskin, G.I. 1996. In *Advances in sonochemistry*, vol. 4, ed. T. Mason, 101–59. Oxford, UK: JAI Press.
39. Eskin, G.I., V.G. Kudryashov, P.N. Shvetsov, Z.K. Kuzminskaya, I.A. Skotnikov, and A.D. Petrov. 1990. *Izv. Akad. Nauk SSSR, Met.*, no. 1:53–57.
40. Eskin, G.I., and S.I. Borovikova. 1993. *Metalloved. Termich. Obrab. Mater.*, no. 6:18–20.
41. Sinyavsky, V.S., V.D. Valkov, and V.D. Kalinin. 1986. *Corrosion and protection of aluminum alloys*. Moscow: Metallurgiya.
42. Ansyutina, A.N., I.I. Guryev, and G.I. Eskin. 1973. *Tekhnol. Legk. Spl.*, no. 9:34–37.
43. Guryev, I.I., G.I. Eskin, A.B. Karan, A.E. Ansyutina, and T.N. Osokina. 1975. *Tekhnol. Legk. Spl.*, no. 12:25–29.
44. Belov, N.A., A.A. Aksenov, and D.G. Eskin. 2002. *Iron in aluminum alloys: Impurity and alloying element*. London: Taylor and Francis.
45. Kuijpers, N.C.W., J. Tirel, D.N. Hanlon, and S. van der Zwaag. 2002. *Mater. Charact.* 48:379–92.

8 Ultrasonic Melt Processing during Shape Casting of Light Alloys

8.1 TECHNOLOGICAL APPROACHES TO SHAPE CASTING WITH ULTRASONIC PROCESSING

Shape casting using ultrasonic processing was historically the first commercial application of ultrasonic treatment, as witnessed by two monographs published in 1965 [1] and 1967 [2] that described the technological schemes and practical experience in investment casting of aluminum alloys. The advantage of using ultrasound in investment casting lies in the relatively small volumes that are required to be processed. On the other hand, the ultrasonic treatment during solidification is applicable only to small castings; otherwise, issues with structure uniformity may arise. Ultrasonic processing for casting large parts should be performed using one of the ex-mold schemes when the liquid metal rather than the semisolid slurry is treated. Different mechanisms are involved, depending on the technological scheme (see Chapter 5).

Precision investment casting is usually done in ceramic molds, where solidification proceeds at a considerably lower rate than in direct-chill casting. Therefore, the grain structure and properties of shape castings are inferior to those of ingots and billets. However, many parts, such as turbine discs and fans, are frequently fabricated by precision casting that allows one to simultaneously form a bulky hub and thin-walled blades without machining. Therefore, the control of the grain structure and properties of cast metal and the filling conditions for thin channels down to 0.1 mm in cross section was a scientific and practical challenge back in the 1960s. It was demonstrated that this problem could be solved by ultrasonic treatment of the solidifying melt. Some of these approaches can also be extended to other technologies of shape casting.

Two methods were regularly used in precision investment casting: pressurized solidification in autoclaves (for improved density) and vacuum suction of melt in the mold (for improved filling) [2]. Ultrasonic tools were fitted in the existing casting equipment.

One system is shown in Figure 8.1a. This installation was designed for casting under pressure with simultaneous sonication of the solidifying alloy. The chamber was made as an autoclave with a sealed lid and a commercially produced 2.5-kW magnetostrictive transducer working at a frequency of 19–20 kHz with a titanium sonotrode mounted in the bottom part. The prepared and dried mold, with a gating system, was coupled with the sonotrode, as shown in Figure 8.1b. The ultrasonic

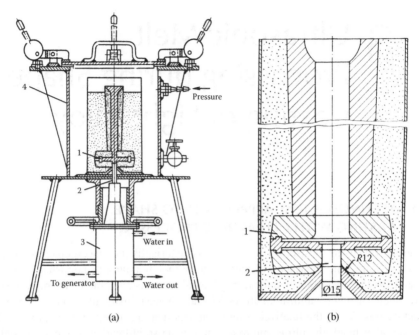

(a)   (b)

FIGURE 8.1   A system for pressurized shape casting (in autoclaves) in ceramic molds with ultrasonic treatment: (a) general view and (b) section view of a ceramic mold for casting a turbine wheel with a sonotrode inserted from beneath: (1) mold, (2) sonotrode, (3) ultrasonic transducer, and (4) autoclave.

transducer was switched on; the melt was poured into the downgate; the lid was rapidly closed; and the autoclave was pressurized. The duration of ultrasonic processing was about 1 min, and the time spent under pressure was 5–6 min. After depressurization, the lid was opened, and the mold and the casting were removed from the autoclave. In order to prevent sticking of the casting onto the sonotrode, the latter was covered with graphite lubricant. Such a system allowed precision casting of parts up to 92 mm in diameter, as shown in Figure 8.2 [3].

FIGURE 8.2   Precision investment castings of aluminum alloys produced using ultrasonic processing during solidification in the mold.

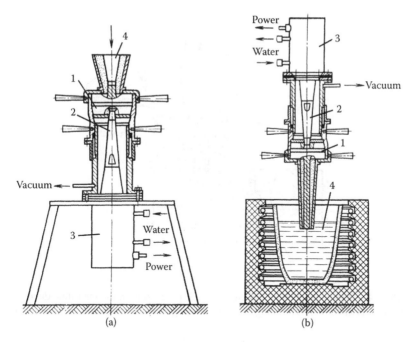

FIGURE 8.3 Schematic diagrams of vacuum suction systems for precision-investment casting with ultrasound introduced into solidifying melt: (a) melt is fed from above and (b) melt is fed from below: (1) mold, (2) sonotrode, (3) ultrasonic transducer, and (4) melt.

Figure 8.3 gives schematic diagrams of vacuum-assisted mold feeding systems, with ceramic casting molds coupled to ultrasonic magnetostrictive transducers in which 50–200-mm-diameter turbine wheels were cast. These systems differ in the method of feeding metal in the mold: either from the top or from the bottom upwards. In the latter case, the assembled mold was held over a crucible containing the melt, and then the transducer was switched on and the spout was lowered into the liquid metal. Once the spout was immersed, the mold was evacuated to about 0.05 MPa, and the melt filled the mold. After 0.5–1.0 min of solidification under ultrasonic field, the sonotrode was removed from the body of the casting, and the assembly was raised and put on a special stand to push the casting out of the mold. Using this approach, the main acting mechanisms are improved filling due to the sonocapillary effect (see Sections 4.1 and 4.2) and enhanced grain refinement due to fragmentation (see Section 5.5). In order to employ other mechanisms such as degassing (see Chapter 3) and activation of substrates (see Sections 5.2–5.4), the ultrasonic processing should be performed before pouring melt into the mold, i.e., in an intermediate volume or in the flow.

Several schemes were suggested for extra-mold ultrasonic processing of the melt. One of the generic ideas is to apply ultrasonic melt treatment (UST) in an intermediate volume, either in a feeder or in a gating system, as shown in Figure 8.4. In this case, the ultrasonic treatment can be combined with introduction of a grain-refining

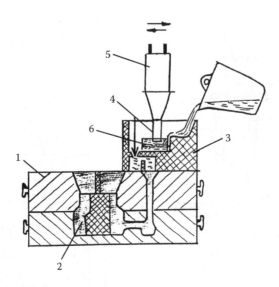

FIGURE 8.4 Principal scheme of ultrasonic melt processing in the melt flow during mold (die) filling: (1) mold box, (2) casting, (3) holding chamber, (4) sonotrode, (5) transducer, and (6) melt.

or -modifying master alloy rod, similar to the ultrasonic melt processing in a launder during DC casting.

An original scheme of ultrasonic melt processing using transversal rather than longitudinal oscillations was suggested by Levy et al. [4]. The setup is presented in Figure 8.5. Casting can be performed in vacuum (as shown), under pressure, and in protective or normal atmosphere. The waveguiding system consists of two half-wavelength supports excited by ultrasonic transducers that are acoustically connected to a sonotrode that also acts as a stopper in a crucible. The transducers excite transversal (bending) oscillations in the sonotrode, with potentially several antinodes of maximum amplitude, where cavitation conditions are met. The crucible is filled with melt and sonicated for a given time to ensure degassing (in the case of reduced or atmospheric pressure) and melt processing. After that, the ultrasound is switched off and the sonotrode is lifted opening the outlet in the bottom of the crucible through which the melt fills the mold. A similar scheme of sonication was recently suggested by Prokic et al. [5, 6], where a specially designed clamp allows the use of a Sialon (ceramic) sonotrode.

An interesting approach was suggested for high-pressure die casting (HPDC) [7]. In this case, the ultrasonic processing is performed in a shot sleeve of an HPDC machine, as shown in Figure 8.6. The ultrasonic system consisted of a 600-W generator, a water-cooled 19.5-kHz transducer, and a Sialon sonotrode 20 mm in diameter. The ultrasonic intensity was about 190 W/cm^2, which corresponded to an amplitude of 30 μm. The sonotrode was preheated to 700°C, and the treatment time in successful trials was 2–10 s for 210 g of melt at a melt temperature of 600–680°C.

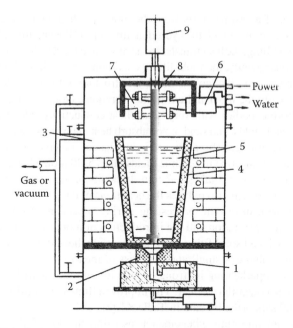

FIGURE 8.5   Principal scheme of a setup with ultrasonic melt processing using transversal oscillations: (1) mold, (2) feeder, (3) chamber, (4) crucible, (5) melt, (6) transducer, (7) waveguide and clamp, (8) sonotrode, and (9) motor for vertical movement.

8.2   EFFECTS OF ULTRASONIC PROCESSING ON SOLIDIFICATION IN MOLDS

The ultrasonic processing during solidification of small-scale castings affects the temperature field, the solidification rate, and the structure of the castings as well as mold filling.

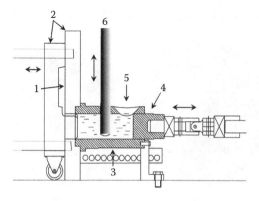

FIGURE 8.6   Diagram showing application of ultrasonic processing during high-pressure die casting: (1) die, (2) frame, (3) shot sleeve with heating, (4) plunger, (5) melt pouring spout, and (6) sonotrode. (Adapted from Khalifa, Tsunekawa, and Okumiya [7].)

The formation of a casting starts from the moment when molten metal enters the mold. When metal fills the mold, the hydrodynamics of the flowing metal has a decisive effect on the mold filling, the heat transfer to the mold, and the density of the solid metal. The flow and, consequently, the mold filling conditions are facilitated by vacuum and ultrasonic vibration. After the mold is filled, the structure of the formation is governed by the heat transfer to the mold and ultrasonic vibrations. Dried ceramic molds are, in fact, capillary-porous bodies with very small heat conductivity. Metal molds absorb heat faster than sand molds, and sand molds absorb heat many times faster than ceramic molds for precision casting. Laboratory and industrial data on precision aluminum castings indicate that solidification in ceramic molds takes 5–10 min, whereas solidification in sand molds of similar size takes 1–2 min. In addition, the ceramic molds are preheated to 80–120°C, which also reduces the heat-transfer rate.

Experiments with ultrasonic processing during precision casting into ceramic molds indicated that cavitation and the introduction of more than 100 W of acoustic power (into the melt and solidifying alloy) substantially heated the metal in the mold. Since the casting volume is small and the mass of melt in the mold hardly exceeds 0.2 kg, ultrasonic superheating is strong enough to improve the filling of thin channels through the sonocapillary effect. The improved thin-section filling is illustrated in Figure 8.7 and supported by data in Table 8.1 [8].

Cavitation is an important phenomenon assisting in mold filling, as shown in Table 8.2 [8]. The filling of thin sections under cavitation conditions depends on the purity of the melt with respect to dissolved hydrogen and insoluble inclusions. The purer the melt, the better is the filling. When the hydrogen concentration is below 0.1 cm^3/100 g (e.g., as a result of ultrasonic degassing), the mold filling increases by a factor of 3.

The analysis of cooling curves recorded in different part of an A356 precision casting indicates that the acoustic energy introduced by ultrasonic treatment slows down melt cooling and the initial stage of solidification by a factor of 6–8, from 120–170 K/min to 14–35 K/min, as seen in Figure 8.8. This slowing down of solidification in the slurry stage of the casting allows for better fragmentation action during ultrasonic cavitation. In HPDC [7], the ultrasonic processing is reported to decrease nucleation undercooling in an ADC12 alloys from 2.9 to 0.4 K.

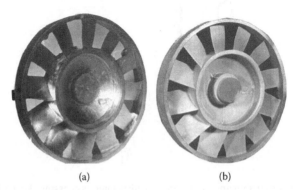

(a)   (b)

FIGURE 8.7 Castings of turbine disk from an A356 alloy produced (a) without ultrasonic processing and (b) with ultrasonic processing.

TABLE 8.1
Effect of Ultrasonic Processing of an Al–Cu–Si Alloy on Filling Thin Casting Channels (%)

	Casting with Ultrasound			Casting without Ultrasound		
Casting Temperature, °C Channel Thickness, mm	720°C	750°C	780°C	720°C	750°C	780°C
1.0	86	95	100	79	86	100
0.7	72	75	88	62	68	84
0.5	35	37	45	22	25	39

TABLE 8.2
Effect of Cavitation on Filling Thin Casting Channels (%)

Cavitation Mode	Precavitation		Threshold	Cavitation	
Amplitude, μm Channel thickness, mm	0	1	3	9	13
1.0	56	58	60	65	70
0.7	38	39	40	50	58
0.5	20	21	25	30	35

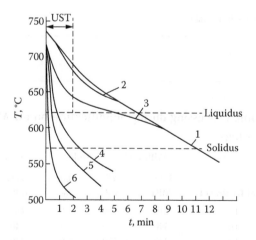

FIGURE 8.8 Cooling curves taken in different sections of an A356 casting similar to that shown in Figure 8.7: (1, 4) center of the casting, (2, 5) blade base, (3, 6) blade body—(1–3) with ultrasonic processing and (4–6) without ultrasonic processing.

8.3 EFFECT OF ULTRASONIC PROCESSING ON STRUCTURE AND PROPERTIES OF SHAPE CASTINGS

Application of ultrasonic processing during solidification of small-scale castings improves the quality of the casting while also providing significant improvements in grain refinement and mechanical properties.

Table 8.3 shows the grain size obtained in castings from different aluminum alloys. The castings are similar to those in Figures 8.2 and 8.7. The nondendritic grain structure can be obtained in casting sections close to the sonotrode. This change in the grain structure has an effect on the strength and ductility characteristics of the castings, as illustrated in Table 8.4.

TABLE 8.3
Effect of Ultrasonic Processing during Solidification on the Grain Size in Castings of Aluminum Alloys

	Grain Size, µm	
Sampling Area	Conventional Casting	With UST in the Mold
A332.2-type (Al–3.5% Cu–9.5% Si–0.25% Mg–0.2% Ti–0.2% Mn)		
Central part	710	80[a]
Base of blade	510	140
Blade	500	240
A224-type (Al–4.8% Cu–0.8% Mn–0.2% Ti)		
Central part	800	75[a]
Base of blade	600	80[a]
Blade	530	140

[a] Nondendritic grains.

TABLE 8.4
Effect of Ultrasonic Processing during Solidification on the Mechanical Properties of Castings

		With UST in the Mold			Conventional Casting		
Alloy	Heat Treatment	UTS, MPa	YS, MPa	El, %	UTS, MPa	YS, MPa	El, %
A356	T5	210	180	5.0	180	160	4.0
A332.2	T6	390	360	2.5	340	310	1.0
A224	T5	340	270	5.0	280	220	3.5

Note: UTS = ultimate tensile strength; YS = yield strength; El = elongation.

Two-stage ultrasonic processing, i.e., melt processing for degassing and then melt or slurry processing for grain refinement and structure modification, is one of possible technological approaches to achieve maximum effect. Figure 8.9 gives a statistical distribution of mechanical properties for A356 (7% Si, 0.3% Mg) and RR53C (2.5% Si, 1.5% Cu, 0.5% Mg, 1.1% Fe, 1% Ni, 0.2% Ti) alloys cast with and without ultrasonic melt processing. The maxima of strength and elongation shift to greater values after ultrasonic treatment, especially in the RR53C alloy that contains Ti. Turbine disks are working at very high rotation speeds and are subjected

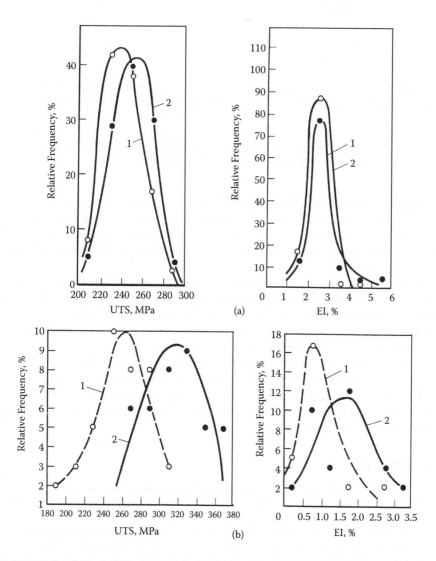

FIGURE 8.9 Statistical distribution of tensile strength and elongation in cast samples from A356 (a) and RR35C (b) alloys subjected to (1) Ar degassing and (2) Ar degassing and ultrasonic melt processing.

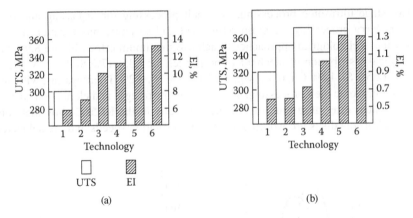

FIGURE 8.10 Effects of inoculation, modification, and ultrasonic treatment of melt in a crucible and during solidification on the mechanical properties of alloys: (a) A356-type and (b) A393-type after casting in metallic molds: (1) without inoculation and ultrasound, (2) without inoculation but with UST in a crucible, (3) same as (2) but with UST during solidification, (4) with inoculation but without ultrasonic treatment, (5) same as (4) but with UST in a crucible, (6) same as (5) but with additional UST during solidification.

to huge centrifugal forces. The quality of the casting and high mechanical properties are a must. As a result of combined ultrasonic treatment—melt degassing and pressurized casting with ultrasonic processing in the mold—the tensile strength of turbine disks from an RR53C alloy increased from 253 to 320 MPa and the elongation from 1.1% to 2%.

The ultrasonic melt processing can be combined with addition of grain refiners and structure modifiers. This was demonstrated in casting of A356-type (7% Si, 0.3% Mg) and A393-type (21% Si, 2.5% Cu, 2.4% Ni, 0.3% Mg, 0.2% Ti) alloys using a variety of processing technologies [9]. Ultrasonic treatment was combined with the addition of complex inoculants (Ti + Zr + B or Na + P + S) to the alloys, applied in both a crucible before casting and during solidification in a preheated metallic mold. The results are given in Figure 8.10 as a variation of mechanical properties.

In all studied castings, ultrasonic treatment reduced the size of grains and eutectic colonies and improved the ductility of the metal. This effect was especially strong when combined with grain refiner/modifier addition.

REFERENCES

1. Eskin, G.I. 1965. *Ultrasonic treatment of molten aluminum.* Moscow: Metallurgiya.
2. Eskin, G.I., V.I. Slotin, and S.Sh. Katsman. 1967. *Precision casting of parts for aviation assemblies from aluminum alloys.* Moscow: Mashinostroenie.
3. Ryzhenkova, M.P., G.I. Eskin, and M.B. Altman. 1974. In *Application of new physical methods for intensification of metallurgical processes, Proc. MISiS,* no. 77:170–73.
4. Levy, L.I., L.B. Maslan, Yu.I. Kitaygorodsky, et al. 1972. Russian invention certificate no. 343765, 07.07.1972.

5. Prokic,] M. 2001. European patent application EP 1238715A1.
6. Puga, H., J. Barbosa, J. Gabriel, E. Seabra, S. Ribeiro, and M. Prokic. 2011. *J. Mater. Proc. Technol.* 211:1026–33.
7. Khalifa, W., Y. Tsunekawa, and M. Okumiya. 2010. *J. Mater. Proc. Technol.* 210:2178–87.
8. Eskin, G.I. 1979. *Ultrasonic melt treatment in processes of shape and continuous casting of light alloys.* Moscow: Mashinostroenie.
9. Altman, M.B., G.I. Eskin, I.S. Gotsev, and N.P. Stromskaya. 1981. In *Alloying and processing of light alloys*, 30–35. Moscow: Nauka.

9 Ultrasonic Processing of Composite and Immiscible Alloys

9.1 ULTRASOUND-ASSISTED INTRODUCTION OF NONMETALLIC PARTICLES AND FIBERS TO LIQUID METALS

Metal-matrix composite (MMC) materials (in recent years also metal-matrix nano-composite materials [MMNC]) have attracted the attention of researchers and industry since the 1950s. Composite materials with micron-scale particles demonstrate improved properties based on the mixture rule, i.e., increased hardness, elastic modulus, strength, size, and thermal stability. The amount of reinforcement introduced varies typically between 5 and 20 vol% or wt%. Nanosized particles have the advantage of a lesser required load (<1 wt%) and additional strengthening on the microscopic level, e.g., by Orowan hardening and the pinning of grain boundaries.

There are many technological ways to produce a composite material, e.g., powder metallurgy, spray deposition, and preform impregnation. Some of these have reached the pilot or industrial scale. At the same time, all of them have limitations in the size and/or shape of components. A liquid-metal route, where a slurry of particles dispersed in the liquid phase can be cast in any shape or in large billets, is a promising technological way of composite manufacturing. This technological approach, however, is hindered by fundamental issues of wettability and agglomeration of particles, with such issues becoming more serious as the particles become finer. Ultrasonic processing found a notable role in overcoming the challenges of liquid-metal composite manufacturing, though it has not yet reached the industrial level. The first attempts to make composite materials using ultrasonic waves and cavitation date back to the 1950s, when tungsten carbide particles were dispersed in liquid lead using 3-kW 21.5-kHz transducer [1]. Thorough investigations on dispersion of high-melting metals and ceramic particles in liquid metals, including aluminum, were performed in the 1960s [2]. These works have reported some innovative ideas on techniques to introduce particles in the melt (see Figure 1.4).

The production of composite materials with particles (in most cases ceramic compounds) comprises several steps:

Preliminary stage:
1. Preparation of particles with the aim of improving the access of the melt to their surface (drying, surface treatment, coating, etc.)

2. Selection of the matrix alloy (additions for surface tension control, grain refining)
3. Selection of the method of feeding the particles into the melt

Main stage:
4. Introduction of particles (combination of mixing techniques, ultrasound parameters, melt temperature, duration of processing, size and geometry of the processed volume, protective atmosphere)
5. Dispersion of particles (combination of mixing techniques and their parameters, melt temperature, duration of processing)

Final stage:
6. Casting and solidification (melt/slurry transfer to the mold, cooling rate)

We will touch upon these aspects in this section, though the focus will be on ultrasonic processing.

The first two aspects are general and are important for any introduction of particles into the liquid phase. As wetting is the major issue, any measures taken to decrease the surface tension and increase the wetting (in addition to ultrasound-induced wetting) are beneficial. The preliminary treatment of particles such as coating them with metals (e.g., Cu, Ni), oxidation, degreasing, and drying is recommended in all cases when it is possible (see reviews [3–5]). The surface of the particles should preferably be free of adsorbed gas and moisture and, hence, relatively smooth. The degassing and degreasing facilitate the access of the liquid phase to the solid particles. The coating improves wettability by reacting with the liquid phase through either melting or forming a new phase with a better wettability. On the side of the liquid phase, elements that are known to decrease the surface tension are frequently added, especially in liquid aluminum, which has high surface tension. The following elements lower the surface tension of liquid aluminum (in descending order per at.%): Bi, Pb, Sb, Li, Ca, Mg, Sn [6]. Out of these elements, addition of Mg is most frequently used, as it does not have the drawbacks of the other additions. In addition, Mg dissolved in liquid aluminum would react with alumina film on the surface of many particles to form spinel and through this reaction improve wetting. Solute titanium may improve wettability of carbon particles and fibers [7]. Among the elements that decrease the surface tension of liquid Mg are Bi, Sb, and Pb [6]. However, liquid magnesium has lower surface tension than aluminum, and additional alloying is not normally used for the purposes of wetting improvement.

The next step in composite production is the selection of a means of delivering the particles into the melt. Let us look at the techniques reported for ultrasonic processing of composites. In many cases, the primary introduction of the particles is done by mechanical mixing using an impeller with subsequent or simultaneous sonication. The impeller design should follow some basic rules [8–11]. In particular, it is recommended to keep the ratio between the impeller diameter d and the mixing vessel diameter D close to 0.3; otherwise the porosity in the composite increases significantly. Similarly, the ratio of the impeller height to its diameter should be kept within

0.35, which allows one to keep the rotation speed low and reduce air entrapment. The distance of the impeller from the bottom of the mixing vessel should be about 0.2–0.3 of the liquid height level. These values are recommended for forming solid suspensions in the liquid phase, provided that the particles are at least partially wetted by the liquid phase. In this case, axial movement of the liquid is preferred. When the particles are poorly wetted, some shearing is advisable; hence a radial component in the liquid flow is useful. In this case, an increase in the previously mentioned ratios is possible. There are also schemes where the impeller is inclined to allow greater freedom in the use of additional devices, e.g., a feeding tube or sonotrode.

The simplest technique of introducing the particles is spraying them onto the surface of the melt using a trough or a tube. These particles are then drawn into the bulk of the melt by vortex (in the case of impeller) or by gravity and surface flows [12–16]. This technique works quite well in magnesium MMCs but have limitations in aluminum MMCs due to the strong oxide film at the melt surface. A combination of the impeller and a protective atmosphere is required.

Particles can be wrapped in a metallic foil (e.g., in aluminum foil for Al MMCs) to form sort of a compact rod that is then slowly fed into the cavitation region, where the foil dissolves to release the particles, thereby exposing them to cavitation [17, 18]. A double wrap in aluminum foils of different thicknesses has been suggested as a means of controlling the release rate [19]. A special feeder device can be used to deliver particles into the cavitation region. Such a system using a worm-type feeder and Ar-atmosphere protection of particles was developed and tried in Mg alloys [20]. The selection of material for the feeding tube is important and may be cumbersome, especially for aluminum. For Mg alloys, steel can be used.

The delivery of particles in the cavitation region can also be achieved using a perforated container (e.g., from Nb) placed underneath the sonotrode [21]. In this case, the geometry of the container as well as the number and size of holes control the release rate.

A next logical step would be to use a kind of master alloy containing a metallic matrix with a large concentration of particles, similar to grain-refining rods. The particles can be spatially distributed in such a composite master alloy and wetted by the matrix. The master alloy can be produced by a powder-metallurgy route, using mechanical alloying for mixing and hot extrusion for consolidation and better particle distribution. Such a scheme was first suggested in the 1960s and implied the direct contact of a sintered aluminum powder (SAP) rod with the sonotrode [2] (see Figure 1.4b). This approach was later used in grain-refining practice (see Section 5.3) and tried in MMNC processing [14, 22]. The application of hollow sonotrodes has also been suggested for introduction of the particles into the melt and in the cavitation region [2, 23, 24] (see Figure 1.4a).

All of these techniques have been tried on the laboratory scale and, obviously, have advantages and disadvantages. The finer the particles, the more complicated their introduction to the melt is. In the case of microscopic (5–100 μm) particles, mechanical or electromagnetic mixing with subsequent or simultaneous ultrasonic processing is sufficient [13, 25]. For example, SiC particles (<20 μm) were introduced into the melt of hypoeutectic (7% Si) and hypereutectic (12.5% Si) alloys by mechanical mixing close to the liquidus temperature [13, 26]. Then the suspension

(2 kg) was heated to 750°C–760°C and subjected to electromagnetic stirring with simultaneous ultrasonic processing (18 kHz, 1 kW acoustic power) for 3–5 min. Then the composite material was cast into a graphite mold. Examination of the structure showed that the grain size decreased 20 times as compared with the composite materials produced by only mechanical stirring. The distances between the SiC particles varied between 2.5 and 9 μm. The agglomerates typical of mechanically stirred composite were completely eliminated. Liquid processing of nanocomposite materials is much more complicated and specific to the nature of the particles.

The processing of a liquid–solid particle mixture is at the heart of the composite technology. Let us look at some examples when ultrasonic cavitation was successfully used for production of metal-matrix composites based on aluminum and magnesium. It is worth noting here that the ultrasonic processing should be performed in a cavitation regime that creates conditions for particle wetting, deagglomeration, and dispersion, as described in Sections 4.1 and 5.3. It was shown that ultrasonic cavitation processing ensures good wetting of 20–40-μm particles of oxides, carbides, and borides in the matrices of aluminum casting alloys [27].

The main parameters of any processing are the volume and the time. This is true for ultrasonic processing as well. The wetting of particles and their dispersion through the liquid volume improve with the time of processing. The following relationship between the size of particles and the duration of ultrasonic processing was established for homogeneous distribution of 1 wt% of γAl_2O_3 in 1 kg of 99.99% pure aluminum at 700°C–720°C at an acoustic power of 500 W (the cavitation threshold is approximately 10 W) [18]:

Particle diameter, μm	80–100	10–20	1–10	0.1–10	0.01–0.1
Processing time, min	1–2	3–5	8–10	12–15	30–45

As one can see, there is a direct correlation between the particle size and the processing time, quite understandable if we recall Figure 5.9 showing that the forces that hold agglomerated particles together increase with particle refinement. In addition, larger specific surface area of fine particles promotes gas adsorption on their surface and makes them float. And finally, this large specific surface makes wetting of the particles much more difficult to achieve, as the difficulties of wetting and liquid access to the particle surface multiply with particle number density.

One of the ways to incorporate small particles into the melt while preventing their flotation and rejection is to make their addition in the semisolid state. The density of the semiliquid metal increases, and the particles can be mixed in using a mechanical impeller. This was demonstrated using AZ91D alloys with addition of 1 to 3 vol% of carbon nanofibers (CNF) (100 nm × 10 μm) [28]. Slurry was prepared by pouring the melt through a cooled launder at 595°C into a crucible. Though the authors did not report the volume of the treated material, it seems that it hardly exceeded 200–300 g. The slurry was then agitated in the crucible using a mechanical turbine at 1750 rpm with simultaneous addition of the reinforcing fibers on the surface. After the end of addition, the mixing was continued for 20 min at 588°C–593°C, where homogeneous distribution of CNF agglomerates was achieved, and then ultrasonic oscillations were applied with an amplitude of 18–20 μm for about 2 min to disperse the agglomerates. Intermittent 3-min processing for 0.5–1.5 min with 15-s resting intervals was shown to be most efficient. When the amplitude was decreased to 16 μm, the processing needed to be continued for 10 min. Finally, the semisolid composite material was solidified by cooling the crucible in water. Another version of this processing route was reported for additions of 6 vol% SiC (25–85 nm, 45 nm av.) to a Mg–18% Zn alloy [29]. The particles were added to a slurry at 575°C using a vortex generated by a rotating blade ($d/D = 0.6$) at 1400 rpm. After this stage, the composite material was heated to the liquid state and particles were dispersed by an ultrasonic sonotrode (20 kHz, null-peak amplitude 30 μm), and the melt was then solidified at a cooling rate of 1 K/s. The treated volume and processing time were not reported, but the volume was rather small (crucible diameter 42.5 mm). A uniform distribution of SiC particles was achieved, with most of the particles spaced at 50–200 nm.

If semisolid processing of composites helps in primary introduction of particles into the matrix, the increased melt temperature can be used to improve wettability and decrease the viscosity of the melt. It is well known that the wettability of many ceramic particles increases with melt temperature [30]. Ultrasonic cavitation, acoustic attenuation, and acoustic flows are facilitated by lower liquid viscosity (see Chapter 2, Equations 2.30, 2.33, 2.34 and Figure. 2.21), which contributes to the efficiency of ultrasonic cavitation processing of composite materials.

Figure 9.1 illustrates the effect of ultrasonic processing on distribution of TiB_2 particles measuring 5–7 μm. A total of 2 wt% of particles was added in 1 kg of an Al–5% Mg–0.05% Ti alloy. The structures shown are characteristic of the three stages of processing: (a) after adding the particles into the vortex during mechanical stirring (400 rpm) for 10 min; (b) after subsequent ultrasonic processing at 720°C for 5 min (frequency 17.5 kHz, null–peak amplitude 30 μm, Nb sonotrode); and (c) after final additional mechanical stirring for 5 min. In all cases, the casting was done in a metallic mold. Although TiB_2 is wetted rather well by liquid aluminum (and the matrix with Mg further facilitates the wetting), mechanical stirring alone results in coarse agglomerates of particles. Ultrasonic cavitation helps in dispersing these agglomerates and in uniform distribution of the particles in the volume. When the ultrasonic processing is finished, particles tend to sediment to the bottom of the crucible, as they are heavier than liquid aluminum. Additional mechanical stirring just before casting helps to overcome this issue.

The practice for nanosized particles is still reserved for small volumes and long processing times. For example, addition of up to 2 wt% SiC particles (<30 nm) to 1 kg of A356 melt using addition of particles through a tube to the melt surface and ultrasonic processing takes about 1 h [14]; addition of 1 wt% Al_2O_3 (25 nm) particles to 250 g of A356 melt using a double-foil technique and cavitation processing takes 45 min [19]. With magnesium alloys, the times can be shorter, as wetting is much easier in liquid magnesium: 15 min is sufficient for adding up to 2.7 wt% TiB_2 particles (25 nm) to 130 g of AZ91D melt using a special feeder [20] or for adding 5 wt% Al_4C_3 particles (<44 μm) to 2 kg of AM60B melt using a Nb cage [21].

Figure 9.2 demonstrates some examples of aluminum- and magnesium-based nanocomposites produced using ultrasonic processing in the liquid state [15, 16].

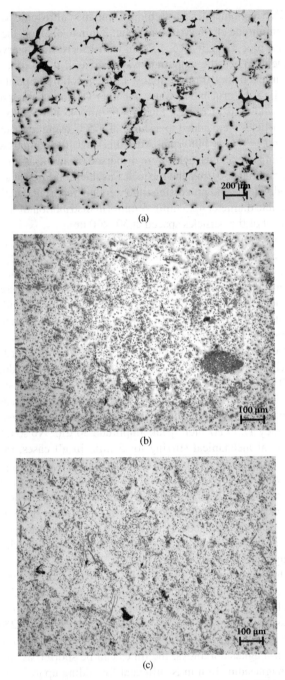

FIGURE 9.1 Microstructure of an Al-based MMC with 2 wt% TiB$_2$ with particle size 5–7 μm: (a) after adding the particles into the vortex during mechanical stirring; (b) after subsequent ultrasonic processing at 720°C for 5 min; and (c) after final additional mechanical stirring for 5 min. Casting in a metallic mold. (Courtesy Vadakke Madam, Brunel University.)

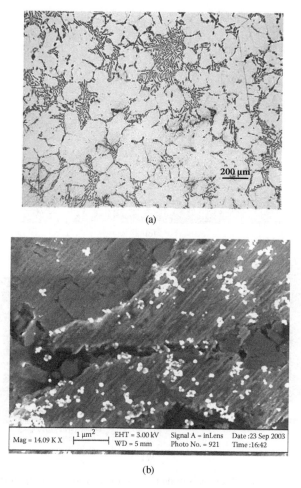

(a)

(b)

FIGURE 9.2 Examples of nanocomposites produced using ultrasonic processing: (a, b) A356+2 wt% SiC (30 nm). (After Yang, Lan, and Li [16].) (*continued*)

Despite relative success and good response of mechanical properties, a tendency for particle clustering persists. Also, most of the particles are pushed to the grain boundaries, whereas the target is to distribute the particles uniformly within the body of the grain.

There is not much data on the effect of ultrasonic parameters on particle distribution. Most of the successful experiments were performed under conditions of developed cavitation with amplitudes of 10 to 40 μm at frequencies 17–22 kHz. It seems that there is a consensus that cavitation is the basis of successful ultrasonic processing of composites, especially at the stage of wetting and deagglomeration. Physical modeling using transparent solutions and mixtures and variable ultrasonic parameters (17–20 kHz, 1.4–4 kW) demonstrated that a better distribution of particles in

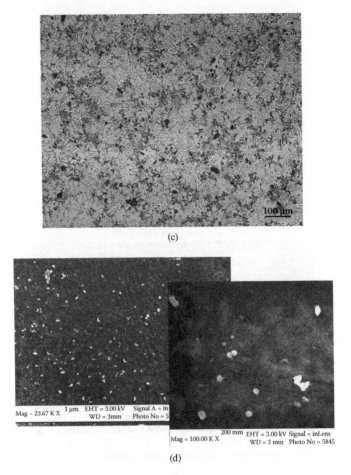

(c)

(d)

FIGURE 9.2 (*continued*) Examples of nanocomposites produced using ultrasonic processing: (c, d) Mg–2% Al–1% Si+2 wt% SiC (50 nm). (After Cao, Konishi, and Li [15].) General view (a, c) and nanoparticles (b, d).

the volume could be achieved at a higher frequency and lower power [31]. In this case, the reagglomeration is prevented and particles are well distributed by acoustic flows in the volume. The results were validated by producing AZ91D–0.5% SiC (40 nm) composite by ultrasonic processing at 20 kHz, 1.4 kW for 6 min.

The next step in producing a nanocomposite is solidification, where a uniform particle distribution in the liquid phase should be retained in the solid matrix. The general rule for capturing the particles during solidification is that the solidification front should move at a velocity exceeding a certain critical value. This critical velocity can be conditionally written as [32]:

$$v_{cr} = \frac{A}{d_p^{(1.5-2)}},$$ (9.1)

where A is a constant dependent on the particle and matrix properties and d_p is the particle diameter.

Or it can be written as [33]:

$$v_{cr} = \frac{0.157}{\mu} \Delta\sigma_{cls}^{2/3} \sigma_{sl}^{1/3} \left(\frac{a}{r_p}\right)^{4/3},$$ (9.2)

where μ is the dynamic viscosity of the liquid; σ_{cls} and σ_{sl} are the interfacial energies between ceramic particle (c), solid metal (s) and liquid metal (l); a is the atomic diameter of atoms in the liquid; and r_p is the particle radius.

Or it can be written as [34]:

$$v_{cr} = \frac{H_{sys}}{36\pi r_p \mu_{eff} L_{cr}},$$ (9.3)

where H_{sys} is the Hamaker constant that describes the van der Waals force between a nanoparticle and the solid phase; μ_{eff} is the effective viscosity of the liquid between the particle and the solid phase; and L_{cr} is the critical distance between the particle and the solid phase.

The analysis performed Xu et al. [34] shows that the critical velocities for 10-nm TiB$_2$ particles vary between 220 and 10^5 µm/s (depending on the model), while the solidification front progresses at a rate of several µm/s in a normal casting process, which is sufficient for much larger particles (>100 µm) [33]. The Al–TiB$_2$ and Mg–TiB$_2$ systems have Hamaker constants of −0.6 and 0.960, respectively, which makes them suitable as a base for MMNC. In contrast, the Al–Al$_2$O$_3$ system has a Hamaker constant of −8.68, with the consequence of a critical velocity that is 14.5 times greater (3200 vs. 220 µm/s for 10 nm). Three possible nanoparticle capture processes were suggested [34]:

Viscous capture when the solidification front advances at a velocity larger than a certain critical value. Extremely high cooling rates are generally needed for nanoparticle capture in most metal–nanoparticle systems, making this mechanism impractical. To overcome this issue, it is possible to affect other parameters in Equation (9.3), e.g., to increase the viscosity of the system so that the nanoparticle can be successfully captured. This can be done by decreasing the melt temperature or increasing the loading of nanoparticles. For example, the increase in volume fraction of 10-nm nanoparticles from 5% to 7% increases the apparent viscosity by almost two orders of magnitude. Another promising way involves the synthesis of new nanoparticles with a metallic shell to increase the Hamaker constants of the nanoparticles with respect to the liquid matrix.

Brownian capture requires the nanoparticles to overcome an energy barrier before being captured. In this case, the selection of particles with a Hamaker constant lower than that of the liquid matrix but close to it (e.g.,

TiB$_2$, SiC in Al, ZrO$_2$ in Mg) or a decrease of the nanoparticle size (e.g., Al$_2$O$_3$ smaller than 2 nm in Al) can be beneficial.

Finally, when the van der Waals potential is negative, *spontaneous capture* may occur, e.g., in the Mg–SiC or Mg–TiB$_2$ systems. In this case, the Hamaker constant of the particles should be higher than that of the liquid matrix. For Al matrices, coating of particles with Cr, Co, Ni, and Cu may help to solve the problem.

9.2 MECHANICAL PROPERTIES OF COMPOSITE MATERIALS PRODUCED USING ULTRASONIC PROCESSING

Early results on metal-matrix composite materials produced using ultrasonic processing showed improved mechanical properties. Vickers hardness of an Al–1% Al$_2$O$_3$ (1 µm) composite produced using the setup in Figure 1.4b showed a 200% increase at 100°C and a 90% increase at 200°C as compared to pure aluminum [2]. Mechanical properties of Al–Al$_2$O$_3$ composites produced using ultrasonic processing showed dependence on the amount of particles and their fineness, as illustrated in Figure 9.3 [12, 18]. Particles of either α or γAl$_2$O$_3$ were wrapped in aluminum foil and added to molten Al (99.99%) into the cavitation zone through an alumina tube with simultaneous stirring with a Nb impeller. The processed volume was 2.5 kg, and a 2.5-kW magnetostrictive transducer working at 20 kHz was used. The introduction of finer particles has an advantage for tensile strength and elongation. It was noted that the introduction of γAl$_2$O$_3$ into the liquid phase was easier due its better wettability, while the improvement of properties was better in the case of αAl$_2$O$_3$, as it was less hydroscopic and the composites had less porosity.

Ultrasound-aided addition of 10%–15% SiC (20–70 µm) to an Al–11% Si–5% Cu–0.15% Ti alloy increased the tensile strength at room temperature from 270 to 320 MPa and at 200°C from 230 to 300 MPa [17]. In a Mg alloy AM60B (Mg–Al–Mn), an addition of 5% of Al$_4$C$_3$ (<44 µm) resulted in a 23%–26% increase in yield strength and hardness and in a 12% improvement in elongation at room temperature, while at 177°C the yield strength increases by 11% and elongation by 28% [21]. The addition of Al$_4$C$_3$ microparticles induced fourfold grain refinement in the magnesium alloy.

The interest in composite materials has shifted in the twenty-first century to nanocomposites. Liquid-metal processing routes typically include ultrasonic processing as a main or supportive stage of introduction and dispersion of nanoparticles. The advantage of nanocomposite materials is in the much lower weight load of particles that is required to obtain good properties. In fact, the best results are obtained when 0.5–1 wt% of nanoparticles is added.

There is a lot of experimental data currently available on the mechanical properties of MMNC based on Al and Mg. Some of the results are compiled in Table 9.1. Note that in most cases, the treated volumes were rather small, ranging from 100 g to 2 kg.

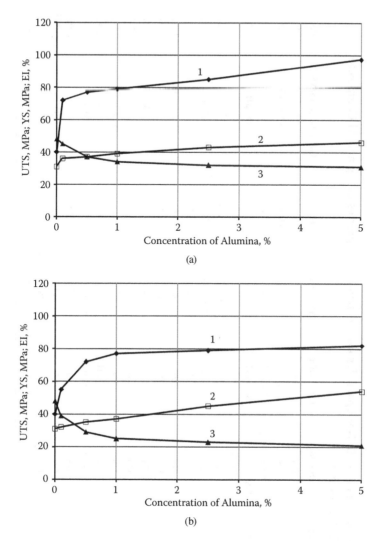

FIGURE 9.3 Effect of αAl₂O₃ concentration on mechanical properties of Al–Al₂O₃ composite materials produced using ultrasonic melt processing: (a) 0.01-μm particles and (b) 1-μm particles: (1) ultimate tensile strength (UTS), (2) yield strength (YS), and (3) elongation (El).

9.3 ULTRASONIC PROCESSING OF IMMISCIBLE ALLOYS AND METALLIC EMULSIONS

The emulsification of immiscible liquids under the action of ultrasound (US) has been known since the 1920s and was demonstrated for water and oil as well as water and mercury [43, 44]. The decisive role of cavitation in the process of emulsification was soon recognized [45]. Already in the mid-1930s, the first successful

TABLE 9.1

Reference Data on the Effect of Reinforcement with Nanoparticles on Mechanical Properties of Mg and Al Composite Materials Produced Using Ultrasonic Processing

Matrix	Particle	Amount, wt%	Technology	UTS, MPa	YS, MPa	El, %	Gain, % (UTS/YS/El)	Ref.
Mg	...	...	US	90	20	14	...	[35]
Mg	SiC, 50 nm	0.5/1	US	121/124	28/30	15.5/14	38;50;10	[35]
Mg	...	...	Friction stirring	370 µH	...	...	...	[36]
Mg	Graphene, 15 nm × 14 µm	1.2 vol%	US + friction stirring	660 µH	...	...	78 (µH)	[36]
AZ91	...	...	US	133	65	N/A	...	[37]
AZ91	SiC, 40 nm	0.5/1	US	180/190	140/140	N/A	43;115	[37]
AZ91	...	...	US	127	70	2	...	[38]
AZ91	SiC, 60 nm	1 vol%	US	220	90	8	73;28;300	[38]
Mg-2Al-1Si	...	...	US	138	49	9	...	[15]
Mg-2Al-1Si	SiC, 50 nm	2	US	157	74	7.5	14;50;-17	[15]
Mg-4% Zn	...	...	US	110	45	5	...	[39]
Mg-4% Zn	SiC, 50 nm	1.5	US	200	67	10	82;49;100	[39]
Mg-18% Zn	...	...	US	760 µH	N/A	N/A	...	[29]
Mg-18% Zn	SiC, 45 nm	6 vol%	SS+US	1830 µH	600 (est.)	N/A	140 (µH)	[29]
AZ91	...	...	US	160	100	3.5	...	[40]a
AZ91	AlN	1	US	185	145	3.0	15;35;-17	[40]a
AZ91D	...	...	US	160	88	2.9	...	[20]
AZ91D	TiB$_2$, 25 nm	1/2.7	US feeder	180/190	104/107	3.3/4.3	19; 21;48	[20]

Alloy	Particle		Processing	UTS	HV	El		Reference
Al	...	...	Reference	60	51 HV	46	...	[41]
Al	Al_2O_3, 10 nm	2	US through crucible	92	102 HV	34	53;100;–26	[41]
A356+ 0.5Cu								
A356+ 0.5Cu	...	...	US	150	97	3	...	[19]
A356+ 0.5Cu	γAl_2O_3, 25 nm	1	US, double foil	198	110	8	32;13;170	[19]
A356	...	...	US	270	110	2.5	...	[16]
A356	SiC, 30 nm	2	US	300	160	3.5	11;45;43	[16]
A356	...	...	US	N/A	N/A	5	...	[14]
A356	SiC, 30 nm	1/1.5	US	N/A	N/A	2.3/3	60;100;–40	[14]
Al–20% Si	...	...	US	85	137	1.1	...	[42]
Al–20% Si	γAl_2O_3, 50 nm	0.5	US, double foil	100	155	1.8	18;13;64	[42]

Note: UTS = ultimate tensile strength; YS = yield strength; El = elongation.

[a] Also reported 20%–25% improvement of tensile properties at 200°C.

experiments on ultrasonic (10 kHz) introduction of Pb in liquid Al and Cd in Al–Si melt were reported [46], and the possibility of producing emulsions of 7%–10% Pb in Al and Zn that were stable even upon remelting was demonstrated [47].

The physics of ultrasonic emulsification is considered elsewhere [48–52]. Some basic factors controlling the process can be summarized as follows. The size of droplets in the emulsion decreases with increasing ultrasonic frequency. At the same time, the higher frequency requires a greater sound intensity. With the increasing intensity and processing time, the emulsion concentration increases up to a certain value when saturation occurs. This saturation is a result of the equilibrium reached between the processes of emulsification (dispersion) and coagulation. A running sound wave is more efficient than a standing wave, with coagulation processes prevailing in the latter case. Precavitation sound processing results in deemulsification. Therefore, cavitation is the important requirement of the process. A low viscosity and a lesser difference in viscosities between the components facilitate emulsification. Additions of surfactants (decreasing the surface tension at the interface) and stabilizers (coating of droplet surface) promote the stability of the emulsion and allow for higher concentrations. In general, the emulsification occurs through local disturbances at the interface between two immiscible liquids, with typically only one liquid undergoing dispersion while the other liquid acts as the source of cavitation bubbles. The disturbance occurs during the expansion phase of the bubble oscillation, while the dispersion happens upon bubble collapse. Figure 9.4 illustrates the formation of a wave disturbance in liquid B caused by an expanding bubble in liquid A. When this bubble starts to contract, it draws the crest of the wave of liquid B upwards, and this crest extends with acceleration following the accelerated contraction of the bubble. When the bubble collapses, the crest disintegrates to form a droplet.

With respect to liquid metals and alloys, ultrasonic emulsification has practical value for the manufacture of free-machining and bearing alloys. These alloys contain additions of low-melting soft elements such as Pb, Bi, and Sn that have either a miscibility gap with Al that causes stratification (Al–Pb, Al–Bi) or have a very large solidification range that triggers gravity segregation (Al–Sn).

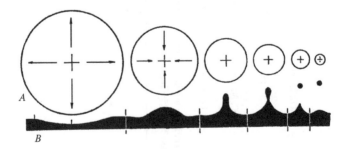

FIGURE 9.4   Schematic showing the mechanism of cavitation emulsification in two immiscible liquids.

Bearing alloys containing up to 10% Pb (additionally up to 10% Sn and up to 4% Sb) can be produced by the following route [32]. The melt is superheated to 1100°C–1200°C, which allows for dissolution of 18%–30% Pb in liquid aluminum. The melt is then poured through a water-cooled ultrasonic funnel (magnetostrictive transducer arranged around the pouring channel). The ultrasonic processing then occurs simultaneously with melt cooling. This creates conditions for nucleation of Pb droplets under intensive mixing that prevents sedimentation. The process ends with direct-chill (DC) casting of billets, where a high cooling rate helps to preserve the emulsion in the solid state. Lead particles 5–40 μm in size can be uniformly distributed in the billet volume. Figure 9.5 shows the

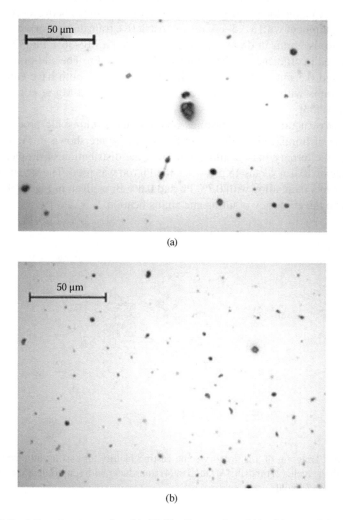

(a)

(b)

FIGURE 9.5 Microstructures of an Al–1% Pb alloy produced (a) without and (b) with ultrasonic processing in a crucible. (Courtesy P. Camean Queijo, Brunel University.)

results of a lab-scale experiment where the ultrasonic treatment of an Al–1% Pb alloy was done in a 0.5-kg crucible. This figure demonstrates the dispersion of the lead particles.

Another technological approach was suggested by G.I. Eskin and tested under laboratory conditions (casting of a 6XXX-series alloy in a metallic mold 95 mm in diameter, 300 mm in height, processed volume 5 kg). The idea was to avoid addition of the low-melting and immiscible components in the furnace, preventing contamination and eliminating the need for high melt superheat. Relatively small (up to 6%) additions of such elements were done using a master alloy or pure-metal rod with ultrasonic processing in the launder, as depicted schematically in Figure 9.6. Two ways of ultrasonic treatment were tried: from the bottom of the launder and from the top of the melt flow, the latter technique being more practical. Standard equipment with a 3.5-kW magnetostrictive transducer working at 18 kHz was used. Rods 9 mm in diameter were made from a Mg–36% Pb–31% Bi alloy and compared in efficiency with 6-mm rods from pure Pb. The addition rates were 1.4 mm/s and 0.5 mm/s, respectively. Lead and bismuth both have eutectic-type phase diagrams with Mg, and are well dissolved in liquid Mg at normal casting temperatures [53].

The microstructures of the castings obtained using ultrasonic processing with introduction of immiscible additions in the melt flow are shown in Figures 9.7a,b. The use of the ternary master alloys improves the distribution of droplets with an average size of 10 μm versus 15 μm when the Pb rod was used. The size distribution of lead droplets in an alloy with 0.7% Pb and 0.6% Bi is given in Figure 9.7c. These data confirm the efficiency of ultrasonic emulsification.

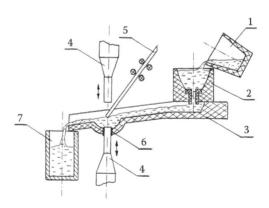

FIGURE 9.6 Diagram of a laboratory-scale casting facility for casting immiscible alloys: (1) ladle with the melt, (2) tundish, (3) launder, (4) transducer with sonotrode, (5) alloying rod, (6) sealing, and (7) mold.

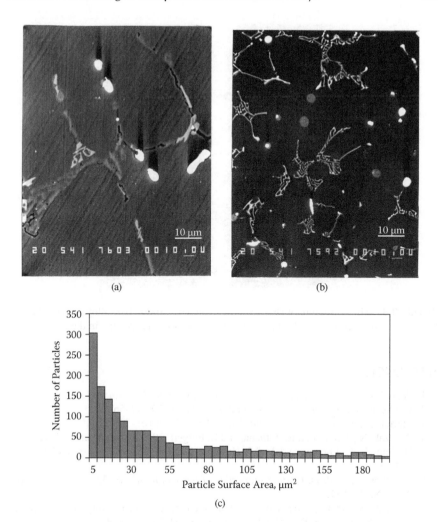

FIGURE 9.7 Microstructure of a 6XXX-series alloy with immiscible additions produced using ultrasonic processing according to scheme in Figure 9.6: (a) addition of 0.6% Pb with a Pb rod, (b) addition of 0.7% Pb and 0.6% Bi with a ternary Mg–Pb–Bi rod, and (c) size distribution of Pb droplets in the alloy (b).

The same approach was used in DC casting of a 6XXX-series alloy with addition of Pb in the melt flow with simultaneous ultrasonic processing of the melt. Figure 9.8 demonstrates good distribution of 10–20-μm lead droplets in the structure of a 6061-type alloy with 6% Pb.

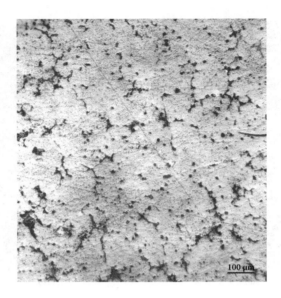

FIGURE 9.8 Microstructure of DC-cast billet from an AA6061-type alloy with 6% Pb produced with introduction of lead using a Pb rod and ultrasonic processing in the melt flow.

REFERENCES

1. Pogodin-Alekseev, I.G., and V.V. Zaboleev-Zotov. 1958. *Lit. Proizvod.*, no. 7:35–36.
2. von Seemann, H.J., and H. Staats. 1968. *Z. Metallde.* 39:347–56.
3. Mortensen, A., and I. Jin. 1992. *Int. Mater. Rev.* 37 (3): 101–28.
4. Sobczak, N., J. Sobczak, R. Asthana, and R. Purgert. 2010. *China Foundry* 7: 425–37.
5. Hashim, J., L. Looney, and M.S.J. Hashimi. 2001. *J. Mater. Process. Technol.* 119: 324–28.
6. Korolkov, A.M. 1960. *Casting properties of metals and alloys*, 32–75. Moscow: Izd. Akad. Nauk SSSR.
7. Sobczak, N. 2004. *Ceram. Trans.*, vol. 146, ed. K. Ewsuk, K. Nogi, M. Reiterer, A. Tomsia, S.J. Glass, R. Waesche, K. Uematsu, and M. Naito, 83–90.
8. Hentrich, P. (ed.). 2000. *Handbook of mixing technology fundamentals, process design, impellers and agitation systems, mechanical design, detail design, sealing technology, reliability and safety, fields of application.* Schopfheim, Germany: Ekato.
9. Hemrajani, R.R., and G.B. Tatterson. 2004. Mechanically stirred vessels. In *Handbook of industrial mixing: Science and practice*, ed. E.L. Paul, V.A. Atiemo-Obeng, and S.M. Kresta, 345–90. Hoboken, NJ: John Wiley & Sons.
10. Atiemo-Obeng, V.A., W.R. Penney, and P. Armenante. 2004. Solid–liquid mixing. In *Handbook of industrial mixing: Science and practice*, ed. E.L. Paul, V.A. Atiemo-Obeng, and S.M. Kresta, 543–84. Hoboken, NJ: John Wiley & Sons.
11. Aniban, N., R.M. Pillai, and B.C. Pai. 2002. *Mater. Design* 23:553–56.
12. Kolmakov, A.I., V.I. Silaeva, T.V. Solov'eva, and G.I. Eskin. 1972. In *Application of ultrasound in industry*, 21–24. Alma-Ata, Kazakhstan: NTO Mashprom.
13. Eskin, G.I., and D.G. Eskin. 2003. *Ultrason. Sonochem.* 10:297–301.
14. Yang, Y., and X. Li. 2007. *J. Manuf. Sci. Eng.* 129: 497–501.
15. Cao, G., H. Konishi, and X. Li. 2008. *Mater. Sci. Eng. A* 486:357–62.
16. Yang, Y., J. Lan, and X. Li. 2004. *Mater. Sci. Eng. A* 380: 378–83.

17. Sidorin, I.I., V.I. Silaeva, T.V. Solov'eva, V.I. Slotin, and G.I. Eskin. 1971. *Metalloved. Termich. Obrab. Met.*, no. 8:23–26.
18. Eskin, G.I. 1974. *Tekhnol. Legk. Spl.*, no. 11:21–25.
19. Choi, H., M. Jones, H. Konishi, and X. Li. 2012. *Metall. Mater. Trans. A* 43A: 738–46.
20. Choi, H., Y. Sun, B.P. Slater, H. Konishi, and X. Li. 2012. *Adv. Eng. Mater.* 14:291–95.
21. Nimityuongskul, S., M. Jones, H. Choi, R. Lakes, S. Kou, and X. Li. 2010. *Mater. Sci. Eng. A* 527:2104–11.
22. Wang, D., M.P. De Cicco, and X. Li. 2012. *Mater. Sci. Eng. A* 532:396–400.
23. von Seemann, H.J., and H. Staats. 1962. German patent DE1126074(B), publ. 22.03.1962.
24. Eskin, G.I., and P.N. Shvetsov. 1969. USSR invention certificate 246060, publ. 01.01.1969.
25. Tsunekawa, Y., H. Sizuki, and Y. Genma. 2001. *Mater. Design* 22:467–72.
26. Eskin, G.I., B.I Semenov, and D.N. Lobkov. 2001. *Lit. Proizvod.*, no. 9:2–8.
27. Sidorin, I.I., V.I. Slotin, T.V. Solov'eva, and G.I. Eskin. 1972. *Izv. Vyssh. Uchebn. Zaved., Mashinostr.*, no. 8:116–20.
28. Mussi, R.G.S., T. Motegi, F. Tanabe, H. Kawamura, K. Anzai, D. Shiba, and M. Suganuma. 2006. *Solid State Phenom.* 116–117:392–96.
29. Chen, L.Y., J.-Y. Peng, J.-Q. Xu, H. Choi, and X.-C. Li. 2013. *Scr. Mater.* 69:634–37.
30. Naidich, Ju.V. 1981. *Progr. Surf. Membr. Sci.* 14:353–484.
31. Shiying, L., G. Feipeng, Z. Qiongyuan, and L. Wenzhen. 2009. *Mater. Sci. Forum* 618–619:433–36.
32. Abramov, O.V. 2000. *Influence of powerful ultrasound on liquid and solid metals*, 142–63. Moscow: Nauka.
33. Kaptay, G. 2001. *Metall. Mater. Trans. A* 32A:993–1005.
34. Xu, J.Q., L.Y. Chen, H. Choi, and X.C. Li. 2012. *J. Phys.: Condens. Matter* 24:paper 255304.
35. Cao, G., H. Konishi, and X. Li. 2008. *J. Manuf. Sci. Eng.* 130:paper 031105.
36. Chen, L.-Y., H. Konishi, A. Fehrenbaher, C. Ma, J.-Q. Xu, H. Choi, H.-F. Xu, F.E. Pfefferkorn, and X.-C. Li. 2012. *Scr. Mater.* 67:29–32.
37. Jia, X.Y., S.Y. Liu, F.P. Gao, Q.Y. Zhang, and W.Z. Li. 2009. *Int. J. Cast Metals Res.* 22 (1–4): 196–99.
38. Nie, K.B., X.J. Wang, X.S. Hu, L. Xu, K. Wu, and M.Y. Zheng. 2011. *Mater. Sci. Eng. A* 528:5278–82.
39. de Cicco, M., H. Konishi, G. Cao, H.S. Choi, L.-S. Turng, J.H. Perepezko, S. Kou, R. Lakes, and X. Li. 2009. *Metall. Mater. Trans. A* 40A:3038–45.
40. Cao, G., H. Choi, J. Oportus, H. Konishi, and X. Li. 2008. *Mater. Sci. Eng. A* 494:127–31.
41. Mula, S., P. Padhi, S.C. Panigrahim, S.K. Pabi, and S. Gosh. 2009. *Mater. Res. Bull.* 44:1154–60.
42. Choi, H., H. Konishi, and X. Li. 2012. *Mater. Sci. Eng. A* 541:159–65.
43. Wood, R.W., and A.L. Loomis. 1927. *Phil. Mag.* 4:417–25.
44. Richards, W.T. 1929. *J. Am. Chem. Soc.* 51:1724–29.
45. Bondy, C., and K. Söllner. 1935. *Trans. Faraday Soc.* 31:835–42.
46. Schmid, G., and L. Ehret. 1937. *Z. Elektrochem.* 43:869–74.
47. Bekker, V. 1941. *Novosti Tekhniki* 10 (4): 25–26.
48. Neduzhy, S.A. 1961. *Akust. Zh.* 7:275–94.
49. Neduzhy, S.A. 1964. *Akust. Zh.* 10:456–64.
50. Abismaıl, B., J.P. Canselier, A.M. Wilhelm, H. Delmas, and C. Gourdon. 1999. *Ultrason. Sonochem.* 6:75–83.
51. Behrend, O., K. Ax, and H. Schubert. 2000. *Ultrason. Sonochem.* 7:77–85.
52. Behrend, O., and H. Schubert. 2001. *Ultrason. Sonochem.* 8:271–76.
53. Fedorov, P.I., V.I. Shachnev, and A.M. Dolgopolova. 1962. *Izv. Vyssh. Uchebn. Zaved., Tsvetn. Metall.*, no. 2:58–64.

10 Rapid Ultrasonic Solidification of Aluminum Alloys

Formation of nondendritic structure requires solidification conditions where numerous solidification centers can form and grow without dendrite branches. This was discussed in Chapter 5, in particular in Section 5.6. Figure 5.20 demonstrates that the dendrite arm spacing and, indeed, the nondendritic grain size similarly depends on the cooling rate. Ultrasonic melt processing is one of the ways to form such nondendritic grain structures; another is rapid solidification [1, 2].

Rapid solidification allows for very high thermal undercooling with potential homogeneous nucleation and spontaneous nondendritic solidification when the energy required for the nucleation becomes less than the energy required for branching. Not all alloys can be easily undercooled to such a degree. Nondendritic structure due to undercooling as a result of high cooling rate was obtained in Ni alloys upon melt atomization with cooling rates above 10^6 K/s [3]. Combination of rapid solidification with ultrasonic melt processing has been investigated as a potentially efficient approach since the 1980s [4]. Atomization of liquids by ultrasonic oscillations has been reported in the first works of Wood and Loomis. The physics of this process is well covered elsewhere [5, 6]. Ultrasonic-aided atomization when the Ni-based melt was dispersed by an oscillating Ar or He flow at 16–20 kHz allowed the formation of nondendritic granules 50–100 µm in size at cooling rates about 10^6 K/s [7].

It is difficult to obtain a sufficiently high thermal undercooling in aluminum alloys; therefore, the formation of nondendritic structure upon rapid solidification/ atomization relies on the enhanced heterogeneous solidification and requires cavitation melt processing before or during atomization.

10.1 BASIC SONICATION SCHEMES FOR RAPID SOLIDIFICATION

Ultrasonic atomization of liquids using an oscillating gas jet (so-called liquid whistles or Hartmann whistles) is one of the ways to obtain fine solid particles and is widely used in various industries. This method can be used for atomizing metallic melts, but it is difficult to achieve the required multiplication of solidification substrates via undercooling for aluminum alloys so that a nondendritic grain structure can be formed. Therefore, atomization for rapid solidification and producing small granules, powder, or fibers should be combined with ultrasonic melt processing to generate numerous solidification substrates and formation of nondendritic structures.

Here we will focus on the atomization when the liquid flow is either preprocessed by ultrasonic cavitation or put in direct contact with an oscillating sonotrode.

Rapidly solidified granules or flakes can be obtained using thin-layer ultrasonic spray casting installations, as those shown in Figures 10.1a,b. The melt (up to 10 kg), superheated 100°C–150°C above the liquidus temperature, is fed through a distribution funnel with a 4–6-mm opening onto the working end of a sonotrode that can be round or made in the form of an oscillating plate. The stream of melt droplets (maximum size 2 mm) passes through a flow of cooling inert gas (helium and/or argon) into a collecting bin containing liquid nitrogen. If the droplets come directly to this bin, then the final product is nearly spherical pellets of 50–150 μm diameter (Figure 10.2a). Alternatively, if the droplets arrive in the collecting bin after colliding with a rotating water-cooled drum, then the final product is platelets of the same size.

Thin strips up to 4 mm and fibers up to 0.2 mm in cross section can be cast in installations shown in Figures 10.1c,d, where a superheated melt (up to 10 kg) is subjected to ultrasonic treatment (18 kHz, null-peak amplitude above 20 μm) right before rapid solidification. Experiments using these installations were conducted with Al–6%

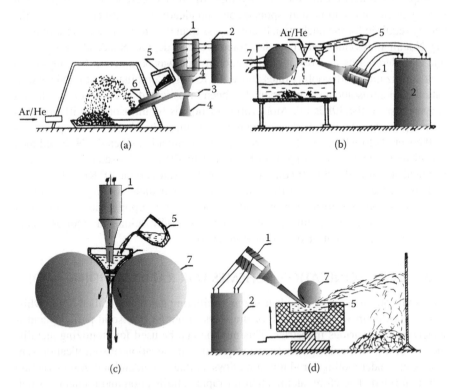

FIGURE 10.1 Schematics of setups for the rapid solidification of aluminum alloys with ultrasonic melt processing: (a, b) for making granules, (c) for making strips, and (d) for making fibers: (1) ultrasonic transducer, (2) generator, (3) flat sonotrode, (4) sonotrode, (5) melt, (6) feeder/distributor, and (7) water-cooled roll or drum.

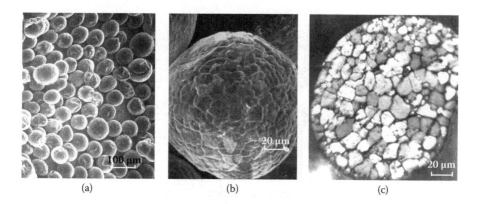

(a)   (b)   (c)

FIGURE 10.2   Structure of granules from an Al–Cu–Zr alloy produced by ultrasonic atomizing in Ar (Figure 10.1a): (a) general appearance, ×100; (b) a single granule, ×500, and nondendritic grain structure, ×500.

Cu–(0.4–3.5)% Zr and Al–(8–8.2)% Zn–(2–2.6)% Cu–(2.4–3)% Mg–(0.4–1.4)% Zr alloys. The nondendritic structure was formed in granules (Figure 10.2) and flakes from aluminum alloys subject to the Zr concentration and completeness of cooling. Similar setups are described by Grant [8] and Pohlman, Heisler, and Cichos [9].

Figure 10.3 shows a prototype of an atomizing plant where the melt is treated by ultrasound (US) immediately before being atomized by a gas jet.

10.2   FORMATION OF NONDENDRITIC STRUCTURE IN RAPID SOLIDIFICATION

Although a nondendritic structure can be formed in granules, uneven cooling in inert gas sometimes results in the formation of both types of structures, as shown in Figure 10.4.

An increase in the Zr concentration in the studied alloys, given the same cooling rate and spraying parameters, facilitates the formation of nondendritic grain structure (Figure 10.5). The small-size fractions of granules have a greater proportion of nondendritic grains than larger-size fractions. This reflects the fact that the equilibrium peritectic point shifts as the cooling rate increases. For Al–Zn–Mg–Cu–Zr alloys, the saturation point corresponds to 1% Zr, whereas for the Al–Cu–Zr alloys, the saturation point is 1.3% Zr. At these Zr concentrations, the number of primary intermetallic particles increases sharply, which, in combination with cavitation, results in the greatly increased amount of solidification substrates for aluminum. Additional cooling of the solidifying droplets in the Ar/He flow extends the range of particle sizes that can act as solidification sites. Consequently, the proportion of nondendritic grains rises.

The nondendritic structure with the grain size smaller than 5 μm was also obtained in strips from Al–Zn–Mg–Cu–Zr alloys and in thin (<200 μm) fibers from Al–Cu–Zr alloys (Figure 10.6). These results confirm the dependence of the nondendritic grain size on the cooling rate, as shown in Figure 5.20.

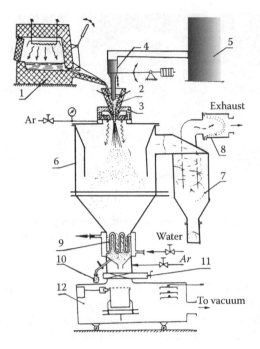

FIGURE 10.3 A technological scheme for pelletizing aluminum alloys by cavitation treatment of the melt in an intermediate container and acoustic atomization in an argon atmosphere: (1) furnace containing the melt, (2) intermediate melt container and casting spout, (3) ultrasonic gas atomizers (liquid whistle), (4) magnetostrictive transducer, (5) ultrasonic generator, (6) atomizing chamber, (7) granule-separating cyclone, (8) filter, (9) water-cooling system, (10) sampling tap, (11) shutter, and (12) granule collector.

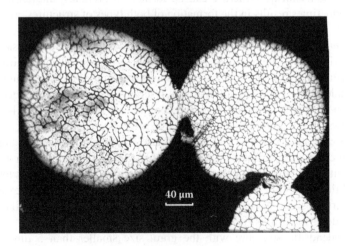

FIGURE 10.4 Microstructure of two fused granules produced by ultrasonic atomizing in Ar showing coexistence of dendritic (left) and nondendritic (right) grain structures due to uneven cooling.

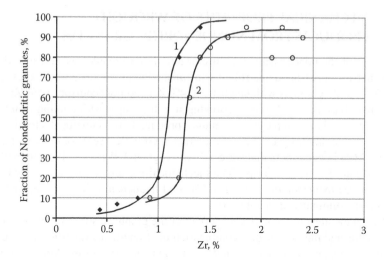

FIGURE 10.5 Effect of Zr concentration in an alloy on the fraction of granules (≤50 µm) with nondendritic structure produced by ultrasonic atomization: (1) Al–Cu–Zr alloys and (2) Al–Zn–Cu–Mg–Zr alloys.

FIGURE 10.6 Fibers (0.2-mm diameter) of an Al–Cu–Zr alloy with nondendritic grain size ≤5 µm obtained by spinning with ultrasonic melt treatment (see Figure 10.1d).

10.3 MECHANICAL PROPERTIES OF WROUGHT PRODUCTS FROM NONDENDRITIC GRANULES

The effect of the nondendritic structure of aluminum granules was estimated from the properties of extruded and rolled semifinished items produced from these granules. Al–Cu–Zr granules with dendritic (0.4% Zr) and nondendritic (1.8% Zr) grain structures obtained by ultrasonic spraying in a thin layer with cooling in nitrogen were selected. For comparison, granules produced by centrifugal atomizing in water—those containing 1.5% Zr— were taken. Granules were compacted at 420°C and 760 MPa in containers that were 320 mm high and 95 mm in diameter and then extruded into 5 × 60-mm strips. The mechanical properties of the extruded strips after heat treatment (solution treatment 550°C, 1 h; quenching in water; aging 170°C, 6 hours) are summarized in Table 10.1.

These data attest to the noticeable effect of the granules' structure on the properties of deformed metal. The best combination of strength and ductility was found in extruded strips produced from granules with nondendritic grains. After rolling, the sheets made from nondendritic granules demonstrate superplasticity facilitated by the very fine recrystallized grain size (annealing at 530°C), as shown in Table 10.2.

TABLE 10.1

Mechanical Properties of Al–Cu–Zr Strips Produced from Different Granules

Granules Produced by	Longitudinal Direction				Transverse Direction			
	UTS, MPa	YS, MPa	El, %	RA, %	UTS, MPa	YS, MPa	El, %	RA, %
US atomizing (dendritic)	365	300	17.0	25.5	460	295	27.0	25.0
US atomizing (nondendritic)	485	335	15.5	24.0	460	360	15.0	25.0
Centrifugal atomizing (dendritic)	445	300	11.7	19.5	335	335	4.5	12.5

Note: UTS = ultimate tensile strength; YS = yield strength; El = elongation; RA = reduction in area.

TABLE 10.2

Structure and Ductility of Sheets from an Al–Cu–Zr Alloy Produced from Different Granules

Granules Produced by	Grain Size, μm		El, %
	Longitudinal	Transversal	
US atomizing (dendritic)	500	32	7
US atomizing (nondendritic)	10	8	164–200
Centrifugal atomizing (dendritic)	100	25	50

We can conclude that ultrasonic processing in the cavitation mode promotes the formation of a nondendritic structure in rapidly solidified granules and flakes. Wrought items produced from such rapidly solidified products have an advantage in mechanical properties and are capable of superplastic deformation.

REFERENCES

1. Matyia, H., B.C. Giessen, and N.J. Grant. 1968. *J. Inst. Met.* 96:30–32.
2. Grant, N.J. 1992. *Metall. Trans. A* 23:1083–93.
3. Patterson, R.I., A.R. Cox, and E.C. van Reuth. 1980. *JOM* 32 (9): 34–39.
4. Savage, S.J., and F.H. Froes. 1984. *JOM* 36 (4): 20–33.
5. Abramov, O.V. 1998. *High-intensity ultrasonics*, 237–63. Amsterdam: Gordon and Breach OPA.
6. Ensminger, D., and L.J. Bond. 2011. *Ultrasonics: Fundamentals, technologies, and applications*, 533–40. Boca Raton, FL: CRC Press.
7. Eskin, G.I. 2006. In *Promising technologies of light and special alloys*, 194–212. Moscow: Fizmatlit.
8. Grant, N.J. 1983. *JOM* 35 (1): 20–27.
9. Pohlman, R., K. Heisler, and M. Cichos. 1974. *Ultrasonics* 12:11–15.

11 Zone Refining of Aluminum with Ultrasonic Melt Processing

11.1 BASICS OF ZONE REFINING

Modern electronics is inconceivable without extra-high-purity metals, including aluminum. More often than not, this material is cleaned of solute impurities by a method known as *floating zone refining*.

This refining procedure, which has been theoretically suggested by Petrov [1] and realized by Pfann [2], is the main industrial method used in the production of extra-high-purity metals and semiconductors. The principle of the method lies in the partitioning of solute atoms between the liquid and the solid phases, with solid being much purer in certain solute elements than the liquid. In this method, a narrow zone of liquid metal is made to travel slowly through the ingot, thus moving impurities by partitioning and diffusion from the refined part of the ingot to the untreated part. In the general case, the efficiency of this method is characterized by the equilibrium partition coefficient k_0, which is defined as the ratio of the solute impurity concentrations in the solid and liquid phases, taken according to the phase diagram,

$$k_0 = \frac{C_s}{C_l}, \qquad (11.1)$$

where C_s and C_l are, respectively, the concentrations of the solute element in the solid and liquid phases in equilibrium at the same temperature. The partition coefficient is smaller or greater than unity, depending on whether the liquidus and solidus of the alloy decrease (eutectic systems) or increase (peritectic systems) with increasing the amount of the solute impurity.

In real solidification, the distribution of impurities between the solid and liquid phases, and in the solid phase, depends not only on the partition coefficient, but also on the diffusion that does not occur instantaneously. As a result, the velocity of the solidification front (solidification rate) and the cooling rate determine the completeness of diffusion equilibrium of composition in the liquid and solid phases. However slow the solidification front advances, the solid phase rejects solute faster than it can diffuse to the bulk liquid and, as result, there is always accumulation of impurity-rich liquid at the solidification front, as shown in Figure 11.1. The existence of this solute-rich layer was recently observed during in-situ solidification of Al–Cu

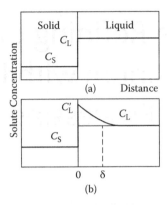

FIGURE 11.1 Distribution of a solute impurity at the solidification front: (a) in equilibrium and (b) in real solidification (floating-zone refining).

alloys [3]. The thickness of this layer and its impurity content define the conditions under which the impurity is transported by diffusion from the liquid into the solid part of the ingot. Figure 11.1b illustrates how the diffusion layer is enriched with impurities if $k_0 < 1$. When $k_0 > 1$, the solidification front is preceded by a depleted layer, and the diffusion transport is from the ingot to the melted zone. In the most typical case of eutectic-type phase diagram of Al with an impurity ($k_0 < 1$), the liquid layer enriched in the impurity ahead of the solidification front will control the concentration of this impurity in the solidifying ingot. In this case, one may speak of the effective impurity partition coefficient $k_{eff} = \frac{C_s}{C_1'}$ in which the difference between C_1' and C_1 depends on the conditions of zone refining, namely, the zone travel rate v, the diffusion coefficient D in liquid, and the thickness of the diffusion layer δ. This dependence can be formalized as follows [4]:

$$k_{eff} = \frac{1}{1 + \left[\left(\frac{1}{k_0}\right) - 1\right]\exp\left(-\frac{v\delta}{D}\right)}.$$

$$(11.2)$$

This equation suggests that for the same values of k_0, v, and D, the transport of impurities from the enriched layer into the liquid bulk is controlled by the layer thickness δ alone. In the general case, the dimensionless quantity $v\delta/D$ represents the normalized rate of ingot growth in zone melting, since it includes the main parameters controlling the efficiency of the zone refining process. For most common alloying elements alloys, D ranges from 10^{-9} to 10^{-8} m²/s, e.g., see for Cu and Fe [5], but for many impurities the diffusion coefficient D is unknown.

The solidification rate or the rate of molten zone travel v along the ingot is typically within 10 mm/s; the depth of the diffusion layer δ is 1–0.01 mm and largely depends on the hydrodynamics of the liquid metal in the pool. The velocity of any liquid flow is known to vanish near a solid wall when the turbulent flow becomes laminar. Therefore, mixing in the bulk of the liquid pool has a lesser effect on the thickness of the diffusion layer than the liquid flow in the immediate vicinity of the solidification front, being directed normal to this front and responsible for the direct transport of the solute impurity from the diffusion layer into the molten zone. Attempts have been made to intensify the transport of impurity from the diffusion later to the liquid pool by forced convection, e.g., by electromagnetic or mechanical stirring in this zone. However, these external fields mainly affect the melt in the bulk of the molten zone, far from the diffusion zone, and have little effect on the solidification front.

The effect of ultrasound on zone melting was first studied on naphthalene ($C_{10}H_8$) purposefully enriched with impurities of azobenzene (C_6H_5N), which turns naphthalene bright orange [6]. The important outcome of this work was that the diffusion layer should be sonicated in the precavitation mode with generated acoustic streams. If the ultrasound power exceeds the cavitation threshold, the cavitation destroys the solidification front, thus interfering with the zone cleaning process. These first experiments were conducted by introducing the ultrasound in the liquid zone, which appeared to be inappropriate for most metals purified by zone remelting, including aluminum, because of the possible dissolution and destruction of the radiating horn in the melt, (Figure 11.2a). Even when using the Nb-based alloys, minuscule

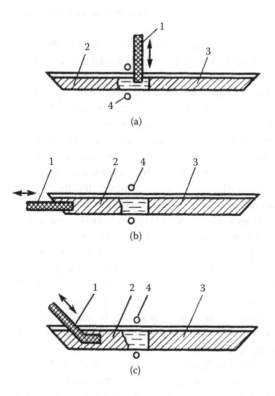

FIGURE 11.2 Arrangement of ultrasonic processing for zone remelting: (a) sonication in the liquid pool, (b) sonication by longitudinal oscillations through the solid ingot, and (c) sonication by interfered oscillations through the solid ingot: (1) waveguide/sonotrode, (2) solid ingot, (3) remelted ingot, and (4) heater.

amounts of dissolved Nb that can be neglected in normal casting and solidification may be beyond the level of allowable impurities in very high-purity aluminum.

The solution was found by sonication through the solid part of the ingot, when the solidification front acts as a radiating surface, as seen in Figures 11.2b,c [7, 8]. In this technique, the impact of ultrasound is absolutely "sterile."

11.2 ULTRASONIC ZONE REFINING OF ALUMINUM

Introduction of ultrasound through the solid part of an ingot is a clean but by no means simple process. The distance between the point of ultrasound entry and the solidification front that radiates the sound into the melt is constantly changing as the zone remelting proceeds. As a result, it is difficult to maintain similar acoustic conditions when the sonotrode is arranged horizontally (longitudinal oscillations, Figure 11.2b) or normally (transversal oscillations) to the ingot. It was found that the arrangement of the sonotrode at an angle to the ingot (Figure 11.2c) results in the dynamic imbalance of the sonotrode–ingot system when complex interference of transversal and longitudinal waves of different frequencies produces almost constant amplitude at the solidification front. An angle of 45° between the axes of the sonotrode and the ingot was shown to be the optimum.

In the setup shown in Figure 11.3, a commercial ultrasonic magnetostrictive transducer operating at 44 kHz was coupled with a special waveguiding system with a titanium sonotrode frozen into the ingot. This arrangement provided for three operating modes at the solidification front: a precavitation mode with amplitude $A > 0.5 \ \mu m$, a threshold mode ($A = 0.75 \ \mu m$), and a developed cavitation mode ($A \geq 1.5 \ \mu m$). The operating conditions of the oscillation system, the transfer of oscillations along the ingot including the liquid pool, and the cavitation onset were controlled by piezoelectric probes fixed by epoxy resin to the system elements using special wire probes that were frozen in the ingot. The magnitude and waveform of signals were monitored with an oscilloscope and a voltmeter, and the frequency was controlled with a digital frequency meter (see Figure 2.18). For comparison of the results, a setup allowed two ingots to be refined simultaneously: one in an ultrasonic field and the other without ultrasonic treatment (Figure 11.3b).

Ingots were prepared by pouring molten metal from a graphite crucible into a graphite mold measuring $18 \times 20 \times 400$ mm. The starting material was aluminum of 99.99% and 99.995% purity. The ingots were then put in a refining mold made from high-purity graphite and zone-refined in an Ar atmosphere at a rate of 1.0 mm/min. A 40-mm-wide remelting zone was generated by an electric resistance heater. The number of remelting runs was varied from 1 to 12, with the heater moving rate and zone size being kept constant.

A spectrum analysis of Si and Fe impurities in the ingot (Figure 11.4) indicated that the 99.99% pure aluminum specimen can be purified in one run in the absence of cavitation. When the conditions for cavitation are reached ($A \geq 1.5 \ \mu m$), the effect of zone refinement disappears. At a threshold amplitude of 0.75 μm, the additional effect of zone purification is insignificant. However, the zone refining efficiency increases significantly under precavitation mode ($A < 0.5 \ \mu m$). It should be noted

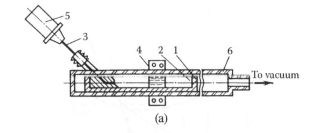

(a)

(b)

FIGURE 11.3   Zone refining with ultrasound introduced through the solid part of the ingot: (a) a schematic and (b) a photograph of the aluminum zone melting setup with (1) graphite mold with aluminum, (2) aluminum ingot, (3) waveguide, (4) heater, (5) transducer, and (6) quartz ampoule.

that equal acoustic powers introduced directly into the liquid pool (contact method) and through the solid part of the ingot (contactless method) produce similar purification effects.

The efficiency of ultrasonic processing during the floating zone refining is illustrated in Figure 11.5, which gives the variation of Si and Cu concentrations over the length of a 99.99% pure Al ingot. The single run with sonication produces better purification than three runs without ultrasound.

The efficiency of ultrasonic zone refining with respect to the partition coefficient or the type of phase diagram is demonstrated in Table 11.1. The ultrasonic zone refining is most efficient for purification from impurities with the eutectic phase diagram with Al and with very low solid solubility ($k_0 \ll 0.1$). The purification effect

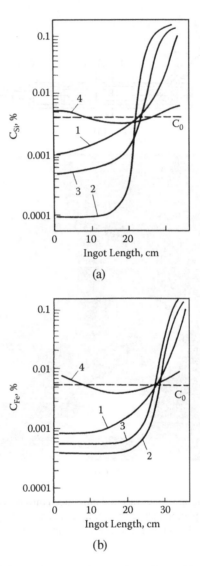

FIGURE 11.4 Effect of ultrasonic processing mode on the purification of 99.99% pure aluminum from (a) Si and (b) Fe after a single refining stage: (1) without sonication, (2) $A = 0.5$ μm (precavitation), (3) $A = 0.75$ μm (cavitation threshold), (4) $A = 1.5$ μm (cavitation).

is still noticeable for impurities with higher solid solubility ($0.1 < k_0 < 1.0$) and for monotectic-type alloying systems.

For impurities that have peritectic-type solidification reaction with Al ($k_0 > 1.0$), ultrasound still helps in purification, perhaps due to the intense acoustic streams that propagate through the entire liquid pool and affect the solidification front from the other side of the liquid zone.

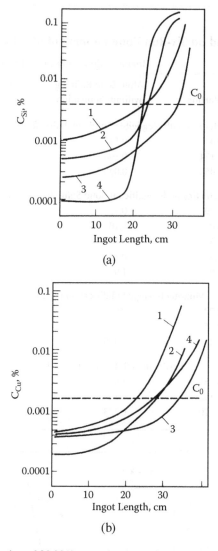

FIGURE 11.5 Purification of 99.99% pure aluminum from (a) silicon and (b) copper after: (1–3) one to three refining runs without sonication, respectively; and (4) after a single refining run with sonication.

It was found that the hydrogen concentration also changed considerably after zone refining, most notably with sonication, as shown in Figure 11.6. The zone refining reduced the hydrogen content from the initial 0.13–0.17 cm³/100 g to the almost-equilibrium solid solubility of 0.036 cm³/100 g, but the degassing rate and the refining effect during sonication were noticeably higher, even after a single run, than in the conventional zone refining after three runs. Note that the Al–H phase diagram is of the eutectic type.

TABLE 11.1

Effect of Ultrasound on Six-Run Zone Refining of 99.99% Pure Aluminum

| | Impurity Concentration, 10^{-4} wt% | | |
Impurities	Starting	After Zone Refining without Sonication	After Zone Refining with Sonication
Eutectics with Negligible Solid Solubility ($k_0 < 0.1$)			
Fe (0.02)	16.4	0.143	0.041
Ni (0.007)	0.174	0.100	0.100
Ca (0.001)	0.450	0.090	0.060
Eutectics with Significant Solid Solubility ($k_0 < 1$)			
Mn (0.94)	0.29	0.22	0.08
Mg (0.51)	36.0	8.10	2.70
Cu (0.17)	11.5	0.46	0.11
Si (0.11)	16.5	0.33	0.11
Zn (0.8)	4.9	1.94	0.73
Monotectic-Type Solidification ($k_0 \leq 0.1$)			
Na (0.0015)	1.70	0.017	0.006
Tl (<0.0005)	0.55	0.022	0.012
Pb (0.11)	0.45	0.300	0.200
Peritectic Solidification ($k_0 > 1$)			
Ti (9.0)	0.531	1.060	0.530
Cr (2.0)	0.192	0.134	0.032
V (4.0)	0.170	0.223	0.130

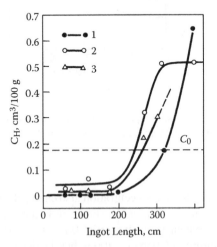

FIGURE 11.6 Effect of ultrasonic processing and the number of remelting runs on the hydrogen concentration in 99.99% pure aluminum: (1) single remelting pass with sonication of $A = 0.5$ μm; (2, 3) conventional zone refining with two and three runs, respectively.

TABLE 11.2

Effect of Ultrasound (US) on Zone Refining of 99.99% and 99.995% Pure Aluminum as a Function of Refining Runs

| Starting Purity, % | No. of Passes | Residual Resistance at a Temperature of Liquid Helium $\rho_{4.2K}$, nΩ-cm | |
		Zone Refining with US	Zone Refining without US
99.99	6	200–300	800–1200
99.99	9	200	600–820
99.995	6	5–70	80–160
99.995	9	5	100

The effect of the number of runs and sonication on the purification efficiency for 99.99% and 99.995% pure aluminum is shown in Table 11.2, where the purity is estimated from the residual electrical resistance at the temperature of liquid helium. The lower the resistance, the purer the metal.

The specific concentrations corresponding to six-run zone remelting of 99.99% and 99.995% pure aluminum are given in Table 11.3. These data prove that ultrasound intensifies the zone refining of aluminum and ensures a more efficient removal of impurities.

The effect of ultrasound on the impurity content of aluminum after zone remelting is illustrated by Table 11.4. A six-run ultrasonic zone remelting of 99.995% pure aluminum boosts its purity above 99.99997%.

TABLE 11.3

Effect of Sonication and Starting Purity on the Composition of Zone-Remelted Aluminum After Six Runs

| Sonication | Impurity Content, 10^{-4} at.% | | | | | | | | |
	Total	Fe	Si	Cu	Zn	Mg	Ti	Se	Mn
				99.99% Al					
Yes	7.2	1.5	0.9	1.2	0.9	0.8	0.9	0.8	0.2
No	11.5	3.0	2.1	1.5	1.8	1.1	1.0	0.8	0.2
				99.995% Al					
Yes	0.66	0.03	0.008	0.2	0.008	0.03	0.2	0.02	0.02
No	3.44	0.09	1.6	0.4	0.008	0.20	0.2	0.02	0.04

TABLE 11.4

**Achieved Purity After Zone Refining of 99.99% and 99.995%
Purity Aluminum (Six Runs at a Remelting Rate of 1 mm/min)**

Sonication	Total Impurity Content, 10^{-4} wt%	Final Purity, wt%
Starting 99.99% Al	85.97	99.991403
No	12.43	99.998717
Yes	3.94	99.999605
Starting 99.995% Al	21.49	99.997250
No	0.89	99.999941
Yes	0.27	99.999971

11.3 MECHANISMS OF SONICATION EFFECTS ON ZONE REFINING

The mechanism of sonication effects on the process of floating-zone refinement is not completely understood. Pioneering studies in this field revealed that acoustic cavitation has an adverse effect, as it disturbs the solidification front and admixes impurity-rich melt into the solid, purified part of the ingot [6, 9]. Conversely, there is an opinion that noncavitating bubbles that pulsate without collapsing may be useful as stirring agents, because of the acoustic streams induced near such bubbles. These considerations should be taken into account, since 44-kHz ultrasound produces hydrogen bubbles of approximately 50–100 μm in aluminum melt. Pfann [2] found that the thickness of the diffusion zone at the solidification front is 1000 μm under stationary conditions and 10 μm under intense stirring. Hence, the contribution of microstreams due to the pulsating bubbles may be substantial.

Acoustic flows are known to affect heat transfer in the boundary layer close to the solid/liquid interface [10, 11]. Here, the sound pulse undergoes sharp changes and causes vortices. In a sufficiently thin boundary layer, the generated forces considerably exceed those that appear due to attenuation in an infinite sound field. The acoustic flows close to the surface of pulsating bubbles and those in the boundary layer (that pulsates with the frequency of the sound field) at the solidification front compare on the same scale with the thickness of the diffusion layer where the impurities are accumulated.

Dedicated experiments were set up to study the true streaming pattern under the conditions of zone refining with sonication. In earlier experiments with organic models and metals, ultrasound was conveyed directly to the liquid pool far from the solidification front, so that the induced acoustic streams stirred the liquid mainly within the liquid pool. When ultrasound was introduced through the solid part, acoustic streaming occurred first of all in the diffusion layer and then in the liquid zone, producing a more pronounced effect on the zone refining process.

The ultrasonic treatment intensifies at least two processes: (1) the transport of the impurity-rich liquid phase from the solidification front to the bulk of the liquid zone, thereby reducing the thickness of the diffusion layer and increasing the impurity

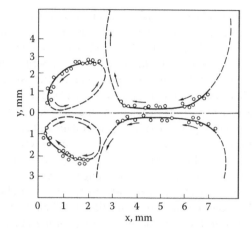

FIGURE 11.7 Pattern of acoustic streaming in the sonicated liquid pool as a function of distance from the solid/liquid interface ($x = 0$) in the zone refining model. Circles represent experimental results; dashed segments show imaginary paths.

concentration in the liquid bulk, and (2) the increase of the temperature gradient at the solidification front, which flattens the solidification surface.

The model setup consisted of a transparent polycarbonate block with metallic insert coupled with the sonotrode. Part of the block was machined to simulate the liquid pool that was filled with glycerin. The dimensions of the solid block and the liquid pool were identical to those during zone refining. The sonication process was filmed at a rate of ten frames per second, covering almost the entire liquid pool. The air bubbles were marked and traced on the film frames and, using frame-to-frame variations, the stream velocities were measured as a function of coordinates for different oscillatory velocities of the sonotrode, that is, at different acoustic powers introduced into the liquid pool.

Figure 11.7 shows the traces of two air bubbles in glycerin measured every ten frames at an interval of 1 s. An analysis of more than 1000 frames gave the characteristics for two opposite streams outlined in Table 11.5.

TABLE 11.5

Effect of Ultrasound (44 kHz) on the Velocity of Acoustic Streams

Acoustic Power, W	Amplitude, μm	Cavitation in the Pool	Acoustic Stream Velocity, mm/s	
			x-axis [a]	y-axis [a]
15	0.5	Absent	0.1	0.05
30	0.5–1.0	Threshold	0.1–0.2	0.1
100	1.2	Developed	0.25–0.35	0.2

[a] See Figure 11.7.

A comparison of the velocities of acoustic streams in glycerin with those of the solidification front under real zone-remelting conditions (1 mm/min = 0.017 mm/s) suggests that the former is always faster than the latter. Consequently, ultrasound is a rather efficient tool that enhances the transport of the impurity-rich part of the melt from the solidification front deep into the melt without the assistance of cavitation phenomena. As the intensity of ultrasound increases and the acoustic pressure exceeds the cavitation threshold and excites cavitation, the velocity of the acoustic stream increases, but the introduced pressure spikes from collapsing bubbles disturb the solidification front and the regular flow pattern, thus lowering the efficiency of refinement.

We can summarize as follows. At low ultrasonic powers, acoustic streaming is formed only at the solidification surface in a range not exceeding 10 × 1 mm. The streaming produces two countercurrent vortices symmetrical about the ingot axis (see Figure 11.7). No flows occur in the bulk liquid. When the power increases but stays below the cavitation level, the accelerated streaming intensifies the effect of ultrasound on the zone refining process. These acoustic streams equalize the temperature field at the solidification front, reduce the width of the diffusion layer, and accelerate the transportation of impurity-rich material away toward the interior of the liquid pool. These quantitative estimates of the effect of ultrasound on aluminum zone refining prove the active role of acoustic streaming in the mechanism of precavitation acoustic zone refining.

11.4 STRUCTURE AND CHARACTERISTICS OF DEFORMED ACOUSTICALLY REFINED ALUMINUM

High-purity aluminum is used primarily in electrical and electronic applications. Ingots from starting at 99.99% and 99.995% pure aluminum before and after zone refining (with and without ultrasonic processing) were separated in the pure and impure parts, cold extruded into 10-mm bars, and then cold drawn, without intermediate annealing, into wires of the following range of diameters: 6.0, 4.0, 1.97, 0.5, 0.3, and 0.2 mm. The structure and properties of these wires were analyzed to estimate the effect of ultrasonic processing on the zone refining.

The study included structural analysis, recrystallization analysis, mechanical tests, corrosion-resistance tests, and the performance of leads of integrated circuits made from the purified aluminum. X-ray analysis indicated that ultrasonic zone refining results in a considerably higher degree of recrystallization [8]. Table 11.6 shows that the wire cold-drawn from aluminum purified with ultrasound demonstrates fully recrystallized structure, while other wires have nonrecrystallized or partially recrystallized structure. The degree of recrystallization can be used as a quality index indicating the presence of insoluble impurities hindering recrystallization. The data in Table 11.7 prove that the recrystallization in high-purity aluminum starts at room temperature, while lesser purity aluminum obtained without sonication finishes recrystallization only above 250–400°C (1-h annealing), which agrees with the values reported in the literature for pure aluminum.

Electrical leads from refined aluminum for electronic devices must be annealed to improve their ductility prior to connection to a silicon or germanium base by

TABLE 11.6

Effect of the Initial Al Purity and Zone Refining with and without Sonication on the Degree of Recrystallization in Cold-Drawn 4.0-mm Wires (Six-Zone Remelting Runs)

	Degree of Recrystallization	
Zone Refining	Pure Part	Impure Part
Starting 99.99% pure Al	nr	nr
After US zone refining	r	pr
After conventional zone refining	pr	nr
Starting 99.995% pure Al	nr	nr
After US zone refining	r	nr
After conventional zone refining	pr	nr

Note: r = recrystallized; pr = partially recrystallized; nr = nonrecrystallized.

ultrasonic or compression welding. Mechanical tests of wires cold-drawn or annealed at 500°C for 1 hour revealed that ultrasonic zone refining markedly improves the ductility of purified aluminum, as illustrated in Tables 11.8 and 11.9.

Specimens of cold-worked aluminum wire were tested in aggressive corrosive media (20% solution of hydrochloric acid). Results of these tests, summarized in Table 11.10, demonstrate that ultrasound applied in zone refining considerably improves the corrosion resistance of purified aluminum.

TABLE 11.7

Degree of Recrystallization as a Function of the Purity and Recrystallization Temperature of Cold-Drawn Aluminum Wires

		Annealing Temperature, °C							
Zone Refining	Al, %	20	50	100	150	200	250	300	400
Starting 99.99% Pure Aluminum									
Not refined	99.991	nr	nr	nr	nr	nr	nr	nr	r
US zone refining	99.9996	r	r	r	r	r	r	r	r
Conventional zone refining	99.9983	pr	pr	pr	pr	pr	pr	pr	pr
Starting 99.995% Pure Aluminum									
Not refined	99.9973	nr	nr	nr	nr	nr	r	r	r
US zone refining	99.99999	r	r	r	r	r	r	r	r
Conventional zone refining	99.99994	pr	pr	pr	pr	pr	r	r	r

Note: r = recrystallized; pr = partially recrystallized; nr = nonrecrystallized.

TABLE 11.8
Tensile Properties of Cold-Drawn Aluminum Wires in Dependence on the Wire Diameter and Aluminum Purity

Diameter, mm	99.99 Al		99.995 Al		99.9999 Al[a]		99.99999 Al[a]	
	YS, MPa	El, %	YS, MPa	El, %	YS, MPa	El, %	YS, MPa	El, %
6.0	55	9.7	28	45.5	19	39.2	18	49.4
4.0	100	11.8	54	9.4	24	16.5	26	28.6
2.0	118	3.3	54	9.2	38	9.3	23	33.0
0.5	148	0.2	89	0.6	38	4.3	27	15.5
0.3	148	0.4	88	0.45	39	8.5	32	15.1

Note: YS = yield strength; El = elongation.
[a] Sonication during zone refining (six runs).

TABLE 11.9
Tensile Properties of 6-mm Cold-Drawn Wire from Ingots of Different Processing History (Zone Remelting with Six Runs)

Zone Refining	UTS, MPa	YS, MPa	El, %	RA, %
	Starting 99.99% Pure Aluminum			
Not refined	108	55	9.7	88.8
Conventional zone refining				
Pure part	61	49	18.6	91.8
Impure part	128	88	8.2	74.7
US zone refining				
Pure part	38	19	39.2	92.8
Impure part	82	74	8.2	85.8
	Starting 99.995% Pure Aluminum			
Not refined	40	28	45.5	91.0
Conventional zone refining				
Pure part	38	19	43.9	86.4
Impure part	42	32	38.9	89.9
US zone refining				
Pure part	39	18	49.4	92.8
Impure part	117	88	9.7	85.5

Note: UTS = ultimate tensile strength; YS = yield strength; El = elongation; RA = reduction in area.

TABLE 11.10

Effect of the Aluminum Purity on Corrosion Resistance

Nominal Purity, %	Al, %	Purification Method	Corrosion Rate, mm/year
99.97	99.97	As received	0.1850
99.99	99.991	As received	0.0022
99.995	99.997	As received	0.0013
99.999+	99.9996	US zone refining	0.0007
99.999+	99.99997	US zone refining	0.00029

A critical operation in manufacturing semiconductors and large-scale integrated circuits is ultrasonic welding of aluminum leads measuring 0.1–0.5 mm in diameter. Commonly used aluminum wire of 99.995% purity has a relatively high strength and a low ductility margin, which makes a very narrow range of optimum processing parameters. This problem has been solved by using high-purity aluminum obtained by zone refining with ultrasound. For comparison, two wire specimens of 0.3-mm diameter and with aluminum purity of 99.997% (standard purity) and 99.99997% (ultrasonic zone refining) were taken for testing. They were ultrasonically welded to a 0.8-μm film of 99.995% pure aluminum. The welding parameters were 12-W power, 55–65-kHz frequency, 0.3–3.0-s duration, and 0.8-N compression force. In ten welded joints, we determined the tensile and shear strength as well as the hardness on the joint and affected zone. These tests demonstrated that ultrasonically refined aluminum improves the quality and hardness of the welded joint.

It is worth noting that the use of 99.99997% pure aluminum improves the appearance of the welded joint, which otherwise becomes dull. Wires made from such ductile aluminum have little springback, and this quality also improves the performance of the ultrasonic welding process.

REFERENCES

1. Petrov, D.A. 1947. *Zh. Fiz. Khim* 21 (12): 29–31.
2. Pfann, W.G. 1966. *Zone melting*. 2nd ed. New York: John Wiley & Sons.
3. Mathiesen, R.H., and L. Arnberg. 2005. *Acta Mater.* 53:947–56.
4. Burton, J.A., R.C. Prim, and W.P. Slichter. 1953. *J. Chem. Phys.* 21:1987–91.
5. Lee, N., and J. Cahoon. 2011. *J. Phase Equilib. Diff.* 32:226–34.
6. Abramov, O.V., I.I. Teumin, V.A. Filonenko, and G.I. Eskin. 1967. *Akust. Zh.* 13 (2): 161–69.
7. Agranat, B.A., G.I. Eskin, E.N. Ozerenskaya, S.A. Afanas'ev, K.B. Yurkevich, and A.V. Stamov-Vitkovsky. 1976. Invention certificate SU 512791 A1, publ. 05.05.1976.
8. Eskin, G.I. 1988. *Ultrasonic treatment of molten aluminum*. Moscow: Metallurgiya.
9. Eskin, G.I., B.A. Agranat, and E.N. Ozerenskaya. 1980. In *Physical and physico-chemical methods of intensification of technological processes, Proc. Mosc. Inst. Steel and Alloys*, vol. 124, 109–118. Moscow: Metallurgiya.
10. Zarembo, L.K. 1970. In *Physics and technology of powerful ultrasound*, vol. 2, ed. L.D. Rozenberg, 87–128. Moscow: Nauka.
11. Abramov, O.V., Yu.S. Astashkin, and V.S. Stepanov. 1979. *Akust. Zh.* 25:180–86.

TABLE 11.10

Effect of the Aluminum Purity on Corrosion Resistance

REFERENCES

12 Semisolid Deformation of Billets with Nondendritic Structure

12.1 INTRODUCTION TO SEMISOLID DEFORMATION

The technology of semisolid deformation or thixotropic deformation has rapidly developed in the last 40 years. This is evidenced by regular biennial conferences on semisolid processing, where leading research centers from Europe, the United States, Japan, China, and Republic of Korea present the results of fundamental studies and technological developments.

The basis of this technology is the ability of a semisolid material with globular (nondendritic) grains to deform under low loads, almost like highly viscous liquid, and easily fill dies and molds. The semisolid materials with such a structure behave like quasi-liquid, i.e., they demonstrate thixotropic* behavior. In other words, the load (especially shear) distributes uniformly over the volume, and the apparent viscosity drops to the viscosity of oil. One can imagine that the globular, nondendritic grains slide in the liquid phase, easily rotating, accommodating the deformation and filling a die at relatively low loads. One of the main conditions for thixotropic behavior of metallic slurries is the nondendritic structure of metal.

The origins of semisolid deformation date back to the research and development at Massachusetts Institute of Technology under the supervision of Prof. M. Flemings [1]. The development of the science and technology of semisolid processing of metals is covered in several good reviews [2, 3]. Electromagnetic or magnetohydrodynamic stirring and mechanical deformation in the semisolid state were initially the main means of producing the desirable nondendritic structure. Since then, new approaches have been suggested and tried.

One of the main advantages of semisolid deformation is the replacement of shape casting with forging. As a result, the combination of good mold filling and advanced process automation with improved properties makes the technology commercially and industrially attractive, especially in automotive and electrotechnical industries.

Semisolid processing allows one to improve the mechanical properties of the final forging up to the level of a wrought alloy, and the increase in tensile strength can be as much as 30% without sacrificing ductility, as demonstrated in Table 12.1.

---

* *Thixix* (Greek) is "touch"; *tropos* means "changing, turning."

TABLE 12.1

Some Characteristics and Properties of Automotive Wheels Produced by Shape Casting and Semisolid Deformation (Alloy A357 T5)

Technology	Weight After Forming, kg	Weight of the Finished part, kg	Production Rate, pcs/h	UTS, MPa	YS, MPa	El, %
Shape casting	11.1	8.6	12	221	152	8
Semisolid forging	7.5	5.1	90	290	214	10

Note: UTS = ultimate tensile strength; YS = yield strength; El = elongation.
Source: Davis [4].

Among the very important advantages are high production rates and near-final shape forming with little material wasted.

Forgings from high-strength aluminum alloys (2XXX and 7XXX series) produced by semisolid deformation have a very attractive combination of high strength and ductility; for example, a forging from an AA7075T6 alloy shows a tensile strength of 495 MPa at an elongation of 7% [4].

In contrast to rheocasting, where the structure suitable for semisolid deformation is formed continuously in the same process with casting/forging, thixoforming assumes that the billet with a required structure is produced before the semisolid deformation proper. The billets cast by direct-chill (DC) casting have fine yet dendritic grain structure. Therefore, these billets require further processing, e.g., holding in the semisolid state to allow for dendrite coarsening to large, globular grains with a grain size two to three times that of the dendrite arms spacing.

Ultrasonic melt processing allows one to achieve nondendritic structure directly in the DC-cast billet (see Chapter 5). This gave the idea of using billets with nondendritic structure (grain size close to the dendrite arm spacing) as a feedstock for semisolid deformation [5, 6]. One of the authors (GE) had an opportunity to discuss this approach with Prof. Flemings during the conference on semisolid processing in Sheffield (1996), who very positively assessed the idea and specifically mentioned ultrasonic melt processing as one of the promising technologies for semisolid processing in his concluding remarks at the conference.

Several other technologies for the formation of nondendritic feedstock for rheocasting have been suggested. For example, "new rheocasting" involves formation of excess nuclei as a result of controlled solidification in a laminar flow along the cooling slope of a crucible or a launder at a temperature close to the liquidus [7]. The control of metal temperature and cooling rate allows the formation of the required fraction of the solid phase inside the crucible with subsequent transport of the slurry to a die. Another interesting technology involves the processing of slurry in a barrel with one or two (intermeshed) screws [3]. Intensive shear deformation produces fine and globular grains that can then be delivered into the die. The mechanism of such intensive shearing is not fully understood and may involve mechanical fragmentation of dendrites, their coarsening under forced convective flows, and hydrodynamic cavitation resulting in phenomena similar to those occurring upon ultrasonic cavitation.

12.2 EFFECT OF NONDENDRITIC STRUCTURE OF BILLETS ON THIXOTROPIC BEHAVIOR OF ALUMINUM ALLOYS

Initially, 285-mm billets from an AA7055-type alloy were selected for studies [5, 6]. Billets with dendritic structure were obtained by regular DC casting. The nondendritic structure was produced by DC casting with ultrasonic melt processing in the mold. The size of the nondendritic grains was 60–70 μm, and the dendritic grains were 1500–2000 μm.

In the solid state, the nondendritic structure is advantageous, mainly because of higher plasticity and a notably higher crack resistance. This enhanced plasticity is retained at temperatures of hot deformation in the solid state, somewhat increasing the allowable strain rate. However, the resistance to plastic deformation (yield stress) remains almost invariable.

For tests in the semisolid state, 35 × 35 × 12-mm test pieces were prepared from longitudinal sections of the billets. These specimens were upset on smooth block heads of a hydraulic press under isothermal conditions, which halved the specimens' height to 6 mm. The upsetting was carried out at 420°C, 525°C, 550°C, 570°C, and 590°C. The proportion and distribution of the liquid phase in specimens before deformation in the semisolid state could be estimated from the microstructure of the samples quenched from the deformation temperature.

Upsetting stress–strain curves taken in the semisolid state over the entire tested temperature range revealed a noticeable reduction of the deformation load for the metal with nondendritic grains. Quantitative data of these tests are summarized in Table 12.2. The maximum upsetting stress for samples with nondendritic grains is significantly lower than that for samples with dendritic grains. This difference increases with the

TABLE 12.2
Effect of the Grain Structure and Testing Temperature on the True Stress σ Required for 50% Upset Ratio at a Strain Rate of 0.1 mm/s in Solid (420°C) and Semisolid States

Upsetting Temperature, °C	σ, MPa		σ_{nd}/σ_d
	Nondendritic (nd)	Dendritic (d)	
420	103.4	111.1	0.93
525	35.8	50.4	0.71
550	21.4	30.5	0.70
570	19.3	30.9	0.62
590	11.7	19.6	0.60

FIGURE 12.1   Microstructures of AA7055 samples after upsetting at 550°C and quenching at 470°C produced from different feedstock: (a) billet with nondendritic structure and (b) billet with dendritic structure.

liquid phase fraction, and at 570°C–590°C (50% liquid) the difference reaches 60%. It is noteworthy that the effect of the nondendritic structure on the upsetting in the solid state (420°C) is markedly lower.

Figure 12.1 shows the grain structure of billets quenched after upsetting at 550°C and cooling to the solidus. The polyhedral structure with nondendritic grains readily deforms in upsetting with 50% of the liquid phase: nondendritic globules slip over one another, especially at the beginning of deformation. The dendritic grains behave differently: their shape radically changes during upsetting, with signs of deformation of the grains proper. As a result, the resistance to deformation increases.

Table 12.3 gives the tensile strength and elongation of solid samples produced by semisolid deformation and subjected to a regular heat treatment. It is clear that the deformation in the semisolid state provides for remarkably high mechanical properties, while the nondendritic structure not only favors the deformation process, but also noticeably increases the ductility of the metal. The strength level complies with the specifications for small forgings produced by deformation in the solid state.

Further development of the technology of nondendritic billets produced by DC casting for semisolid deformation was done using a casting scheme with ultrasonic processing in the melt flow (see Section 5.8). Two alloys were selected: low-alloyed AA6007 and high-strength AA7050. The former alloy was cast as 145-mm and

TABLE 12.3

Effect of the Initial Billet Structure on the Mechanical Properties of Samples Produced by Semisolid Deformation (An AA7055-Type Alloy with Deformation Temperature 570°C)

Billet Grain Structure	UTS, MPa	El, %
Nondendritic	670	10.9
Dendritic	650	7.8

178-mm billets with the composition: 0.6%–0.9% Mg; 0.9%–1.4% Si, 0.12%–0.14% Zr, 0.5%–0.7% Fe, and up to 0.1% of (Cu + Zn + Cr + Mn). The latter alloy was cast as 178-mm billets with the composition: 6.2% Zn, 2.5% Cu, 2.5% Mg, and 0.15% Zr. An Al5Ti1B master alloy rod was introduced in the melt flow with simultaneous ultrasonic cavitation treatment in the launder during DC casting. The size of nondendritic grain (after ultrasonic processing) was 50 μm, and the size of dendritic grains (without ultrasonic processing) was 200–250 μm for AA6007 and 300–400 μm for AA7050. The obtained billets were cut in pieces up to 300 mm long for semisolid forging and rheological testing.

The dependence of the solid fraction on temperature in the semisolid state can be estimated using either thermal analysis or thermodynamic calculations. The results of the latter are shown in Figure 12.2. The solidification range of an AA6007

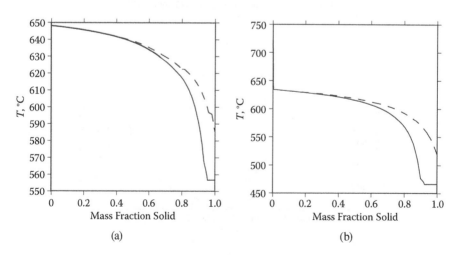

(a)   (b)

FIGURE 12.2 Thermodynamic calculations showing the evolution of solid fraction during solidification of (a) AA6007 and (b) AA7050 alloys. Solid lines correspond to Scheil (nonequilibrium) solidification, and dashed lines correspond to equilibrium solidification.

alloy extends from 575°C to 650°C and that of AA7050 ranges from 465°C (524°C equilibrium) to 635°C.

Physical modeling of semisolid deformation was performed using cylindrical samples (Ø10 × 13 mm for AA6007 and Ø60 × 16 mm for AA7050) that were upset in open dies under isothermal conditions. It was shown that thixotropic properties start to appear at about 20% of the liquid phase and are most pronounced at 30%–40% liquid. Experiments with a higher content of liquid are difficult due to disfiguration of the sample because of tensile stresses, so extrapolation was used.

Figure 12.3 demonstrates the development of the liquid fraction and the compressive yield strength for both studied alloys. The possibility of semisolid deformation was assessed by isothermal tests of AA7050 in the temperature range 560–600°C, which corresponded to the liquid phase content of 8–25%. Figure 12.4 gives the results obtained at 590°C. It is clear that the nondendritic structure results in a deformation force about four times less than that required for the dendritic structure, and this force is independent of the deformation rate [8].

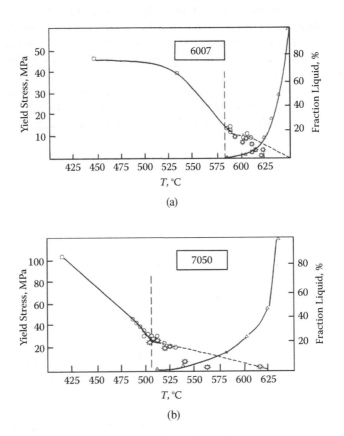

FIGURE 12.3 Dependence of yield stress and fraction liquid on the temperature for (a) AA6007 and (b) AA7050 alloys: o = samples without cracks; ✿ = samples with cracks and extrapolated values (dashed line); Δ = fraction liquid.

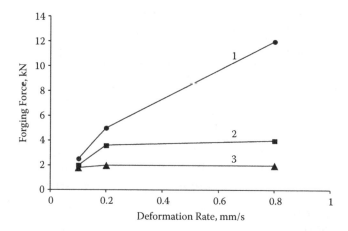

FIGURE 12.4 Effect of deformation rate in the semisolid state on the forging force at 590°C for different feedstock: (1) dendritic structure, grain size 2 mm; (2) dendritic structure, grain size 400 μm; and (3) nondendritic structure, grain size 50 μm.

12.3 RHEOLOGICAL PARAMETERS OF SEMISOLID ALLOYS

Optimization of process technology usually requires computer simulation, which is not possible without having a reliable and validated constitutive model of the material.

The rheological parameters of the selected alloys were studied experimentally [9, 10]. As the tensile strength of semisolid material is very small, compression tests were used. Cylindrical samples (Ø10 × 13 mm) were upset in a hydraulic press in the temperature range 420°C–610°C for AA7050 and at 400°C–630°C for AA6007 at a strain rate of 0.1 s⁻¹. The samples tend to fracture from the surface, even at small fractions liquid, due to the tensile component of stress, this effect being more severe at high deformation rates. Therefore, extrapolation of the results was used to predict the properties in the semisolid state [11]. The extrapolation technique includes simultaneous measurements of the upsetting response and the indentation (see Figure 12.5). The results of these tests allow one to determine a similarity criterion, which is later used to recalculate the yield stress upon semisolid deformation from the indentation data obtained in the semisolid state.

Using the results of mechanical testing (upsetting and indentation) and reference data [12], the dependencies of the yield stress on the temperature, deformation rate, and accumulated strain were determined for the AA7050 and AA6007 alloys. The temperature dependence was demonstrated in Figure 12.3, as the effects of the deformation rate and accumulated deformation were relatively small [9].

12.4 COMPUTER SIMULATION OF SEMISOLID FORGING

The measured rheological characteristics along with thermophysical properties of the alloys were used in finite-element modeling with Q-Form software [9].

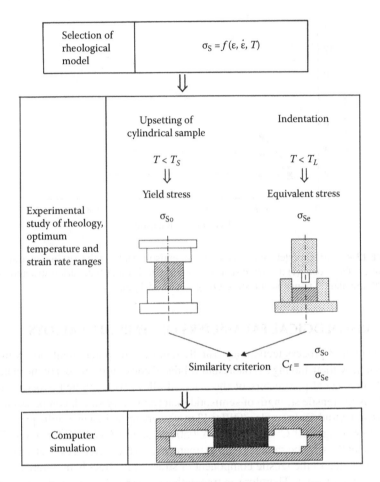

FIGURE 12.5 Principle of computer simulation of semisolid deformation.

A viscoplastic model was used for liquid fractions up to 40%. The die temperature and geometry were the variables, and the metal temperature, accumulated deformation, and average stress were the outputs of the simulations. The piston velocity was 30 mm/s (approx. 1 s⁻¹) or 300 mm/s (10 s⁻¹), and the initial metal temperature was 610°C (approx. 40% liquid). The simulation was performed for an AA7050 alloy with a nondendritic grain size of 50 μm. The results are given in Table 12.4.

The accumulated deformation is not uniformly distributed in the cross section, especially in the die with right-angle sides (1). In addition, the upper surface of the part in the right-angle dies experiences folding, with the maximum accumulated strain up to 4.5. The sloped sides of the dies improve die filling and decrease accumulated stresses and strains. The best combination is 3 in Table 12.4. The average stresses after the end of semisolid forging are compressive, with tensile stresses only at the surface. The minimum tensile stress was

TABLE 12.4

Distribution of Temperature, Accumulated Strain, and Stress Upon Semisolid Forging of a Cylindrical Part in Dies with Right (1) and Sloped (2–4) Sides (AA7050 Alloy, Nondendritic Grain Size 50 μm, Deformation Temperature 620°C)

No.	Die Temp., °C	Def. Rate, s^{-1}	Fraction Liquid at Start, %	Fraction Liquid at End, %	T_{max}, °C	T_{min}, °C	ΔT, °C	ε_{max}	ε_{min}	σ_{max}, MPa	σ_{min}, MPa
1	400	1	40	18	579	574	5	4.94	0.125	−7.3	−118
2	400	1	40	18	580	574	6	3.84	0.11	−2.3	−115
3	600	1	40	23	587	584	3	3.79	0.12	−7.8	−112
4	400	10	40	22	586	578	8	3.87	0.11	−1.0	−125

found for combination 3 (Table 12.4), 5.4 MPa, which is well below the yield stress for the AA7050 alloy at this temperature and strain rate (14 MPa). The temperature distribution during deformation is quite uniform in all studied cases. At the same time, combination 3 is again preferable, as it ensures the minimum temperature difference across the section (3°C) and the maximum liquid fraction at the end of the deformation (23%).

The computer simulation results were validated by experiments [8]. The computer simulations allow us to conclude that the semisolid deformation of nondendritic billets should be performed (a) in the temperature range corresponding to 18%–40% of the liquid phase at a deformation rate not less than 1 s^{-1}; (b) the geometry of the die should prevent folding and tensile stresses (e.g., by using sloping faces); and (c) the variation of temperature throughout the cross section of the forged part should not exceed 10°C.

12.5 SEMISOLID DEFORMATION OF CASTING ALLOYS WITH NONDENDRITIC GRAIN STRUCTURE

One of the advantages of semisolid deformation is the replacement of shape casting using the same type of casting alloys. The near-net shape combines here with improved mechanical properties. Casting alloys, however, contain a substantial amount of eutectics surrounding the primary aluminum grains. The semisolid deformation of such alloys requires some optimization.

We prepared about 5 kg of an A357-type alloy (7.4% Si, 0.6% Mg, 0.25% Fe, 0.22% Ti). After melting and degassing at 800°C, the melt was held for 1 h in a graphite crucible and then cooled to 720°C [13, 14].

The nondendritic structure was formed as follows: The melt was poured through a water-cooled launder into a thin-walled titanium mold (Ø65 × 100 mm) preheated to 300°C–350°C. During melt flow in the launder, the melt cooled to a subliquidus temperature of 630°C, which ensured the formation of a suspension of small crystals in the lower portion of the melt (close to the launder bottom). Ultrasonic treatment (22 kHz, 1 kW) was then applied to the slurry just before the entry to the mold, which resulted in multiplication of solid grains. Samples without ultrasonic processing were produced as well. The mold was well heat-insulated and was covered with an insulating cover after the end of filling.

Samples were quenched at 575°C in water, and the structure was evaluated using the following parameters [14, 15]:

- Average grain size
- Shape factor F_g

$$F_g = L_{ge}^2/4\pi S_g, \tag{12.1}$$

where L_{ge} is the length of the boundary between grains and eutectics and S_g is the average grain area

TABLE 12.5

Effect of Ultrasonic Processing on the Structure Parameters and Thixotropic Propensity of Billet from an A357-Type Alloy

Ultrasonic Processing	D_g, µm	F_g	C_g	K
No	79	2.39	0.21	0.50
Yes	67	1.89	0.07	0.132

- Branching coefficient C_g

$$C_g = 2L_{gg}/(2L_{gg} + L_{gc}),\qquad(12.2)$$

where L_{gg} is the length of the boundary between directly connected grains.

The quality criterion was suggested for evaluating the thixotropic propensity of the structure:

$$K = F_g C_g.\qquad(12.3)$$

Table 12.5 gives the results of structure evaluation [14, 15]. The application of ultrasonic processing results in grain refinement, decreases the branching of grains, makes grains more equiaxed, and decreases the quality factor, which indicates that the structure approaches nondendritic grain type.

The produced billets with nondendritic structure were machined to a diameter of 56 mm and a height of 30 mm for subsequent semisolid deformation. The temperature range of deformation was selected as 565°C–590°C [16]. The billets were upset at a deformation rate of 5 mm/s. Figure 12.6 shows the schematic of final parts produced at two temperatures. Examination of the forgings shows that the higher temperature (590°C) was beneficial for the quality, in particular for the

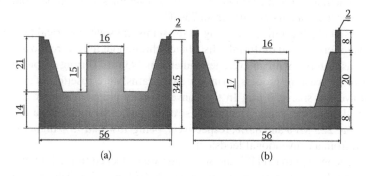

FIGURE 12.6 Shape and dimensions of a part after semisolid deformation performed with the starting temperature (a) 565°C and (b) 590°C.

formation of a defect-free thin wall. Computer simulation of the process showed that the die should be preheated to the temperature of the semisolid deformation, thereby ensuring uniform temperature distribution and preventing tensile stresses in thin sections.

12.6   APPLICATION OF ULTRASONIC PROCESSING TO RHEOCASTING

The difference between thixoforming and rheocasting is in the continuity of the process and in the forming technology. In the former case, the billet with nondendritic structure is prepared and solidified before the forming stage and is later heated to the semisolid state and formed by forging or extrusion. In the latter case, the nondendritic structure is achieved by processing the semisolid slurry immediately before the forming stage, which then more resembles casting, e.g., high-pressure (HPDC) or low-pressure die casting (LPDC).

A common way of achieving the nondendritic structure is by holding the slurry at a certain temperature to allow for grain coarsening and globularization. Recent research shows that the application of ultrasonic processing to the slurry could be beneficial for the formation of nondendritic grain structure [17–19]. Two approaches were suggested: (a) treatment by direct introduction of a sonotrode into the slurry and (b) treatment by indirect sonication through the bottom of a crucible.

In the first (direct) approach [17, 18], an AA5052 alloy was melted at 720–750°C and degassed with Ar. The melt was cooled to a pouring temperature of 670–680°C, and about 500 g of melt was poured into a metallic crucible preheated to 550°C, followed by application of ultrasound (20 kHz, 1.2 kW, intermittent processing using a piezoceramic transducer with a Ti sonotrode, with 1-s treatment followed by 1-s resting). The sonotrode was immersed into the melt 15–20 mm from the surface. Total processing time was about 120 s. In order to prevent the melt from oxidation, an Ar protective atmosphere was used during the slurry preparation process. In the process of ultrasonic processing at about 645°C (4–5°C below the liquidus), the melt was converted into a semisolid slurry with a certain solid fraction. The semisolid slurry was immediately poured into a permanent steel mold or into the shot sleeve of a HPDC machine to produce standard tensile test samples. For comparison, samples were cast by the same techniques from 730°C.

As a result of ultrasonic processing in the semisolid state, spherical primary (Al) grains were formed and uniformly distributed throughout the entire cross section of the rheocast sample, with an average diameter of 143 μm. During secondary solidification in the casting process, primary (Al) grains were rosettelike or spherical, with an average size <50 μm. In the conventionally prepared sample, (Al) grains had typical dendritic shape. The tensile strength and elongation of the rheo-HPDC sample were 225 MPa and 8.6%, respectively, being 15% and 75% higher than those produced from a conventional feedstock.

In the second approach [19], the sonotrode was attached to the bottom of a steel cup (50-mm diameter) containing the melt (AA5083 or A356). The cup was pushed from the top to maintain good acoustic contact. Ultrasonic processing was performed using a 2.6-kW generator and a piezoceramic transducer at 20 kHz. Intermittent treatment with 1-s sonication followed by a 1.5-s resting time was used. The alloys were melted at 740–750°C, degassed with Ar, cooled to a temperature of 650–660°C, and poured into the metallic cup preheated to 570°C. About 450 g of the melt was then treated by ultrasound at a temperature starting from 2°C above the liquidus, i.e., at 632°C (AA5083) or 617°C (A356). The melt was cooled down from the liquidus temperature at a rate of about 6°C/min. The processing time varied between 20 and 70 s. Samples were taken by a quartz tube and quenched in water to observe the in-situ structure. Part of the slurry was solidified in a ceramic mold and part in a HPDC machine. The injection speed of die-casting was 4 m/s, and the specific pressure of injection was 40 MPa. The die was preheated to 200°C before processing. Samples produced without ultrasonication were also cast using the same techniques for comparison at a pouring temperature of 730°C.

The increase in the treatment time (accompanied by a decrease in temperature and increase in the solid fraction) resulted in progressive spheroidization of primary (Al) grains. After 50 s of processing, the number of primary grains also increased while their shape remained mainly globular (Figure 12.7b), which is in clear contrast to dendritic grains in samples without sonication (Figure 12.7a). Finally, after 60–70 s of processing, the size of grains started to increase, probably due to agglomeration. The measurements indicated that the solid fractions of the AA5083 slurry treated

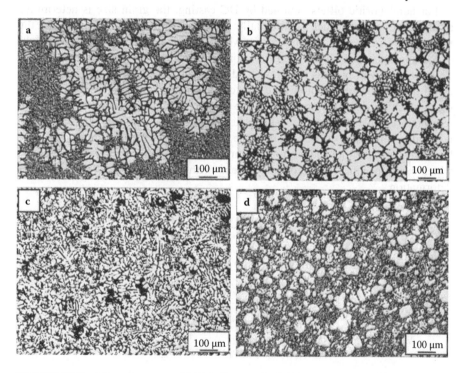

FIGURE 12.7 Microstructure of A356-alloy semisolid slurry (a, b) and HPDC samples (c, d) produced by conventional casting without ultrasonication (a, c) and with 50-s ultrasonic processing of a slurry (b, d). (After Lu et al. [19].)

by ultrasound were 10%, 16%, 22%, 25%, and 27% after 20, 30, 40, 50, and 60 s of processing, respectively; and for the A356 slurry: 14.5%, 19.5%, 22%, and 27% for 20, 40, 50, and 60 s of processing, respectively. The structure analysis showed that 50 s was sufficient to produce spherical grains with an average size of 60–70 μm and a shape factor of about 0.5.

The structure differences between HPDC samples made from sonicated and conventional feedstock are illustrated in Figures 12.7c,d. Samples produced by HPDC showed a clear advantage of their globular semisolid structure, the result of ultrasonic processing of the slurry, as illustrated in Table 12.6 [18, 19]. This type of structure also ensured lower porosity.

The results presented in this chapter indicate that billets with nondendritic structure—produced using ultrasonic processing of the melt or slurry—have advantages in semisolid deformation over billets with dendritic structure, which is in general agreement with earlier findings [2]. However, the available literature does not give a sufficient base for comparing the structural changes caused by both deformation technologies, so there is still a need for further study of this issue. Topics to be investigated include (a) the effect of the initial grain size on the process of metal flow in the solid–liquid state and (b) the optimization of the deformation temperature range to ensure the stability of metal flow in the die.

For nondendritic billets obtained by DC casting, the grain size is determined by the cooling rate, which is essentially a function of the casting speed and the billet size. Grains may become larger when the metal is heated above the solidus temperature and conditioned for a long time at the deformation temperature. This grain-growth process requires a dedicated study. The effect of grain size on metal flow in the solid–liquid state also depends on the dimensions and shape of the product, the deformation load, and the cooling schedule. Even a small change in temperature could result in a considerable change in the solid/liquid ratio with corresponding shrinkage effects. An increase in deformation temperature increases the heat transfer to the tool and requires an improved thermal insulation of the process, which would benefit from automatic systems with thorough control of heating and deformation.

TABLE 12.6
Effect of Rheocasting on Tensile Properties of Aluminum Alloys

Alloy	Processing	UTS, MPa	El, %
AA5083	Conventional HPDC	254	6.8
	Rheo HPDC (50-s UST)	283	9.0
A356	Conventional HPDC	223	4.3
	Rheo HPDC (50-s UST)	254	7.5

Source: Wu et al. [18] and Lu et al. [19].

REFERENCES

1. Spencer, D.B., R. Mehrabian, and M.C. Flemings. 1972. *Metall. Trans.* 3:1925–32.
2. Flemings, M.C. 1991. *Metall. Trans. A* 22A:957–80.
3. Fan, Z. 2002. *Int. Mater. Rev.* 47 (2): 49–85.
4. Davis, J.R. (ed.) 1993. *Aluminum and aluminum alloys: ASM specialty handbook*, 102–3. Materials Park, OH: ASM International.
5. Dobatkin, V.I., and G.I. Eskin. 1996. *Tsvetn. Met.* 2:68–71.
6. Dobatkin, V.I., and G.I. Eskin. 1996. In *Proc. 4th Int. Conf. on semi-solid processing of alloys and composites*, ed. D.H. Kirkwood, and P. Kapranos, 193–96. Sheffield, UK: University of Sheffield.
7. Kaufmann, H., H. Wabusseg, and P.J. Uggowitzer. 2000. *Aluminium* 76 (1–2): 70–75.
8. Eskin, G.I., S.G. Bochvar, and V.V. Belotserkovets. 1998. *Tekhnol. Legk. Spl.*, no. 3:23–28.
9. Eskin, G.I., and V.N. Serebryany. 2000. In *Proc. 6th Int. Conf. on semi-solid processing of alloys and composites*, ed. G.L. Chiametta, and M. Rosso, 361–66. Brescia, Italy: Edimet.
10. Eskin, G.I., and V.N. Serebryany. 2001. *Tekhnol. Legk. Spl.*, no. 5–6:32–38.
11. Kopp, R., D. Neudenberger, M. Wimmer, and G. Winning. 1998. In *Proc. 5th Int. Conf. on semi-solid processing of alloys and composites*, ed. A. K. Bhasin, J.J. Moore, K.P. Young, and S. Midson, 165–72. Golden, CO: Colorado School of Mines.
12. Kong, G.G., and Y.H. Moon. 1998. In *Proc. 5th Int. Conf. on semi-solid processing of alloys and composites*, ed. A. K. Bhasin, J.J. Moore, K.P. Young, and S. Midson, 573–80. Golden, CO: Colorado School of Mines.
13. Eskin, G.I., M.V. Moiseeva, and V.N. Serebryany. 2003. *Tekhnol. Legk. Spl.*, no. 4:3–7.
14. Eskin, G.I., B.I. Semenov, V.N. Serebryany, and Yu.P. Kirdeev. 2002. In *Proc. 7th Int. Conf. on advanced semi-solid processing of alloys and composites*, ed. Y. Tsutsui, M. Kiuchi, and K. Ichikawa, 397–402. Tsukuba, Japan: Nat. Inst. Adv. Industr. Sci. Technol.
15. Eskin, G.I., B.I. Semenov, V.N. Serebryany, and Yu.P. Kirdeev. 2003. *Metallurg. Machinostr.*, no. 2:41–45.
16. Apelian, D. 2000. In *Proc. 6th Int. Conf. on semi-solid processing of alloys and composites*, ed. G.L. Chiametta and M. Rosso, 47–54. Brescia, Italy: Edimet.
17. Lü, S., S. Wu, Z. Zhu, P. An, and Y. Mao. 2010. *Trans. Nonferrous Met. Soc. China* 20:758–62.
18. Wu, S., S. Lü, P. An, and H. Nakae. 2012. *Mater. Lett.* 73:150–53.
19. Lü, S., S. Wu, C. Lin, Z. Hu, and P. An. 2011. *Mater. Sci. Eng. A* 528:8635–40.

13 Ultrasonic Melt Processing: Practical Issues

13.1 ULTRASONIC EQUIPMENT

We have discussed the main types of ultrasonic sources (transducers) in Chapter 1. The development of technology related to both the materials and the electronics used gives the modern user a wide array of equipment for ultrasonic generation.

Piezoceramic transducers are now designed to work in the desirable low-frequency range of 18–25 kHz. Several manufacturers (e.g., Hielscher [Germany] and Mecasonic [France]) now produce transducers, which are powerful (up to 5 kW), efficient (up to 90%), and compact, as seen in Figure 13.1a. Their inherent weaknesses for high-temperature applications are (a) the requirement for additional cooling of the transducer and (b) a narrow working range of frequencies. In reality, the first problem can be overcome by either forced-air flow cooling or by a water-cooled jacket around the upper part of the waveguide. This can solve the overheating problem for laboratory-scale research, but it will be very challenging at the industrial scale. The second problem is more severe. Modern piezoceramic systems are designed and built in such a way that the main parameter to maintain is the working amplitude (typically 10 μm peak to peak). The power capacity is spent on maintaining the amplitude under different acoustic loads; for example, it may take 1 kW to maintain the maximum amplitude of a sonotrode in air and 2 kW to maintain the same amplitude in liquid aluminum. The set amplitude can be changed in fractions of the maximum amplitude. However, the facility for tuning the frequency of the transducer is very limited. Operation at high temperatures accompanied by thermal expansion results in geometrical distortions of the initial sonotrode shape and dimensions, which inevitably leads to a change in the resonance frequency of the sonotrode. A piezoceramic transducer will not be able to respond to these changes by tuning its frequency beyond a certain narrow range and will overload. The result is either automatic switch-off or failure and fracture of the piezoceramic assembly, as shown in Figure 13.2.

Magnetostrictive systems went a long way toward development that was more concerned with the electronics of a generator, while the design of a magnetostrictive transducer (MSC-5-18-type) remaining practically unchanged since the 1950s. However, powerful generators changed from water-cooled-tube electronic schemes (1950s) to air-cooled thyristors (1970s) and then to air-cooled transistor-based

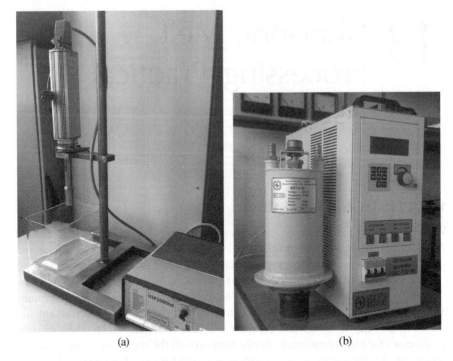

(a)   (b)

FIGURE 13.1 Examples of modern ultrasonic equipment: (a) with piezoceramic transducer and (b) with magnetostrictive transducer.

FIGURE 13.2 A piezoceramic element fractured due to overloading (arrow points at fractured fragments).

electronics [1–4]. Modern magnetostrictive systems are powerful (up to 5 kW), low-frequency (17 to 25 kHz), compact, versatile, and reliable (see Figure 13.1b), yet they are less energy efficient than their piezoceramic counterparts (approx. 50%). Magnetostrictive transducers are made of a stack of magnetostrictive sheets (Ni, Co, Fe-based alloys) that are water cooled and, therefore, protected from overheating during high-temperature processing such as melt treatment. The electronic control system of a generator provides for a relatively wide range of automatic frequency tuning, typically ±1 kHz (narrowband), which is sufficient to maintain the transducer operational, while the geometry of a sonotrode changes with heating/cooling. Although each magnetostrictive transducer has its own resonance frequency that corresponds to its maximum acoustic energy-conversion ability (and the maximum oscillation amplitude at a given power level), the system will continue working within the given frequency range, even if with a lesser efficiency.

Commercial magnetostrictive systems control themselves in a different way than the piezoceramic systems. Such systems maintain their resonance conditions, i.e., they tune to the maximum output power by changing the working frequency of the transducer to maintain the resonance conditions of the entire acoustic system (transducer, waveguide, sonotrode, load). In such a way, the system produces ultrasonic vibrations that may have different amplitudes in dependence on the working conditions. The input power of the transducer can be changed to increase or decrease the working amplitude. Some of the generators are able to maintain the amplitude level automatically by tuning the input power. In this case, a feedback voltage of the magnetostrictive coil (reverse magnetostriction) is used as an input for tuning.

In general, the low-frequency and less-energy-efficient magnetostrictive technology has become obsolete as a mainstream of ultrasonic equipment manufacturing. However, this technology remains essential for metallurgical applications. There are several producers that supply modern magnetostrictive equipment to metallurgical research and industry, e.g., Reltec (Russia), Afalina (Russia), and Advanced Sonics (USA). Table 13.1 compares some characteristics of magnetostrictive systems, both modern and obsolete.

13.2 DESIGN OF WAVEGUIDE AND SONOTRODE

The acoustic system that transmits ultrasonic oscillations from the transducer to the processed medium (liquid metal) usually consists of several elements, i.e., a waveguide (extension), booster, and a sonotrode or horn, as shown in Figure 13.3. The entire system should be acoustically calculated in such a way that (1) it is in resonance with the transducer at its own frequency; (2) it assures the required level of amplitude at the working frequency; and (3) the joints and connections should be in the points of minimum stress (maximum displacement amplitude).

The basic condition of acoustic calculation for the elements of the waveguide and sonotrode systems is that each element should be in resonance at the resonance frequency of the transducer. This conditions is fulfilled if each element has the length multiple of the half-wavelength of the ultrasonic wave propagating in this element. This

TABLE 13.1

Specification of Ultrasonic Equipment Based on Magnetostrictive Transducers

Parameter	Tube Generator (Obsolete) UZG-6.3	Thyristor Generator (Obsolete) UZG3-4	Transistor Generator		
			Reltec USGC-5-22 MS	Afalina USG5-25	Advanced Sonics PPP-4K
Generator output power, kW	6.3	4	5	5	3.4
Frequency range, kHz	18–25	18–22	7–10; 16–20; 20–24	15–50	17–18 (or sweeping frequency)
Generator output voltage, V	220, 380	360	220, 440	220, 360	220–240
Polarizing current, A	...	5–15	1–25	5–25	1–25
Consumed power (max), kW	12	5	6	5.5	4
Weight generator/ transducer, kg	620/15	200/15	25/15	15/15	...

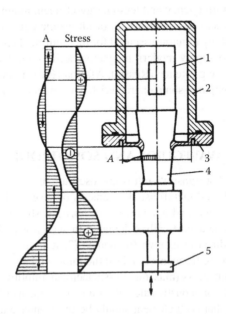

FIGURE 13.3   Acoustic system: (1) ultrasonic transducer, (2) casing, (3) support, (4) booster/ extension, and (5) sonotrode. The graphs on the left show the variation of amplitude (A) and stress along the vertical axis of the acoustic system.

half-wavelength depends on the resonance frequency (transducer parameter), sound velocity and density (material parameter), and geometry (acoustic element parameter).

If we recall Equations (2.3) and (2.4), we can simplify the calculation of the wavelength as

$$\lambda = \frac{c}{f} = \frac{1}{f}\sqrt{\frac{E}{\rho}}, \tag{13.1}$$

where f is the resonance frequency, c is the sound velocity, E is Young's modulus, and ρ is the density of the element material.

A cylindrical element will then have a resonance length calculated using Equation (13.1) and can be $\frac{1}{2}n\lambda$ in actual length. In this case, the end of the element will be characterized by the maximum amplitude, while the top (usually a connection point) is defined by the minimum stress. In order to maintain the longitudinal mode of vibrations in this element, its diameter should fulfill the following condition:

$$0.05 < \frac{d}{\lambda} < 0.5. \tag{13.2}$$

In many cases, amplification of the amplitude is required to achieve a higher cavitation concentration. In this case, a conical or exponential element is used.

The exponential element has the running cross-sectional area S changing with the length x according to an exponential rule: $S = S_0 e^{-\gamma x}$, where γ is the tapering coefficient. The resonance wavelength for an exponential element can be determined as:

$$\lambda = \frac{c}{f}\sqrt{1+\left[\frac{\ln(D_1/D_2)}{\pi}\right]^2}, \tag{13.3}$$

where D_1 and D_2 are the larger and smaller diameters of the exponential element. The ratio D_1/D_2 will be the amplification factor for the amplitude.

The wavelength dependence on the material parameters and element geometry is more complicated in the case of a conical element:

$$\tan\frac{\pi f\lambda}{c} = \frac{\pi f c(D_1 - D_2)^2}{c^2(D_1 - D_2)^2 + (\pi\lambda f)^2 D_1 D_2} \tag{13.4}$$

with the amplification factor:

$$M = \frac{D_1}{D_2}\cos\frac{\pi f\lambda}{c} - \frac{c(D_1 - D_2)}{\pi f\lambda D_2}\sin\frac{\pi f\lambda}{c}. \tag{13.5}$$

The largest amplification factor will be for a so-called stepped design where the element consist of two or more sections each is multiple to $\frac{1}{4}\lambda$. In this case, the amplification will be equal to $(D_1/D_2)^2$.

With the same larger and smaller diameters, the conical element will be a few mm longer than the exponential and stepped ones, while the amplification factor of the stepped element will be much larger than that of the other designs. According to the calculations performed for these three designs (made of an aluminum alloy), the difference in lengths and amplifications is as follows [5]:

Geometry	D_1, mm	D_2, mm	Length, mm	Amplification	Frequency
Stepped	30	10	134	9	20
Exponential	30	10	135	3	20
Conical	30	10	140	2.84	20

As one can see, the conical geometry is close in performance to the exponential one. It is also easier to manufacture. On the other hand, the calculation of the resonance length is quite cumbersome (compare Equations 13.3 and 13.4). In practice, the formula for the exponential geometry is used to calculate the length of the conical sonotrode with a safety factor of 1.1, after which fine-tuning of the length can be performed. For that, the sonotrode (or extension) is either connected to the magnetostrictive transducer with the known resonance frequency or to a special device that can measure the resonance frequency. If the attachment of the sonotrode to the magnetostrictive transducer results in a decreased tuned frequency, then the sonotrode is longer than necessary and should be shortened (see Equation 13.1).

An element of the acoustic system will have some flats, holes, and other added geometrical features, e.g., for connection and tightening. There are simple rules that allow one to adjust the resonance length calculated for a sonotrode of simple geometry to the real geometry. The added element (for example thickening) would result in shortening the sonotrode by a length that can accommodate the mass of the added feature. On the other hand, the cutting of flats or holes in the sonotrode would call for a corresponding extension of its length. The distribution of amplitudes do not change much: the distribution curve stretches in the case of flats and shortens in the case of thickening. The amplification coefficient decreases somewhat.

There are a few other tips on how to make the elements of the acoustic system robust and reliable. These elements work under the conditions of high-frequency fatigue, and any sharp corners, edges, and scratches are potential sites for crack initiation. Figure 13.4a gives an example of a crack that has developed in a sonotrode, apparently originating from the thread in a connection hole. The formation of a crack in a sonotrode manifests itself by the clearly changing pitch of the sound generated by the working ultrasonic system. Instead of a monotonous white noise, whistling of a lower frequency appears. To avoid further damage, the work should be immediately stopped and the sonotrode inspected for damage. A similar change in sound pitch would accompany a loosening connection between acoustic elements,

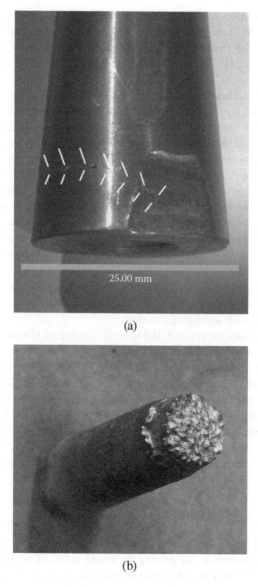

(a)

(b)

FIGURE 13.4 Typical example of sonotrode damage: (a) crack from a stress concentrator (arrows show the crack) and (b) tip erosion.

so the soundness of the sonotrode can be checked by tightening the connections and switching on the transducer. If the whistling sound persists, the damage is obvious, and the failed acoustic element should be changed.

Common practice in designing sonotrodes for high sound intensities requires sufficient filleting at junctions, corners, and flats to prevent premature fatigue failure.

Typical radii of the fillet would be 1–5 mm. For a stepped sonotrode (where the step occurs at the quarter-wave position, i.e., at the node of maximum stress), a rule of thumb is to make the radius of the fillet equal to the difference between the radii of the two cylinders [4]. This may be impractical if the ratio of diameters is much greater than two.

The connection between different elements of an acoustic system is usually made with threaded studs. It is recommended to have a fine thread with a pitch of 1.0–1.5 mm. There are also other types of connections used such as clamps and flanges that should be located in the places of minimum displacement amplitude $\left(\frac{1}{4}\lambda\right)$. The contact surface between two acoustic elements should be smooth and flat to ensure good acoustic contact. A small amount of graphite-based lubricant can be added to the thread and to the contact surface to avoid their (ultrasonic) welding and to ease unscrewing. Two elements should be tightened together with sufficient force; in some cases it is recommended to do the final tightening on a working system. In this case, care should be taken to have vibration-safe grips on the wrench used. In practice, a variety of shapes is used for different applications. Figure 13.5a gives some examples.

Nowadays, finite-element simulations can be used for modeling and calculation of sonotrodes of different shapes. We refer the reader to other sources [3, 4, 6] for more detailed information on acoustic mechanics.

As we have already mentioned, the specific feature of ultrasonic melt processing is high temperature and heating of the sonotrode. The heating, on the one hand, causes expansion and changing resonance conditions; on the other hand, it also calls for cooling of the transducer. There were attempts to design a water-cooled sonotrode that can be used for melt processing, even of high-temperature metals such as copper- and iron-based alloys, as illustrated in Figure 13.6 [7, 8]. In this case, the sonotrode becomes a submerged chill and acts as a source of nucleating crystals dispersed by cavitation and distributed by acoustic streaming. Safety issues related to a water-cooled tool submerged into the liquid metal should be considered.

13.3 SELECTION OF MATERIALS FOR SONOTRODES

The ideal material for a sonotrode working in contact with liquid metals should meet the following requirements: (a) wetting by liquid metal in an ultrasonic field, (b) high density, (c) high melting point, (d) low solubility in liquid metal, (e) high elastic modulus and weak temperature dependence of elastic properties, (f) high fatigue endurance, (g) small thermal expansion coefficient (LTEC), (h) low heat conductivity, and (i) low sensitivity to thermal cycles.

Table 13.2 gives basic physical properties of some materials that are used or can potentially be used as boosters and sonotrodes. They are ranked with respect to the sound velocity, which is the main parameter in transmitting acoustic oscillations (see Equation 13.1) determining the length of an acoustic element. One can see that a ceramic called Sialon tops the table while metallic Nb is in the lower ranks.

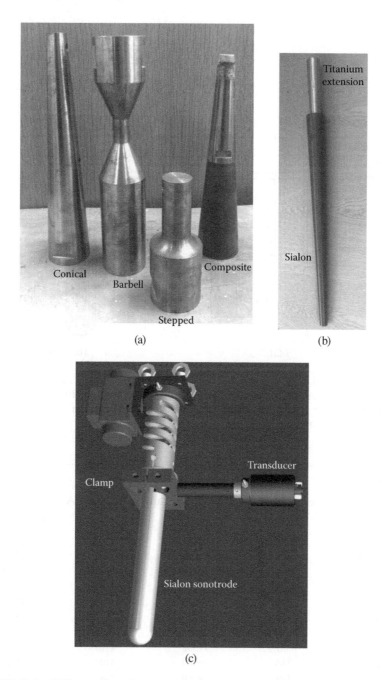

(a)   (b)

(c)

FIGURE 13.5 Different types of sonotrodes: (a) metallic boosters and sonotrodes of various geometries, (b) ceramic sonotrode with a metallic insert (Courtesy of S. Komarov, Nippon Light Metals.), and (c) a scheme of a ceramic sonotrode with a clamp. (Courtesy of M. Prokič, MPI.)

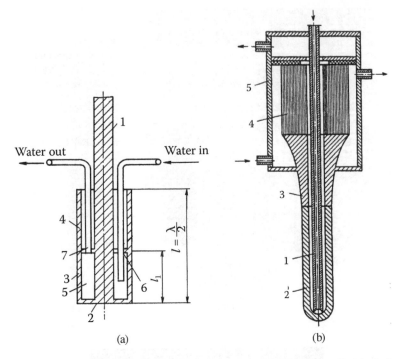

FIGURE 13.6 Design of water-cooled sonotrodes: (a): (1) horn, (2) cylindrical element, (3) lower chamber, (4) upper chamber, (5) water-cooling channel, and (6) partition (After Teumin [7]); and (b): (1) cooling tube, (2) hollow sonotrode, (3) booster, (4) transducer, and (5) water-cooled chamber. (After Abramov et al. [8].)

TABLE 13.2
Physical Properties of Some Materials Relevant to Acoustic Applications

Material	E, GPa	ρ, kg/m³	c, m/s	LTEC, 10^6 K^{-1}
Sialon (SiAlON)	288–306	3230	9440–9730	3.04–3.2
Molybdenum	351.5	10200	5450	4.8
Quartz (SiO₂)	71.7	2210	5370	0.55
Al AA2024	72	2660	5150	24
Steel (low carbon)	209.2	7810	5100–5250	12
Titanium Ti–Al	116.2	4430	5072	8.6
Nickel	210.9	8900	4785	13.4
Tungsten	351.5	19340	4310	4.5
Tantalum	186	16600	4100	6.4
Niobium	104	8560	3480	7.1

Note: E = Young's modulus; ρ = density of the element material; c = sound velocity; LTEC = thermal expansion coefficient.

Numerous experiments with different materials revealed, however, that it is hard to find a candidate to meet all the requirements. At the moment, only metals can satisfy these requirements to a practical extent. Ceramics (e.g., Sialon) do possess positive service characteristics; however, they are very sensitive to thermal cycling, are brittle, and are very difficult to machine; in addition, the connection of a ceramic element to a metallic one is very challenging. There are interesting developments in this area that we will briefly touch upon later on.

At the moment, for the ultrasonic treatment of molten light alloys, the best performance characteristics have been achieved with high-melting metals, specifically, niobium alloys. It is true that their acoustic characteristics are not the best. However, these alloys exhibit elastic characteristics that are insensitive to temperature variations in a liquid-metal (Al, Mg) environment. Young's modulus of Nb changes only slightly in the temperature range 20–1200°C, whereas the elastic moduli of other high-melting metals (titanium, molybdenum, tungsten, steel) monotonically decrease with temperature [9]. The thermal stability of the elastic characteristics of niobium ensures the stability of its acoustic properties when operating in liquid metal.

What is even more important to industrial application, Nb is quite resistant to dissolution and erosion under agitated conditions in liquid Al [10, 11]. When a sonotrode interacts with the liquid metal for a long time, acoustic cavitation in the melt can cause the sonotrode to dissolve in liquid metal, and the collapsing cavitation bubbles can cause so-called cavitation erosion of the radiating face (see Figure 13.4b). As a rule, the first process is rapidly terminated after a thin film of aluminum is formed on the sonotrode surface (aluminization). Therefore, the ultrasonic erosion is of major significance.

Pioneering experimental studies on cavitation stability of various metals in aluminum melts were performed in the 1960s [10]. Sonotrodes of the same geometry made from different metallic materials were submerged by 10 mm in 10 kg of molten 99.85% pure aluminum at 740°C and operated at 18 kHz, 20-μm amplitude for up to 1 h. The chemical analysis of the melt was used as an indicator of sonotrode dissolution and erosion. The results are shown in Figure 13.7 and in Table 13.3. All studied materials suffer from dissolution and erosion; however, Nb appears to be the most stable.

A mechanism of cavitation erosion in liquid metals was suggested and experimentally confirmed in the 1960s for a variety of sonotrode materials and melts [10, 12]. The interaction of a sonotrode material with liquid metal (e.g., aluminum) produces intermetallic compounds Al_xMe_y with a formation rate and hardness specific of each particular interacting system. Intermetallic compounds of all Al–Me systems are as a rule harder and more brittle than the sonotrode metal. As a result of this difference in properties, the surface layer of intermetallic compounds easily breaks off and disperses under the impact of cavitation, leaving behind a pure metal surface where the intermetallic formation process begins anew and the erosion proceeds further. Figure 13.8 illustrates the suggested mechanism. The existence of such an intermetallic layer (10–15 μm thick) was confirmed for a Ti sonotrode in contact with liquid Al [10] and Zn [12].

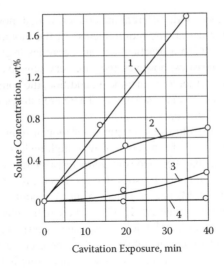

FIGURE 13.7 Dissolution kinetics of different sonotrode materials in liquid aluminum at 740°C and 20-μm 18-Hz vibrations: (1) stainless steel, (2) Ti, (3) Mo, and (4) Nb alloy. (Adapted from Eskin [10].)

The erosion rate depends on the properties of the intermetallic compound and on the kinetics of its formation. For example, a brittle compound Al_3Me forms at 0.002 wt% Fe in the Al–Fe system, at 0.1–0.2 wt% Mo in the Al–Mo system, at 0.15–0.19 wt% Ti in the Al–Ti system, and at 0.18–0.28 wt% Nb in the Al–Nb system. The hardness of these intermetallic compounds also differs, as shown here:

	Al_3Fe	Al_3Ti	Al_3Mo	Al_3Nb
μH, MPa	9600	5950	3660	3750

TABLE 13.3
Kinetics of Dissolution and Erosion of Different Sonotrode Materials in Liquid Aluminum

Exposure to Cavitation, min	Concentration in the Melt, wt%				
	Steel	Ti	Mo	W	Nb
2	0.080	0.018	0.012	0.005	0.0005
4	0.155	0.030	0.024	0.010	0.0026
5	0.22	0.040	0.028	0.012	0.0010
8	0.61	0.048	0.034	0.017	0.0011
10	dissolved	0.053	0.034	0.018	0.0022
12	...	0.061	0.040	0.022	0.0024
15	...	0.062	0.040	0.025	0.0028

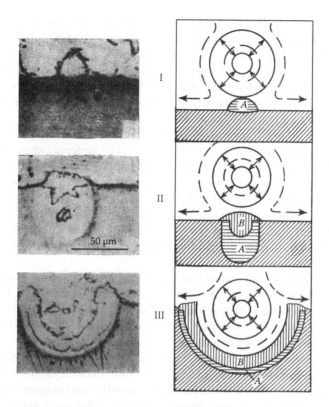

FIGURE 13.8 Mechanism of cavitation erosion of a metallic sonotrode in metallic melt: *I*, starting stage of stresses caused by a cavitation bubble and the formation of a diffusion zone A; *II*, formation of a brittle compound B; and *III*, fracture and breaking away of the inter-metallic compound, exposure of the fresh metal in the pit, and the repetition of the process. On the left: experimental observation of the same process for Ti sonotrode in molten Zn. (Adapted from Grießhammer [12].)

These data indicate why sonotrodes from iron-based steel—where the hardest and brittlest compound is formed—erode at a much higher rate than those made from titanium-, molybdenum-, and especially niobium-based alloys. Tantalum is also a good candidate for sonotrode materials with exceptionally high resistance; however, the use of this metal is prohibitively expensive.

These studies demonstrated that Nb and its alloys are the most suitable metallic materials for sonotrode application in liquid aluminum. The practice showed that sonotrodes made from commercial Nb alloys (C103-type, Nb–Mo–Zr) can with-stand up to 300 hours of continuous operation in liquid aluminum. A note should be made on the quality of the metal used for machining sonotrodes. High-temperature metals (Nb, Ta, W) are produced via a powder metallurgy route followed by elec-tron-beam or arc remelting, forging, and annealing. The stock used for sonotrodes should be of the highest quality with regard to inclusions, porosity, and residual stresses. Therefore, double electron-beam remelting in protective atmosphere, thoroughly controlled forging, and vacuum annealing are recommended.

In liquid Mg, titanium or low-alloy steel sonotrodes can be used, especially in research and in short-exposure foundry applications. Intermediate members of the acoustic system are usually made of Ti alloys, which have a good combination of acoustic properties, thermal stability, low density, and low thermal conductivity. In many cases, low-carbon steel can be used as well.

The idea of using ceramic sonotrodes dates back to the very first applications of ultrasonic processing to liquid metals [13]. Sintered corundum (Al_2O_3), quartz (SiO_2), and hard porcelain (Al_2O_3, SiO_2, $Si_3Al_4O_{12}$) were used to introduce ultrasonic vibration into the melt. A system with hard porcelain sonotrodes was commercially produced by Atlas Werke in the 1940s (see Figure 1.2b), where the porcelain element was mechanically clamped to the transducer. This system was used to cast aluminum ingots (see Figure 1.2a). The use of ceramics at these times proved to be unreliable, with frequent cracking of ceramic elements and bad acoustic contact.

New ceramic materials are currently under scrutiny for sonotrode applications. The acoustic properties of Sialon (Table 13.2) are good; mechanically, it is a rather tough material able to withstand thermal shocks and even be machined to some extent. This material is also almost inert to liquid aluminum up to 1000°C [11]. The bottleneck is the connection of a ceramic sonotrode to metallic parts of the acoustic system, e.g., transducer or booster. Ceramics are usually connected to metals by soldering, but this technology is not well suited for high-frequency applications. One of the ways to overcome the problem is to make a metallic insert in green-body ceramic that, after sintering, can be used for the connection. This insert can be a cylinder with internal thread embedded into the ceramic part or a stud with external thread protruding from the ceramic part. This solution works, and an example is shown in Figure 13.5b. Another solution to the problem is by using clamps. In this case, a transducer can be clamped to the ceramic sonotrode and excite transverse oscillations in the latter, as seen in Figure 13.5c [14, 15]. Sonotrodes made from modern ceramics such as Sialon can potentially extend the application of usual ultrasonic processing schemes to higher-temperature melts such as Cu-based alloys.

13.4 CHARACTERIZATION OF ULTRASONIC PROCESSING

The main means of characterizing the acoustic parameters of ultrasonic processing (amplitude, cavitation threshold, acoustic pressure, etc.) were described in Section 2.6. In this section, we briefly touch upon existing technical solutions for acoustic characterization.

One of the main parameters that can be used for estimating the energy transmitted to the melt is the amplitude of the radiating face of a sonotrode. The direct measurements are possible only in air, with subsequent recalculation to the actual amplitude in the melt by using a feedback voltage of the transducer or calibration curves obtained by other means, e.g., calorimetry. The construction of most piezoceramic systems and some magnetostrictive systems includes a vibrometer or another precalibrated sensor that allows them to display the actual amplitude.

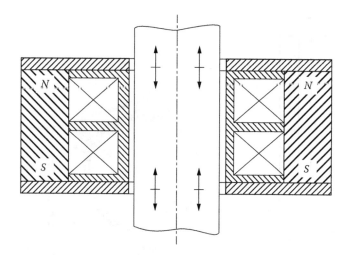

FIGURE 13.9 A scheme of an electrodynamic vibrometer.

One such scheme includes an electrodynamic vibrometer, as depicted in Figure 13.9 [16]. The magnetic circuit includes a circular permanent magnet with two disk-shaped magnetic coils pressed into the magnet. The measured sonotrode is placed through the vibrometer with 0.05–0.1-mm gaps. The use of two gauging coils connected oppositely helps to pick the total electromotive force from both magnetic coils that otherwise would be cancelled. In this setting, each section will measure the amplitude of oscillating velocity in the immediate vicinity, while the total measured signal will give the averaged amplitude in the section between the two magnetic coils. In addition, if the gauging coils are commuted aiding, the signal produced will be proportional to the total mechanical stress at the measured section. A simple switch can be provided to get both values. The absolute sensitivity is constant in the frequency range 15–50 kHz but depends on the material (magnetic properties) of the sonotrode. Alloyed steel will give forty-five times stronger response than a titanium alloy.

In the absence of this useful tool, a researcher needs to have a stand-alone sensor to measure the amplitude of a particular sonotrode at particular conditions, e.g., at a given transducer power. The simplest method uses a mechanical micrometer with a ball or pin probe (Figure 13.10) that is put in contact with the radiating face of an idle sonotrode, and then the measurement is taken after the sonotrode is switched on. The height obtained in the ball-indicator can be recalculated to the null–peak amplitude as [17]:

$$A \ (\mu m) = 0.003 \sqrt{\frac{h \ (cm)}{f \ (kHz)}}. \tag{13.6}$$

In the case of micrometers, the read value corresponds to the peak-to-peak amplitude.

Contactless measurements can be performed using laser interferometry, high-speed laser distance meters, and induction meters. In the last case, the changing

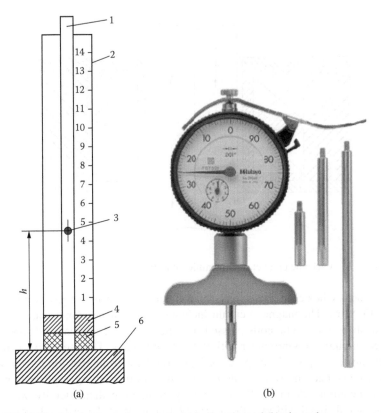

FIGURE 13.10 Mechanical vibrometers: (a) ball probe and (b) pin probe.

distance between the vibrating face and the induction coil in the sensor changes the magnetic conductivity of the gap and, in turn, the inductance of the coil that is transferred to alternating voltage. The signal is then processed to give the amplitude. A reliable setup of this type (VN1-4) is manufactured by Belorussian State University of Informatics and Radioelectronics (BSUIR), as shown in Figure 13.11a. This vibrometer measures the null–peak amplitude, works in the range of amplitudes 0–100 μm at frequencies 15–50 kHz, and is accurate within 2.5 rel.%. An example of measurements obtained using this vibrometer is given in Figure 13.11b. These results also illustrate the dependence of the amplitude on the input power in an MST-5-18 transducer and the amplification with the use of combined conical–cylinder and conical–conical elements of a sonotrode.

Most useful information about acoustic field and cavitation activity in the treated liquid can be obtained using so-called cavitometers or hydrophones. These systems are well developed for room-temperature applications. One can look at a hydrophone as a reverse transducer, where mechanical vibrations excited in the probe are transmitted to a converter (piezoelectric or electrodynamic) that turns the mechanical vibrations into an electrical signal. The magnitude and frequency spectrum

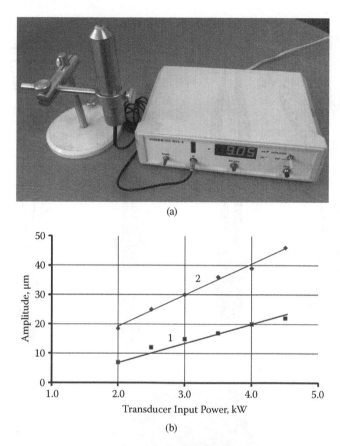

(a)

(b)

FIGURE 13.11   (a) Contactless vibrometer VM1-4 and (b) results on amplitude measurements obtained with this vibrometer: (1) conical Ti booster 65 to 40 mm and cylindrical Nb sonotrode 40 mm in diameter and (2) conical Ti booster 65 to 40 mm and conical Nb sonotrode 40 to 20 mm.

of this signal is then filtered and analyzed to produce valuable information about acoustic pressure, cavitation threshold, and high-frequency harmonics reflecting the cavitation development. The principle of cavitometer operation and examples of its use are given in Section 2.6 (see Figures 2.20–2.22).

Low-temperature cavitometers are usually based on one of the schemes shown in Figure 13.12a [16]. Figure 13.12b gives an example of a handheld room-temperature cavitometer. Unfortunately, these direct schemes, where the sensing element is placed in the head of the probe cannot be used at high temperatures, let alone in liquid metals. For this purpose, a special elongated probe made of tungsten or titanium is designed for submerging into the melt, while the converter is located at a distance. Such a unique high-temperature cavitometer (ICA-3HT) has been developed and manufactured by Belorussian State University of Informatics and Radioelectronics and is shown in Figure 13.12c.

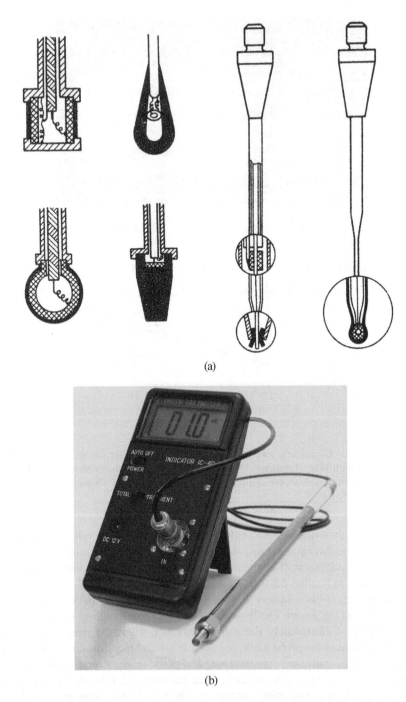

(a)

(b)

FIGURE 13.12   (a) Various schemes of cavitometers (hydrophones) for measuring cavitation activity and acoustic pressure in liquids; (b) room-temperature handheld cavitometer; and (c) high-temperature cavitometer. (Courtesy of N. Dezhkunov, BSUIR.) (*continued*)

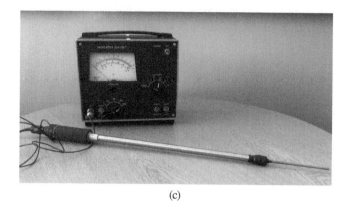

(c)

FIGURE 13.12 (*continued*) (a) Various schemes of cavitometers (hydrophones) for measuring cavitation activity and acoustic pressure in liquids; (b) room-temperature handheld cavitometer; and (c) high-temperature cavitometer. (Courtesy of N. Dezhkunov, BSUIR.)

REFERENCES

1. Mason, W.P. (ed.). 1964. *Physical acoustics: Principles and methods*. Vol. 1: *Methods and devices Part B*. New York: Academic Press.
2. Eskin, G.I. 1988. *Ultrasonic treatment of molten aluminum*. 2nd ed. Moscow: Metallurgiya.
3. Abramov, O.V. 1998. *High-intensity ultrasonics: Theory and industrial applications*. Amsterdam: Gordon and Breach OPA.
4. Ensminger, D., and L.J. Bond. 2011. *Ultrasonics: Fundamentals, technologies, and applications*. 3rd ed. Boca Raton, FL: CRC Press.
5. Patel, D.M., and A.U. Rajurkar. 2011. *Proc. Int. Conf. Comput. Meth. Manufacturing. ICCMM 2011*, 115–21. Assam, India: Indian Institute of Technology Guwahati.
6. Ensminger, D., and F.B. Stulen. 2008. *Ultrasonics: Data, equations and their practical uses*. Boca Raton, FL: CRC Press.
7. Teumin, I.I. 1968. *Introduction of ultrasonic oscillations in processed medium*. Moscow: Machinostroenie.
8. Abramov, V., O. Abramov, V. Bulgakov, and F. Sommer. 1998. *Mater. Lett.* 37:27–34.
9. Ferraro, R.J., and R.B. McLellan. 1979. *Metall. Trans. A* 10A:1699–1702.
10. Eskin, G.I. 1965. *Ultrasonic treatment of molten aluminum*. Moscow: Metallurgiya.
11. Yan, M., and Z. Fan. 2001. *J. Mater. Sci.* 36:285–95.
12. Grießhammer, A. 1967. *Ultrasonics* 5:229–32.
13. Hiedemann, E.A. 1954. *J. Acoust. Soc. Am.* 26:831–842.
14. Puga, H., J. Barbosa, J. Gabriel, E. Seabra, S. Ribeiro, and M. Prokic. 2011. *J. Mater. Process. Technol.* 211:1026–33.
15. Puga, H., J. Barbosa, D. Soares, F. Silva, and S. Ribeiro. 2009. *J. Mater. Process. Technol.* 209:5195–5203.
16. Makarov, L.O. 1983. *Acoustic measurements in processes of ultrasonic technology*. Moscow: Mashinostroeniye.
17. Teumin, I.I. 1961. *Techn. Inform. Bull. Spec. Des. Bur. (OKB) for ultrasonic and high-frequency equipment*, no. 1 (19): 9–12.

Index

For Product Safety Concerns and Information please contact our
EU representative GPSR@taylorandfrancis.com
Taylor & Francis Verlag GmbH, Kaufingerstraße 24, 80331 München, Germany